INTRODUCTION TO THE
GLOBAL STRUCTURE OF SPACE-TIME

THEORETICAL FOUNDATIONS OF COSMOLOGY

INTRODUCTION TO THE
GLOBAL STRUCTURE OF SPACE-TIME

THEORETICAL FOUNDATIONS OF COSMOLOGY

MICHAEL HELLER

Pontifical Academy of Theology
Cracow, Poland
and
Vatican Observatory, Vatican City

World Scientific
Singapore • New Jersey • London • Hong Kong

Published by

World Scientific Publishing Co. Pte. Ltd.
5 Toh Tuck Link, Singapore 596224
USA office: 27 Warren Street, Suite 401-402, Hackensack, NJ 07601
UK office: 57 Shelton Street, Covent Garden, London WC2H 9HE

THEORETICAL FOUNDATIONS OF COSMOLOGY
Introduction to the Global Structure of Space-Time

ISBN 13-978-981-02-0756-4
      10-981-02-0756-5

Typeset by Z. Golda,
supported by the Polish Grant GMEN 141.

Table of Contents

Preface

The aim of the present book is to discuss theoretical foundations of cosmology. And nothing more, but possibly also nothing less. Therefore, it is not a textbook in any sense of this word. One of its principal goals is to study the assumptions upon which relativistic cosmology is founded, and, first of all, those assumptions that are implicit in the mathematical structure of cosmology.

The leading idea of the book may be formulated in the following way: *The Universe is a structure.* This structure can be investigated with the help of either local or non–local (global) mathematical methods. In the first case, we have theoretical physics, in the traditional meaning of this word (i.e. in the sense of "local physics"). In the second case, we have cosmology. In this sense, *cosmology is a non–local physics.* In the spirit of this approach, I use, from the very beginning, global mathematical methods, even in these parts of the material which are usually regarded as only a preparation to cosmology.

It is worth noticing that cosmology, understood as a non–local physics, is something more than a theory of cosmological models. Cosmology in such a narrow sense (as a theory of world models) appears in the book not earlier than in the last two chapters; I think, however, that, when analysing the global structure of space–time in all foregoing chapters, I am studying truly cosmological issues. Both special and general theories of relativity are presented with stress on their respective global aspects.

I assume that the reader is familiar with mathematical methods which are employed throughout the book; although, for his or her convenience, I usually give main definitions with short comments and some examples. In this way, I hope, the reading of the book could be smooth enough without having too often to consult other textbooks or monographs. However, I am aware that more responsible acquaintance with the book would sometimes require the aid of some other publications. *Bibliographical notes*, at the end of every chapter (with the exception of chapters 5 and 6), could assist the reader in this respect.

I would like to stress once more that my intention is not to teach the reader how to do calculations (I assume that he already knows it), but to clarify concepts and show the efficiency of structures.

Definitions, theorems, examples, and so on, are numbered in a continuous manner within each section; the method of numbering is natural and needs no explanation. Only the definitions of those concepts which play a role in the exposition have been numbered, definitions of all other concepts having been incorporated into the text.

The organisation of the material is the following. In the first two chapters foundations and assumptions of the contemporary theory of space–time are discussed. The second chapter contains elements of special relativity. I am especially interested in the existence of the Lorentz structure. Since it is, in principle, a global problem, it conveys — in my opinion — information of cosmological significance. The third chapter is a concise repetition of the fibre bundle method, one of the main theoretical tools used in the book. With the help of it the general theory of relativity is presented in the fourth chapter. The method, on the one hand, makes manifest similarities of this theory with gauge theories of contemporary physics and, on the other hand, stresses the peculiarity of Einstein's theory of gravity as a theory of the frame bundle over space–time. The fifth chapter occupies the central place in the book. It poses, to use Einstein's expression, "the cosmological problem", and explicitly formulates the leading idea of the entire book. Although no single formula appears in this chapter, it is by no means a popular exposition: its full understanding presupposes a considerable mathematical and cosmological culture by the reader. Analyses carried out in this chapter (and a program it outlines) are applied to the simplest world models, the so–called Robertson–Walker cosmology, in the sixth chapter. It goes without saying that the Robertson–Walker cosmology is reconstructed with the help of global methods (the concept of warped product plays a key role here).

The rather extensive Appendix (of the size of an average chapter) is aimed at revealing a beautiful logic inherent in the evolution of space–time theories, starting from the Aristotelian physics and ending with general relativity. The application of the fibre bundle method, helped — if necessary — by some concepts of the mathematical category theory, has allowed us to reconstruct former theories of space–time in the modern geometric framework, and to show a "natural" character of the Einstein theory of relativity within the context of this evolutionary chain. It turns out that general relativity is logically the almost unavoidable "next step" along the path of development. In my opinion, it is a strong, albeit purely esthetic, argument on behalf of Einstein's theory.

At the end of most of the chapters there is a section entitled *Comments and Remarks*. Here I consider some problems more loosely connected with the content of the given chapter, and sometimes I allow myself digressions of a philosophical character.

I am fully aware that the book leaves many problems either completely untouched or non–satisfactorily treated. Such problems as: Bianchi cosmology, physics of the early Universe, further chapters of the global theory of space–time (including the classical singularity problem), many questions of observational cosmology, although well (and sometimes very well) developed as far as their technical aspects are concerned, still await a more careful methodological elaboration. Each of these problems (and many others as well) deserve a separate monograph. The *Bibliographical Essay*, after the sixth chapter, containing a list of books on cosmological topics with short remarks and explanations, is intended to recompense the reader for the fact that the book ends at a point where the most interesting action only begins.

M. H.

March 1985

INTRODUCTION TO THE
GLOBAL STRUCTURE OF SPACE-TIME

THEORETICAL FOUNDATIONS OF COSMOLOGY

Chapter 1

Manifold Model of Space–Time

0. Introduction

Each physical theory contains cosmological elements, usually as tacit assumptions, since each physical theory develops on an "arena", and the theory of such an arena can be thought of as a part of cosmology. Macroscopic physical theories occur in space–time and all of them presuppose a certain structure of it, called *differential manifold structure* or simply *manifold structure*. This structure is common to all macroscopic physical theories which differ from one another only when structures are taken into account that are superimposed on the manifold structure. The problem of a suitable arena for theories of microphysics is still an open question. Quite a common opinion asserts that, in quantum mechanics also, manifold continues to play its role of an arena for physical processes. Even if the opponents of this opinion are right, one has to admit that space–time modelled by differential manifold preserves its importance in all quantum theories: all measurements that constitute the empirical base of such theories are performed within the macroscopic space–time, and they must appear, in this or another way, in the mathematical formalism of these theories (here all discussions have their origin concerning the theoretical status of space–time in theories of microphysics).

To construct the manifold model of space–time is the goal of the present chapter. Its larger parts are of a purely mathematical character. Sec. 1 prepares, so to speak, the mathematical environment for the theory of differential manifolds. It presents, in a more or less informal manner, some information on functional spaces, and in particular on Banach spaces, and their role in modern differential geometry as well as some preliminary remarks concerning the geometric model of space–time. Sec. 2 contains the definition of differential manifold (in the case of a finite number of dimensions) and introduces the most important concepts of the manifold theory. Sec. 3 deals with one of such concepts, namely the concept of manifold orientability. Vector and tensor fields are introduced in sec. 4, and the notions of immersion, embedding and submanifolds appear in sec. 5. Mathematical tools, prepared in this way, enable us to present in sec. 6 a succinct construction of the manifold model of space–time. Sec. 7 contains comments and remarks more loosely connected with the main topic of the chapter. Among others, the attention of the reader is drawn to some difficulties inherent in the very concept of space–time manifold, which made impossible the full implementation into a physical theory the so–called Mach's principle, once postulated by Einstein and broadly discussed till now. At the end of sec. 7 some remarks can be found concerning the status of space–time manifold in theories of the microworld.

Mathematical material presented in this chapter, as well as in a few next chapters, is of a repetitive character, therefore definitions are stressed rather than theorems and their proofs, mutual dependencies between concepts that will be needed later on rather than particular examples. In the "Bibliographical Note" at the end of the chapter, the reader can find information concerning books on the manifold theory. He is encouraged to choose something from that list. Clear mathematical perspective will make further reading of this book easier.

The model of space–time, as a four–dimensional differential manifold, presented in this chapter is very general. It will be gradually enriched in the following chapters.

1. Banach Spaces and Manifolds — Mathematical Environment of the Space–Time Theory

The general concept of function appeared in mathematics not earlier than the beginning of the XIXth century. It was Riemann who, in his famous inaugural lecture *On the Hypotheses That Lie at the Foundations of Geometry*, first noted the necessity to investigate functional spaces. Further development of the function theory consisted mainly in a transition to the study of functions whose domains and ranges are not limited to the axis of real numbers or to spaces with finite numbers of dimensions. Owing to this transition, functional spaces were put into their natural environment, namely the general theory of topological spaces.

Dwelling within the region of elementary mathematics creates a dangerous illusion. One usually deals with "well–behaved" functions, for instance, with continuous or differentiable (as many times as needed) functions, with functional spaces having sufficiently smooth structures, etc. This creates an impression that well–behaved functions exist in general, and "pathological cases", such as noncontinuous or nondifferentiable functions, are rare exceptions. An instant of reflection (for which seldom a time is left in a spate of computations) is enough to see that things are just the opposite. There are well–behaved functions that constitute exceptional regions (of "measure zero") within the realm of all possible functions.

Non–maliciousness of God, of which Einstein spoke so often, reveals itself in that in his architectural plans the Creator has used well–behaved functions or at least such functions that can be approximated by well–behaved functions. The functions being elements of the Banach space belong to this privileged class.

A linear normed and complete space is said to be a *Banach space*. This definition can be thought of as a natural generalization of the concept of vector field in the Euclidean space. To each element x of a Banach space (to each "vector") one ascribes a real number $\|x\|$, called the *norm* of the element (a generalization of a vector's length). The norm is defined with the help of known axioms that formalise its properties. Linearity of a Banach space ensures that one can add its elements to each other and multiply them by numbers (real or complex). Completeness of a Banach space means that every Cauchy sequence reaches its limit in this space.

The properties expressed in the above definition of Banach space guarantee that, on the one hand, this space is a very general object containing a rich class of special cases often met in the mathematical practice and, on the other hand, it is a well-behaved object eliminating many pathological situations which would require special caution from the mathematician.

In particular, from the definition of Banach space it follows that a metric can be introduced in this space. Indeed, the expression $\|x_1 - x_2\|$, where x_1 and x_2 are elements of a Banach space, satisfies the metric axioms. Therefore, Banach spaces are metrizable.

One of the main advantages of Banach spaces is that they constitute a natural environment for calculus. The generalization of the derivative in the space $\mathbf{R}^n$ (which is a special case of a Banach space) to the derivative in Banach spaces in general is so immediate that many theorems, obtained in this way, look as if they were mechanically repeated from the elementary course of differential calculus. In fact, this procedure is far from being trivial. In the transition process to general Banach spaces, the notion of a derivative of a function is generalized to the concept of a mapping from one Banach space to another. In this way, functional analysis acquires one of its most fundamental tools.

Calculus can also be done in spaces more general than Banach spaces, for instance, in any topological vector space. However, in such a case, the generalization is not that natural (for example, the implicit function theorem can no longer be valid without additional assumptions).

The theory of functional spaces is a powerful tool of modern differential geometry. Differential geometry has grown up from analytic investigations of curves and surfaces in the three–dimensional Euclidean space. The generalization went in two directions: first, from three to any number of dimensions (also infinite) and, second, towards getting rid of the "surrounding" Euclidean space. In this way, the crucial concept of modern differential geometry was born, namely the concept of *differential manifold modelled on a topological vector space of any (also infinite) number of dimensions* or, in particular, *on a Banach space*. In the latter case one speaks about a *Banach manifold*.

The manifold concept is a generalization of the parametric representation of a surface in $\mathbf{R}^3$, i.e. of a $\mathcal{C}^k$ mapping of an open subset of a given subset into $\mathbf{R}^2$ In the case of an infinite–dimensional Banach manifold, the space $\mathbf{R}^2$ is replaced by an infinite–dimensional Banach space.

Let X be any set and $\mathcal{B}$ a Banach space. The set of pairs (U_i, ϕ_i), called *local charts*, such that

(1) the sets U_i $(i \in \mathcal{I})$ cover X,

(2) the mappings ϕ_i are $\mathcal{C}^k$ bijections from open subsets U_i onto open subsets of $\mathcal{B}$,

(3) the sets $\phi_i(U_i \cap U_j)$, for each pair $i, j \in \mathcal{I}$, are open in $\mathcal{B}$, and mappings of the

type

$$\phi_j \circ \phi_i^{-1}: \ \phi_i(U_i \cap U_j) \longrightarrow \phi_j(U_i \cap U_j)$$

are of C^k class,
is called *the C^k atlas on X modelled on the space $\mathcal{B}$*. Two atlases are called *equivalent* if their union is an atlas. The equivalence of atlases is an equivalence relation.

A set X together with an equivalence relation of C^k atlases on X modelled on the Banach space $\mathcal{B}$ is said to be a C^k *differential manifold modelled on the Banach space $\mathcal{B}$*. The dimension of such a differential manifold is, from definition, the dimension of $\mathcal{B}$; it can be infinite. The atlases that belong to the equivalence class appearing in the above manifold definition are called atlases *admissible on X*.

Infinite–dimensional manifolds modelled on Banach spaces, in short also called *Banach manifolds*, are in fact functional spaces; however, one can introduce on them a majority of structures which are natural generalizations of structures known from the elementary course of differential geometry. For instance, tangent and cotangent spaces, and consequently vector and tensor fields as well as differential forms, can be defined on them. In particular, symplectic forms and Hamiltonian vector fields are naturally introduced on Banach manifolds. In close analogy to the finite–dimensional case one can construct various fibre bundles over Banach manifolds and define fields of different geometric objects as cross–sections of the corresponding fibre bundles over these manifolds. Tensor fields and differential forms turn out to be special cases of such constructions. In this way, one can also introduce a metric (the field of a metric tensor) on a Banach manifold, changing it into a *Riemann manifold*. The geometry of such manifolds is further constructed in analogy with the finite–dimensional case.

The spaces $\mathbf{R}^n$ (and $\mathbf{C}^n$) are Banach spaces with the norm defined in a natural way, and differential manifolds modelled on $\mathbf{R}^n$ (or $\mathbf{C}^n$), (their precise definitions will be given in the next section), play a key role in the traditional differential geometry; they are simply called *differential manifolds (real or complex)*. Differential manifolds are locally diffeomorphic to $\mathbf{R}^n$ (or $\mathbf{C}^n$), and, therefore, at each of their points there exists tangent vector space with the tangent vectors defined in an "intrinsic manner", i.e. with no reference to any space in which the manifold would be immersed. On the one hand, differential manifolds preserve well–known properties of surfaces immersed in the Euclidean spaces, and, on the other hand, they are entirely autonomous in the sense that, although modelled on $\mathbf{R}^n$ (or $\mathbf{C}^n$), they do not need to be considered with reference to any "surrounding space".

Owing to these properties, in the contemporary theoretical physics four–dimensional (real) differential manifolds serve as models of space–time.[1] This remains true both with respect to classical physics and with respect to special and general rela-

---

[1] The reader should pay attention to a certain fuzziness of commonly accepted terminology: one speaks of manifolds **modelled on** Banach spaces, and one speaks on the physical space–time **modelled by** a four–dimensional manifold. Modelling in the former sense is determined by the manifold definition; modelling in the latter sense is one of the most fundamental procedures used in theoretical physics, whose methodological aspects are studied by philosophers of science.

tivity. It turns out that space–times of pre–Newtonian theories (those of Aristotle and Galileo) can also be modelled by four–dimensional differential manifolds (see, Appendix). The manifold model of the physical space–time is a subject matter of the present book. In this introductory section, I want only to place this space–time model in its natural mathematical environment, namely in the general theory of differential manifolds modelled on Banach spaces. Real differential manifolds constitute a very special class of Banach manifolds, but nevertheless they are general enough to provide description tools of a rich family of phenomena occurring in the real universe. It seems that a certain tension between generality and peculiarity of mathematical constructions is a key characteristic of the conceptual structure of contemporary physics.

The concept of normed spaces was introduced by Stefan Banach in 1922 in his doctoral thesis. From Banach's work, *Theorie des operations lineaires* (1932) the "mature age of normed spaces" began (Bourbaki 1969). The manifold concept started to crystalise itself in the XIXth century, simultaneously in several branches of mathematics, where the necessity to investigate locally Euclidean spaces arose. In 1854, in his outspoken inaugural lecture, Riemann made some remarks which turned out to be important for the development of the manifold concept (see Riemann 1868). The next step belongs to Poincaré: in 1880, when studying certain topological techniques in the theory of differential equations, he introduced what is to–day called flows on a manifold. A precise definition of a smooth two–dimensional manifold was given by Weyl in 1913, in his book *The Concept of a Riemann Surface* (see Weyl 1955). Understanding of the manifold concept evolved rather laboriously in the first half of the XXth century to become, after 1950, one of the most crucial chapters of modern topology and differential geometry.

2. Differential Manifolds

A topological space X that is homeomorphic to $\mathbf{R}^n$ in the neighbourhood of each of its points is said to be a *topological manifold*. In the following, we shall assume — unless this assumption is directly revoked — that topological manifolds have the *Haussdorf property*, i.e., any two different points of such a topological manifold possess disjoint neighbourhoods. The existence of a local homeomorphism between X and $\mathbf{R}^n$ should be understood in the sense of the following definitions.

A pair (U, ϕ), where U is an open subset of X, and $\phi\colon U \to V$ a homeomorphism onto an open subset $V \subset \mathbf{R}^n$ is said to be a *(local) chart* on the manifold X. By definition n is the *dimension* of the topological manifold. Coordinates $(x^1, \ldots, x^n)$ of the image $\phi(x) \in \mathbf{R}^n$ of a point $x \in X$ are called *(local) coordinates* of the point x in the chart (U, ϕ); this chart is also traditionally called a *(local) coordinate system on* X. It is assumed that X is covered by open sets U_α being domains of charts on X.

In the following, we shall additionally assume that the set $\{U_\alpha\}$ covering X is countable (X as a topological space has a countable base). However, this assumption is not a part of the manifold definition. Because of future applications it is convenient to accept that X is a connected space, or at least that the dimension of X is the same in the neighbourhood of each of its points (this property follows from the connectedness of X).

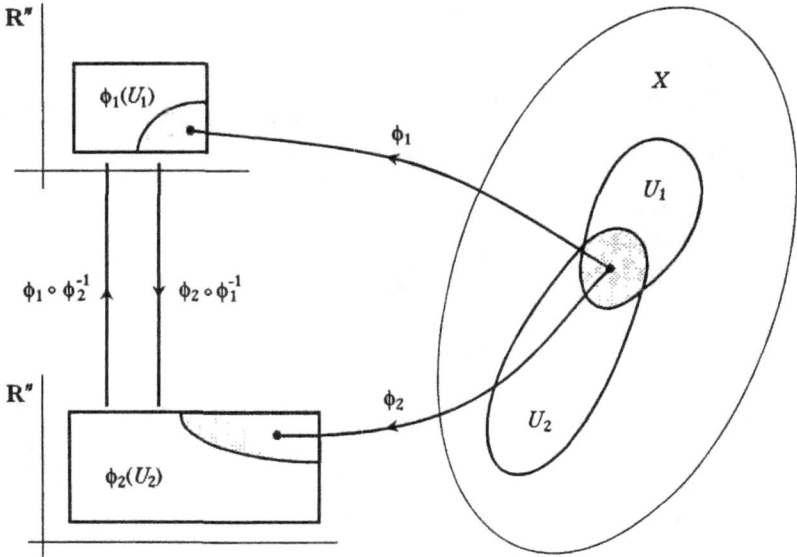

Fig. 1.1. Compatibility of maps on topological manifold.

Two charts (U_α, ϕ_α) and (U_β, ϕ_β) on X are said to be $\mathcal{C}^k$ *consistent* if — assuming that $U_\alpha \cap U_\beta \neq \emptyset$ — the mappings

$$\phi_\beta \circ \phi_\alpha^{-1}: \; \phi_\alpha(U_\alpha \cap U_\beta) \longrightarrow \phi_\beta(U_\alpha \cap U_\beta),$$
$$\phi_\alpha \circ \phi_\beta^{-1}: \; \phi_\beta(U_\alpha \cap U_\beta) \longrightarrow \phi_\alpha(U_\alpha \cap U_\beta)$$

from open subsets of $\mathbf{R}^n$ into open subsets of $\mathbf{R}^n$ are of class $\mathcal{C}^k$ (see Fig. 1.1). The compatibility condition means that there exists a smooth (in the sense of $\mathcal{C}^k$) transition from coordinates (x^i), $i = 1,\ldots,n$, defined on U_α, to coordinates (y^i) defined on U_β, in the region on which U_α and U_β overlap, i.e.,

$$(x^i) \longmapsto (y^i) = f^i(x^i),$$

where f^i are real functions of class $\mathcal{C}^k$ (and analogously the other way round: $(y^i) \mapsto (x^i)$). k may be also infinite. If the mappings $\phi_\beta \circ \phi_\alpha^{-1}$ ($\phi_\alpha \circ \phi_\beta^{-1}$) are analytic, we write $k = \omega$.

The set of charts $\{(U_\alpha, \phi_\alpha)\}$ covering X (i.e. $X = \cup_\alpha U_\alpha$) and pairwise $\mathcal{C}^k$ consistent on overlapping domains, is said to a $\mathcal{C}^k$ *atlas A on X*. Two $\mathcal{C}^k$ atlases are said to be $\mathcal{C}^k$ *equivalent* if their union is a $\mathcal{C}^k$ atlas on X.

2.1. Definition. The pair $\mathcal{M} = (X, \{A\})$, where X is a topological space, and $\{A\}$ a set (an equivalence class, in fact) of equivalent $\mathcal{C}^k$ atlases on X, is said to be a $\mathcal{C}^k$ *differential manifold* (or $\mathcal{C}^k$ manifold, for short). If there is no danger of misunderstanding, we will often omit the indication of the differentiability class $(\mathcal{C}^k)$ in the name of manifolds. If $k = \omega$, a manifold is said to be *analytical*.

In practice, one works with one atlas treating it as a representation of the entire equivalence class. We shall often make use of this simplification.[2]

2.2. Example. The space $X = \mathbf{R}^n$ can be given differential structure by defining an atlas consisting of the one chart (U_α, ϕ), where $U = \mathbf{R}^n$, $\phi = \mathrm{id}$.

2.3. Example. To define a differential structure on a one–dimensional sphere

$$\mathcal{S}^1 = \{(x^1, x^2) \in \mathbf{R}^2 \colon (x^1)^2 + (x^2)^2 = 1\},$$

an atlas is needed consisting of at least of two charts (U_α, ϕ_α), $\alpha = 1, 2$, (see Fig. 1.2)

$$U_1 = \mathcal{S}^1 \setminus \{(-1, 0)\}, \qquad \phi_1 \colon U_1 \longrightarrow \mathcal{D}_1 \subset \mathbf{R}^1,$$
$$U_2 = \mathcal{S}^1 \setminus \{(1, 0)\}, \qquad \phi_2 \colon U_2 \longrightarrow \mathcal{D}_2 \subset \mathbf{R}^1.$$

The mappings ϕ_α can be defined in the following way,

$$\phi_1^{-1} \colon \Theta \longmapsto (\cos \Theta, \sin \Theta), \qquad -\pi < \Theta < \pi,$$
$$\phi_2^{-1} \colon \Theta \longmapsto (\cos \Theta, \sin \Theta), \qquad 0 < \Theta < 2\pi$$

(see, Thirring 1978, pp. 10–11).

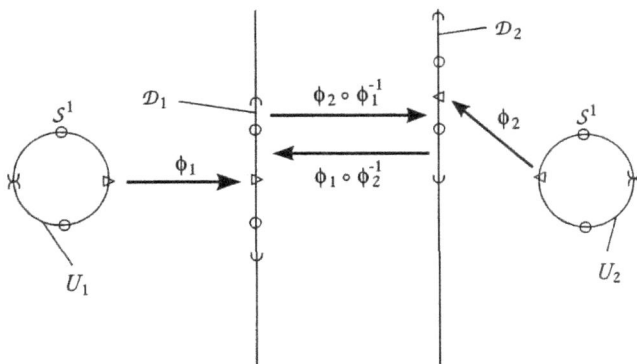

Fig. 1.2. Atlas on $\mathcal{S}^1$ For clarity the set $\mathcal{S}^1$ is shown in two copies. Little circles, triangles, and parentheses indicate which points correspond to each other.

---

[2]Against a more common usage we prefer the term *differential* rather than *differentiable* in the name $\mathcal{C}^k$ manifolds, understanding the difference between these two terms more or less in the same manner as one understands the difference between *metric* and *metrizable*. In this sense, not every topological manifold is differentiable, but if it is $\mathcal{C}^1$ differentiable, it is also $\mathcal{C}^\infty$ differentiable.

If, in the chain of definitions leading to the C^k manifold concept, the space $\mathbf{R}^n$ is replaced by its "upper" ("lower") part, i.e., by the region in $\mathbf{R}^n$ satisfying the condition $u^1 \geq 0$ ($u^1 \leq 0$), where u^1 is a coordinate in $\mathbf{R}^n$ (this region will be denoted by $\frac{1}{2}\mathbf{R}^n$), one obtains the so-called C^k *manifold with boundary*. The *boundary* $\partial\mathcal{M}$ (sometimes denoted also by $\dot{\mathcal{M}}$) of a manifold $\mathcal{M}$ with boundary is defined to be the set of those points of $\mathcal{M}$, the images of which under ϕ_α are situated on the boundary of $\frac{1}{2}\mathbf{R}^n$ in $\mathbf{R}^n$; the mappings ϕ_α define local coordinate systems on $\mathcal{M}$.

2.4. Example. Let $X = [a, b] \subset \mathbf{R}^1$. Any atlas on X must consist of at least two charts (U_α, ϕ_α), $\alpha = 1, 2$; for instance,

$$U_1 = [a, b), \qquad \phi_1 \colon x \longmapsto x - a,$$
$$U_2 = (a, b], \qquad \phi_2 \colon x \longmapsto b - x.$$

Then $\partial\mathcal{M} = \{a\} \cup \{b\}$.

Any C^k manifold can be also constructed in a different way, more clearly showing the role of the modelling space $\mathbf{R}^n$. The construction recipe is the following (see Bishop and Goldberg 1980, pp. 31–33). Prepare a suitable number of disjoint copies of $\mathbf{R}^n$, and from them cut out the open regions that will be modelling local coordinate systems on the manifold. The open regions which are to be overlapping with each other should be cut out of different copies of $\mathbf{R}^n$. Now identify those points of the open regions which are transformed into each other when the transition is made from one local coordinate system to another local coordinate system. To be more precise, on the set $\mathcal{P}$, being the union of disjoint open regions of different copies of $\mathbf{R}^n$, one defines an equivalence relation $\mathcal{E}$, treating as equivalent those points of different open regions which correspond to each other under local coordinate transformations. The manifold is to be defined as the quotient space $\mathcal{M} = \mathcal{P}/\mathcal{E}$. A topology on $\mathcal{M}$ can be introduced by demanding that local coordinate transformations be homeomorphisms. Such a topology is called a *manifold topology*. In short, a manifold can be glued together out of a sufficient number of pieces of $\mathbf{R}^n$. Whitney (1936) has proved that the two ways presented above of defining manifolds are equivalent.

2.5. Definition. Let $\mathcal{M} = (X^n, \{A\})$ and $\mathcal{N} = (Y^p, \{B\})$ be two C^k manifolds, and (U, ϕ) and (W, ψ) their local charts at points $x \in U$ and $y \in W$, respectively. Let us consider a mapping $f \colon X^n \to Y^n$ (Fig. 1.3). The expression $\psi \circ f \circ \phi^{-1}$ *represents the mapping* f in the charts (U, ϕ) and (W, ψ). The mapping f is C^r ($r \leq k$) at x if $\psi \circ f \circ \phi^{-1}$ is C^r at $\phi(x)$. It can be shown that this definition is independent of the choice of a chart. If a mapping f is C^r at each point $x \in X^n$, f is said to be of *class* C^r (or simply to be C^r).

If $Y^p = \mathbf{R}^1$ (Fig. 1.4), f is called a *function* on X^n, and the expression $f \circ \phi^{-1}$ its *representative* in the chart (U, ϕ).

2.6. Definition. If a mapping $f \colon X^n \to Y^p$ (of class C^k) is a bijection, and if both f and its inverse f^{-1} are continuously differentiable (are C^k), f is called (C^k) *diffeomorphism*. Two manifolds (of class C^k) between which there *exists* a (C^k) diffeomorphism are called (C^k) *diffeomorphic*.

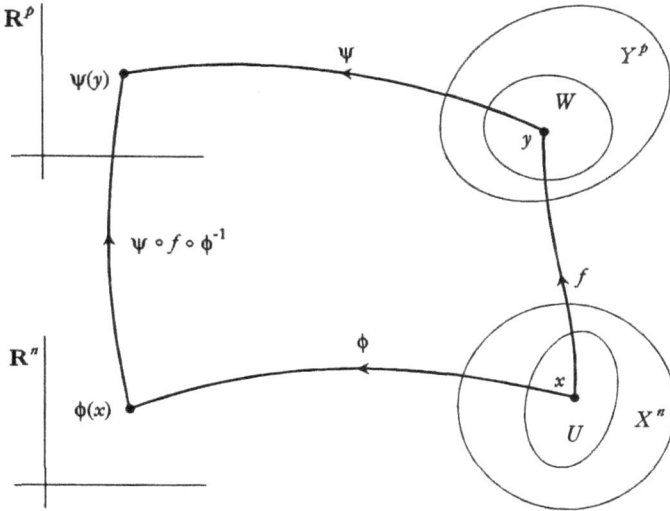

Fig. 1.3. Mapping $f: X^n \to Y^p$.

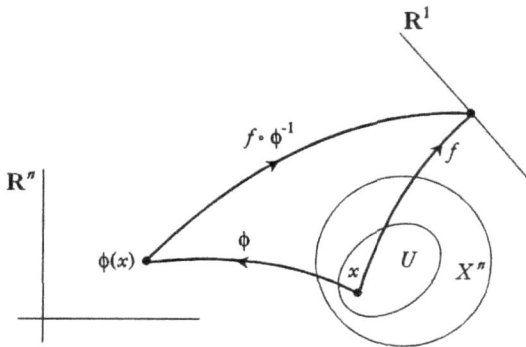

Fig. 1.4. Function f on manifold.

In the theory of manifolds, diffeomorphic manifolds are indistinguishable from each other; in other words, there exists a category, called the *category of C^k manifolds*, the objects of which are C^k manifolds and morphisms are differentiable mappings of class C^k between C^k manifolds. C^k diffeomorphisms are isomorphisms in this category. (See Trautman 1970b, 1973b; Sulanke and Wintgen 1972; concise information on the category theory is given in the Appendix, sec. 1.)

From the theorem proved by Whitney (1936) it follows that the theory of n–dimensional manifolds can be identified, in a sense, with the theory of subsets $\mathbf{R}^m$ with $m \leq 2n + 1$.

3. Orientability of Manifolds

Let (x^i) and (y^i) be local coordinate systems on open subsets of $\mathbf{R}^n$. The coordinate systems are said to have *the same orientation* if the determinant of the Jacobian matrix $\mathcal{J} = \det(\partial x^i / \partial y^j)$ is positive at each point of a given open subset of $\mathbf{R}^n$. The *orientation* of a chart (U, ϕ) of a manifold $\mathcal{M}$ is defined through the orientation of coordinate systems on $\phi(U) \subset \mathbf{R}^n$. A manifold is said to be *orientable* if there is an atlas on it such that any two charts (U_1, ϕ_1) and (U_2, ϕ_2), the domains of which overlap $(U_1 \cap U_2 \neq \emptyset)$, have the same orientation. A manifold defined in terms of such an atlas is said to be *oriented*. A manifold which is not orientable is said to be *nonorientable*.

Two atlases of a given manifold are said to possess *the same orientation* if all their charts have pairwise the same orientations. To be of the same orientation is for the atlases an equivalence relation. An equivalence class of this relation is called an *orientation* of this atlas. Every connected manifold admits exactly two orientations.

The space $\mathbf{R}^n$ considered as a manifold with an atlas consisting of one chart $(\mathbf{R}^n, \mathrm{id})$ (see example 2.2), is an orientable manifold. The Möbius strip can serve as an example of a nonorientable manifold. In the case of two–sided surfaces in the three–dimensional Euclidean space, orientability is equivalent to two–sidedness of a surface (the Möbius strip is a one–sided surface). (For more information see Brickell and Clark 1970, pp. 119–125; Bishop and Goldberg 1980, pp. 162–164).

There exists a visual method of deciding whether a given manifold is orientable or not. Let us associate a *graph* $\mathcal{G}(A)$ with each atlas $A = \{(U_\alpha, \phi_\alpha)\}$ of a manifold $\mathcal{M} = (X, \{A\})$ in the following way. Let the knots of the graph $\mathcal{G}$ correspond to the charts ϕ_α of the atlas A, and the edges of the graph, linking the corresponding knots, to the connected components of the overlapping regions $U_\alpha \cap U_\beta \neq 0$. All these correspondencies are assumed to be one–to–one. Two knots of the graph can be linked by countably many edges. To all edges of the graph let us ascribe the numbers: $+1$ if the Jacobian matrix in the connected component of the region $U_\alpha \cap U_\beta$ that corresponds to that region is positive, and -1 if this Jacobian matrix is negative. In this way, each knot is linked with itself, forming a loop to which a number $+1$ is ascribed. The sequence of edges forming a loop will be called a *cycle*. To the sequences of edges (and consequently, to cycles as well) we ascribe a number, being the product of numbers which correspond to the edges forming this sequence (this cycle). We say that the *change of orientation of a knot* takes place if all numbers, corresponding to edges emanating from this knot, are multiplied by -1. Let us notice that this procedure does not change the number corresponding to the loop emanating from, and returning to, a given knot (this number is twice multiplied by -1). A graph is said to be *orientable* if, by suitably changing the orientation of its knots, one can obtain the situation in which to all edges of the graph $\mathcal{G}$ the numbers $+1$ are ascribed.

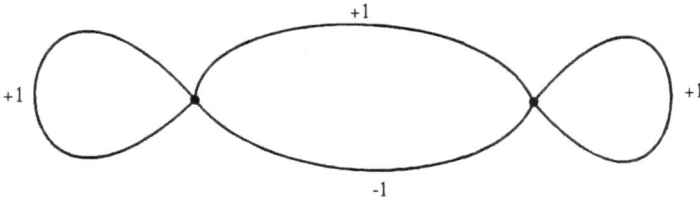

Fig. 1.5. Graph corresponding to the Möbius strip.

Of course, the change of orientation of a knot belonging to a graph $\mathcal{G}$ corresponds to the change of orientation of the chart ϕ_α associated with this knot and belonging to the atlas A of a given manifold $\mathcal{M}$. A manifold $\mathcal{M}$ is orientable if and only if the graph $\mathcal{G}$ corresponding to its atlas A is orientable. It can be shown that a graph $\mathcal{G}$ is oriented, and consequently the manifold corresponding to it is oriented, if $\mathcal{G}$ contains no cycle to which the number -1 is ascribed. (See Sulanke and Wintgen 1972).

4. Vector and Tensor Fields on Manifolds

The fact that manifolds are modelled on the spaces $\mathbf{R}^n$ implies the existence of the tangent space $\mathcal{T}_x(\mathcal{M})$ to a given manifold $\mathcal{M} = (X^n, \{A\})$ at each point $x \in X^n$ that is isomorphic to $\mathbf{R}^n$. (If misunderstanding can be excluded we shall also write $x \in \mathcal{M}$.)

Let us consider two differentiable functions f_1 and f_2 on a manifold $\mathcal{M}$. These functions are said to have the same *germ* at a point $x \in \mathcal{M}$ if there is a neighbourhood of x on which f_1 and f_2 coincide. The *germ of a function* f at x is — from the definition — the equivalence class of functions having the germs at x equal to that of f. Germs of functions at x form a vector space (with respect to addition and multiplication by real numbers). Moreover, since the product of two germs at a point x (defined as the pointwise multiplication of functions) is a germ at x, the germs at x form an algebra.

Finally, let us recall that a linear mapping satisfying the Leibniz rule is called *derivation*. Now, a vector tangent to a manifold $\mathcal{M}$ at a point $x \in \mathcal{M}$ can be defined as a derivation on the algebra of germs of differentiable functions at x. The set of all vectors tangent to $\mathcal{M}$ at x is called the *tangent space* to $\mathcal{M}$ at x and is denoted by $\mathcal{T}_x(\mathcal{M})$. Of course, tangent spaces to a manifold are vector spaces (with respect to addition and multiplication by real numbers).

One could translate the above abstract definitions into a language honoured by the handbook tradition, and say that the tangent vector v_x to a manifold $\mathcal{M}$ at a point $x \in \mathcal{M}$ is a linear mapping from the space $F(\mathcal{N})$ of differentiable functions, defined on a neighbourhood $\mathcal{N}$ of x, into the space $\mathbf{R}^1$, i.e.,

$$v_x(\alpha f + \beta g) = \alpha v_x(f) + \beta v_x(g),$$

where $\alpha, \beta \in \mathbf{R}^1$ and $f, g \in F(\mathcal{N})$; moreover, the mapping is supposed to satisfy the Leibniz rule

$$v_x(fg) = f(x)v_x(g) + g(x)v_x(f).$$

The set of all such tangent vectors to a manifold $\mathcal{M}$ at $x \in X$ is the tangent space $T_x(\mathcal{M})$.

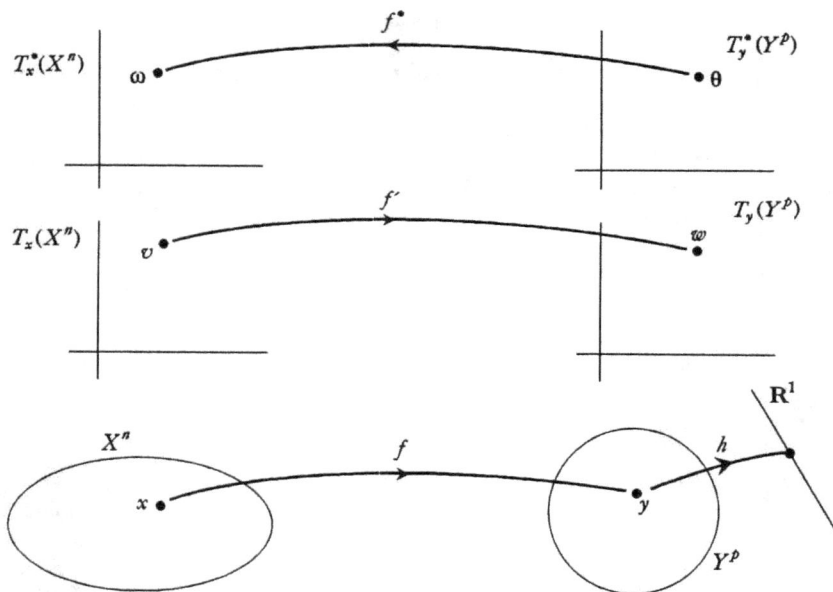

Fig. 1.6. Derivative f' and pull back f^* of a mapping f.

Let us consider the mapping

$$f \colon X^n \longrightarrow Y^p$$

differentiable at $x \in X^n$. The *derivative* of this mapping at x is defined to be a linear mapping

$$f'(x) \colon T_x(X^n) \to T_y(Y^p),$$

transforming $v \in T_x(X^n)$ into $w \in T_y(Y^p)$ in such a way that, for every function h at the point $y = f(x) \in Y^p$, one has $w(h) = v(h \circ f)$ (Fig. 1.6). It can be easily seen that under the mapping $f'(x)$ a tangent vector to a curve C in X at a point x is transformed into the tangent vector to the curve $f(C)$ in Y^p at the point y.

Let $\mathcal{V}$ and $\mathcal{W}$ be vector spaces. The set $\mathbb{L}(\mathcal{V}, \mathcal{W})$ of all linear mappings from $\mathcal{V}$ into $\mathcal{W}$ is a vector space (with addition and multiplication by scalars, defined in a natural way). The vector space $\mathbb{L}(\mathcal{V}, \mathbf{R}^1)$ is called *dual* to $\mathcal{V}$ and is denoted by $\mathcal{V}^*$. The space $\mathcal{T}_x^*(\mathcal{M})$, dual with respect to tangent space $\mathcal{T}_x(\mathcal{M})$, is called the *cotangent space* to the manifold $\mathcal{M}$ at a point x; its elements are called *covectors*. The following names are also used: *contravariant vectors* — for elements of $\mathcal{T}_x(\mathcal{M})$, and *covariant vectors* — for elements of $\mathcal{T}_x^*(\mathcal{M})$.

Let us consider a differentiable mapping

$$f\colon X^n \longrightarrow Y^p$$

and a covariant vector $\theta \in \mathcal{T}_y^*(Y^p)$. We define the *pull back* of θ (induced by the mapping f) as a linear mapping

$$f^*\colon \mathcal{T}_y^*(Y^p) \longrightarrow \mathcal{T}_x^*(X^n),$$

transforming θ into $\omega \in \mathcal{T}_x^*(\mathcal{M})$ in such a way that, for any $v \in \mathcal{T}_x(X^n)$, the functional equality $(f^*\theta)v = \theta(f'v) \circ f$ is satisfied (or the numerical equality $(f^*\theta)_x v_x = \theta_y(f'v)_y$ (see Fig. 1.6)).

And consequently, the *pull back of a function* g at a point $y = f(x) \in Y^p$ (induced by the mapping f) is defined as $f^*(g) = g \circ f$ (or more precisely: $f^*(g) \mapsto (g \circ f)(x)$).

Let $\mathcal{V}_1, \ldots, \mathcal{V}_n$ and $\mathcal{W}$ be vector spaces. A mapping $f\colon \mathcal{V}_1 \times \cdots \times \mathcal{V}_n \to \mathcal{W}$, linear with respect to each variable, is said to be a *multilinear mapping*. Let $\tau_i \in \mathcal{V}_i$, $i = 1, \ldots, n$. We define the multilinear mapping

$$\tau_1 \otimes \cdots \otimes \tau_n\colon \mathcal{V}_1 \times \cdots \times \mathcal{V}_n \longrightarrow \mathbf{R}^1$$

by

$$\tau_1 \otimes \cdots \otimes \tau_n(v_1, \ldots, v_n) = (\tau_1(v_1)), \ldots, (\tau_n(v_n)),$$

where $v_i \in \mathcal{V}_i$. This function is called the *tensor product*.

4.1. Definition. A real multilinear function

$$f\colon \underbrace{\mathcal{V}^* \times \cdots \times \mathcal{V}^*}_{r \text{ times}} \times \underbrace{\mathcal{V} \times \cdots \times \mathcal{V}}_{s \text{ times}} \longrightarrow \mathbf{R}^1$$

is called *a tensor of the type* (r, s) (over the space $\mathcal{V}$). Tensors of the type (r, s) form the vector space

$$\underbrace{\mathcal{V} \otimes \cdots \otimes \mathcal{V}}_{r \text{ times}} \otimes \underbrace{\mathcal{V}^* \otimes \cdots \otimes \mathcal{V}^*}_{s \text{ times}} = T^r{}_s(\mathcal{V})$$

(or in the abbreviated form $\otimes^r \mathcal{V} \otimes^s \mathcal{V}^* = T^r{}_s(\mathcal{V})$). If the dimension of $\mathcal{V}$ is n, $\dim \mathcal{V} = n$, then $\dim T^r{}_s(\mathcal{V}) = n^{r+s}$. Tensors of the type $(0,0)$ are scalars, tensors of the type $(1,0)$ are contravariant vectors, and tensors of the type $(0,1)$ are covariant vectors.

A *tensor of the type* (r,s) *on a manifold* $\mathcal{M}$ is defined to be a real multilinear function on

$$\otimes^r T_x(\mathcal{M}) \otimes^s T_x^*(\mathcal{M}) = T^r{}_s(T_x(\mathcal{M})).$$

A *tensor field* T *of the type* (r,s) *on a manifold* $\mathcal{M}$ is defined to be a mapping $\mathcal{M} \to T^r{}_s(\mathcal{M})$ such that, for every $x \in \mathcal{M}$, one has $T(x) \in T^r{}_s(T_x(\mathcal{M}))$. A tensor field of the type $(1,0)$ on $\mathcal{M}$ is a field of contravariant vectors on $\mathcal{M}$; a tensor field of the type $(0,1)$ (of the class $\mathcal{C}^k$) is also called a *one–form*. A tensor field is called *symmetric* (*antisymmetric* or *skew–symmetric*) if, at each point $x \in \mathcal{M}$, $T(x)$ is a symmetric (antisymmetric) tensor. Totally antisymmetric tensor field of the type $(0,p)$ (of the class $\mathcal{C}^k$) is called a *p–form*.

The pull back mapping can be generalized to any p–form. Let f be a differentiable mapping

$$f\colon X^n \to Y^q,$$

and ω any p–form on Y^q. The *pull back* $f^*\omega$ of the p–form ω (induced by the mapping f) is defined in the following way,

$$f^*(\omega)(v_1,\ldots,v_p) = \omega(f'v_1,\ldots,f'v_p),$$

where $v_i \in T_x(X^n)$, and $f'v_i \in T_{f(x)}(Y^q)$.

5. Submanifolds

Let $f\colon \mathcal{N}^q \to \mathcal{M}^p$ be a differentiable mapping from a differential manifold $\mathcal{N}^q$ to a differential manifold $\mathcal{M}^p$. This mapping is said to be of *rank* r at a point $y \in \mathcal{N}^q$ if the dimension of $f'(T_y(\mathcal{N}^q))$ is r. The following, almost self–evident, relationships hold:

if the rank of f at y is q $(= \dim \mathcal{N}^q)$, the induced mapping f' at y is injective, and $q \leq p$;

if the rank of f at y is p $(= \dim \mathcal{M}^p)$, the mapping f' at y is surjective, and $q \geq p$.

If the rank of the mapping f (of the class $\mathcal{C}^k$) at each point of the manifold $\mathcal{N}^q$ is q, f is called the *immersion* (of the class $\mathcal{C}^k$), and the manifold $\mathcal{N}^q$ is said to be *immersed* in the manifold $\mathcal{M}^p$. Let us note that the set $f(\mathcal{N}^q)$ need not carry the manifold structure since f need not be injective (however, if f is an immersion, f' is always an injection; this immediately follows from the above statements and definitions). Immersion that is also an injection is called *embedding*, and one correspondingly speaks of an *embedded* manifold into one another. If f is an embedding, the set $f(\mathcal{N}^q) \subset \mathcal{M}^p$ with the differential structure induced by f is a manifold, although this differential structure need not be consistent with the original differential structure on

$\mathcal{M}^p$; if it is consistent, f is said to be a *regular embedding*. If f is a regular embedding, $f(\mathcal{N}^p)$ is called[3] a *submanifold* of $\mathcal{M}^p$.

An open subset of a manifold $\mathcal{M}^p$ considered, in a natural way, as a submanifold of $\mathcal{M}^p$, is called an *open submanifold* of $\mathcal{M}^p$

6. Manifold as a Model of Space–Time

All macroscopic theories of contemporary physics — and theories that have played important roles in the history of physics (provided they have been reconstructed in the language of modern mathematics, see Appendix) — presuppose that the arena in which physical processes occur has the structure of a four–dimensional differential manifold. Points of this manifold are called *events*. Events are determined by their four coordinates in local charts. Local charts are treated as local reference frames (see below, sec. 7, A). One of the coordinates in each of such charts is considered to be the time coordinate, the remaining three — space coordinates. This distinction of time and space coordinates has a purely local meaning: it is defined only in a local chart. One should expect that the possibility to cover the entire manifold with a single chart (i.e. the possibility of an atlas consisting of one chart, as it is the case in $\mathbf{R}^n$) is a rare exception. Since from the mathematical point of view time and space coordinates are on equal footing, one often prefers speaking of four coordinates of events or — to be more precise — saying that a four–dimensional differential manifold serves as a *mathematical model of the physical space–time*. With less care, but more easy, one says simply that the space–time is a four–dimensional manifold. Therefore, it can be asserted that space–time is a set of all possible events, where by an event one should mean an "idealized occurrence in the physical world having extension neither in space nor in time" (Geroch 1978, p. 3), for instance, an idealization of an elastic collision of two particles.

The four–dimensionality of space–time is an empirical fact. A physical counterpart of the dimension is a number of the degrees of freedom of a physical system. For instance, three degrees of freedom in a translational motion correspond to the three–dimensionality of space. The possibility to move along only one temporal direction is reflected in the one–dimensionality of time. Let us notice, however, that the "arrow" along this direction (the so–called *arrow of time*) has not been determined by the model. The fact (known from the very personal experience) that physical processes can "move" only from the past to the future has no explanation within the present model. At most, we can, in a purely conventional manner, call one end of the temporal direction the *future*, and the other and — the *past*. In the following, we shall often make use of this convention.

---

[3]One should note the difference in terminology used by different authors (see, for example, Choquet–Bruhat et al. 1982, pp. 239–242; Bishop and Goldberg 1980, pp. 40–43; Kobayashi and Nomizu 1963, chapt. 1, sec. 1; Brickell and Clark 1970, pp. 68–83; O'Neill 1983, pp. 15–21; Hawking and Ellis 1973, pp. 23–24).

Usually, it is assumed that a manifold modelling the physical space–time is orientable, and, additionally, that it is separately time and space orientable (this will be defined in chap. 2, sec. 1; see also below sec. 7, B, and sec. 1. 3 in Sachs and Wu 1977).

The manifold model ascribes to space–time not only the continuity but also (in general) a "sufficiently smooth" continuity. Usually, it is assumed that all mappings appearing in the definition are of class C^k, where k is as large a number as it is required by a given physical situation. Most often, one simply assumes[4] that a modelling manifold is of class C^∞

The postulate of continuity leads to the question of the status of this concept in microphysics. According to quite a common opinion, it is a smooth space–time manifold that continues to play the role of an arena for physical processes in quantum mechanics. However, one could quote convincing reasons supporting the view that a smooth space–time is a macroscopic concept: we coin this concept within the world of macrophysics, in which all measurements are performed providing us with information about the micro–world, and then "projecting it down" into the domain of quantum phenomena. Such a view is corroborated by an analysis of the mathematical structure of quantum mechanics. Unfortunately, this problem goes beyond the limits of the present book: in the following — except for sporadic remarks — we will deal only with space–times of macroscopic physical theories which are always modelled by smooth manifolds.

Finally, one should note an interesting methodological problem: can the manifold model of space–time be defined operationally (or at least quasi–operationally)? And, if the answer is positive, how can it be defined? According to Trautman's (1964) recipe, a space–time manifold can be constructed with the help of "negligibly small" clocks, the relative positions of which are parametrised by triples of real numbers. In the construction proposed by Synge (1965), a fundamental tool is again a clock, and a fundamental operation — a time measurement. However, this author tacitly assumed the existence of signals that would inform the observer on distant events. A construction, free from such gaps, was proposed by Ehlers, Pirani and Schild (1972). In their approach, free fall and light propagation serve as "elementary operations" (see also Ehlers 1973). These operations allow one to construct not only the manifold, but also the metric structure of space–time (see, next chapter and Heller 1977).

7. *Comments and Remarks*

A. Local charts on a manifold correspond to local frames of reference in space–time (although not all coordinate transformations a physicist would consider as changing to new frames of reference (see, Dixon 1978, pp. 2–7). Full discusion of this problem

---

[4]Some authors, when speaking of a C^k space–time, have in mind the fact that the metric defined on this space–time is of class C^4 (see, for instance, Tipler et al. 1980, pp. 155–157).

is possible only after a metric has been introduced on a manifold (see, chapter 2, sec. 5, D). The fact that a manifold admits — on the strength of its definition — all atlases equivalent to a given atlas (see above, sec. 2) shows a certain "relativity" of the space–time model discussed here: the very concept of space–time includes all possible classes of reference frames consistent with it, and none of these classes is *a priori* distinguished in any way. Some classes of reference frames may become privileged only afterwards, when the space–time considered is enriched by additional postulates which should always be physically motivated.

B. The problem of space–time orientability, space orientability and time orientability belongs to the most fundamental problems of physics and cosmology. Having in mind that at the present stage of our analysis we have no "generally valid" method to split space–time into space and time separately, let us refer to interesting remarks of a mathematician, Laurent Schwartz (1967, vol. 2, chapt. 6, sec. 5), who after asking the question

"Is our universe an orientable manifold?" continues: Let us put aside the relativistic point of view that teaches us to look at the universe as a four–dimensional manifold, and let us assume that our world is modelled by a three–dimensional C^∞ manifold. Is it orientable? Do the known laws of physics allow us to equip it with a canonical orientation? To begin, let us assume that our world, considered as a three–dimensional manifold, is not orientable. This means that in the Universe there exist some paths, similar to those in the Möbius strip, such that starting a journey along one of them with a certain orientation, and pursuing the orientation continously during the travel, one comes back to the staring point with the opposite orientation. A human being travelling throughout the Universe along such a path, after coming back to the Earth, would find that its heart is at the right side of its body, and its liver at the left side. Letters in a book, read during the travel, would now run from right to left. All this without any sudden change experienced by the traveller during his trip. Anyway, after his coming back he would judge himself absolutely normal and "invariant" On the other hand, he would think that all phenomena on the Earth had been "reversed" during his absence. Schwartz concludes that all arguments for the world's orientability, quoted in elementary textbooks of physics and mathematics, rest on a very weak basis.

C. From the times of Einstein, there exists in physics a set of ideas called the *doctrine* or the *principle of Mach*. These ideas go back to Leibniz and Berkeley; however, they were read out by Einstein from the works of Ernst Mach in which this author criticised the Newtonian mechanics. Roughly speaking, Mach's principle postulates a causal relationship between the local physics and the global structure of space–time (for an account of Mach's original doctrine, see Heller 1988). In its more maximalistic formulations Mach's principle requires that the global structure of the world should *uniquely* determine the local physics[5]. (For more details see Goenner 1970, Reinhardt 1973, Heller 1975a, Raine and Heller 1981, Raine 1981a, Barbour 1983). As the number of quoted works testifies (and this is only an "unfair sample" of the continuously increasing literature), discussions on the subject are still going on. It seems that the status of Mach's principle in the theory of general relativity

---

[5]In the original Einstein's formulation Mach's principle postulates the "relativity of mass" the mass of a test body should be uniquely determined by the distribution and motions of all masses in the Universe.

has been clarified in the works of Raine (1975, 1981a), but Mach's ideology manifests some tendencies to go beyond this theory and to provide new inspirations in creating new physical theories (see, for example, Barbour 1974, 1983; Barbour and Bertotti 1977; Raine and Heller, chapter 13). I do not want now to go deeper into this set of problems; I would like only to note that the manifold model of space–time imposes substantial restrictions upon the possibilities to implement Mach's doctrine into a physical theory. Any manifold has — from its very definition — local topological and differential properties identical with those of the modelling space R^n. Although this circumstance leaves a certain freedom to manipulate the local structure of space–time with the help of field quations (see, Raine 1975, 1981a), this does not change the fact that space–time must locally be R^n irrespectively of its global structure. This limitation can be removed only by postponing modeling space–time by a differental manifold (see, below, comment E).

D. The manifold concept has been defined in terms of the modelling space R^n. A manifold can be also constructed by suitably gluing copies of R^n (see, above, sec. 2). In this sense, the arithmetical space R^n is *a priori* with respect to the manifold concept. However, it is not a logical necessity, but first the result of our decision on what a manifold is, and second the question of agreement of this decision with empirical data. Under the present precision of empirical data the model can, in principle, be discarded and replaced by a new one. Kant was wrong when asserting that the Euclidean character of space is an *a priori* condition of any experiment.

E. A host of empirical data testifies that, within the domain controlled by macro-physics, there is no need to give up the manifold model of space–time, although it is not impossible — as we have seen — that, in microphysical theories, space–time appears from outside, as it were, namely as a consequence of the fact that an observer performs all his measurements being immersed in the macroscopic space–time. The necessity to go beyond the manifold model of space–time could appear within a domain of quantum gravity. There are strong reasons compelling one to believe that a quantisation of gravity — first of all, at very early stages of the world's evolution, in the vicinity of the initial singularity — could hardly be avoided. Although some other proposals are known, it seems highly probable that the quantisation procedure would require giving up (or rather to generalize) the manifold model of space–time. As an example of this view, let us quote from Trautman's work:

> "The topological and differential structures of space–time do not seem to possess a well–defined operational meaning. Therefore, it is likely that they will have to be abandoned, or rather replaced by another structure which will be more closely related to, and influenced by, physical phenomena than the absolute, locally Euclidean manifold structure of space–time assumed in all current theories. In my opinion, a satisfactory quantum theory of space, time and gravitation will have to do away with the notion of a differentiable manifold as a model of space–time" (Trautman 1973b, p. 183).

Of course, one should expect that the future, long-expected quantum theory of gravity would give, as its limiting case, the present manifold model of space–time. For this reason this model ought to be regarded as a permanent achievement of contemporary physics.

Bibliographical Note. Fundamental information on Banach spaces could be found in the two–volume book of Maurin (1976, 1980), a more extensive study of functional analysis in the classical monograph by Riesz and Nagy (1972), and a very transparent introduction in the book by Roman (1975, vol. 2). Chapter 7A of the book by Choquet–Bruhat et al. (1982) gives a short account of differential manifolds modelled on infinitely–dimensional topological spaces.

The concept of (finitely–dimensional) differential manifold is presented in all handbooks of modern differential geometry. From among many excellent books in this field, the two volumes by Kobayashi and Nomizu (1963) merit our special attention. This book is often quoted by physicists, and many works of differential geometry written after 1963 are based on it and often contain generalizations or applications of its results. More concise, but also very good, accounts of differential geometry can be found in Bishop and Crittenden (1964) or Hicks (1965). The books by Auslander and MacKenzie (1963), and by Brickell and Clark (1970) are devoted entirely to manifolds; the last one contains many useful examples. The reader interested in an introductory exposition should consult Wallace (1968) or O'Neill (1966). The first of these books deals with a topological aspect of manifolds, the second is an elementary, but modern, exposition of differential geometry with a number of computational exercises; the manifold concept is introduced gradually as a generalization of the concept of plane in the Euclidean space. The book by Schutz (1984) is especially recommended because of its didactic values. The reader is introduced to both modern geometric structures and computational techniques. The selection of the material has been made in view of future applications in physics; such applications are also discussed. Conceptual foundations of the manifold theory are very clearly discussed in a little book by Isham (1989).

The textbook of differential geometry written by Sikorski (1972) is exceptional in a certain sense. The exposition of differential geometry is here based on a concept of differential space, introduced by the author. A *differential space* is defined to be the pair $(\mathcal{M}, \mathcal{C})$ where $\mathcal{M}$ is any set, and $\mathcal{C}$ a family of real functions on $\mathcal{M}$, satisfying two conditions that axiomatically define the "smoothness" of functions belonging to $\mathcal{C}$, namely: (1) $\mathcal{C}$ is closed with respect to localization, (2) $\mathcal{C}$ is closed with respect to superposition with Euclidean functions. To be more precise, let $(\mathcal{M}, \tau)$ be any topological space, and $\mathcal{C}$ any non–empty family of functions on $\mathcal{M}$. A function f, defined on $\mathcal{N} \subset \mathcal{M}$, is said to be a *local $\mathcal{C}$-function* if, for every $p \in \mathcal{N}$, there is a neighbourhood $\mathcal{U}$ in the topological space $(\mathcal{N}, \tau_N)$, where τ_N is a topology induced in $\mathcal{N}$ by τ, and a function $g \in \mathcal{C}$ such that $f|_U = g|_U$. The set of local $\mathcal{C}$–functions is denoted by $\mathcal{C}_A$; one obviously has $\mathcal{C} \subset \mathcal{C}_M$. The family of functions $\mathcal{C}$ is said to be *closed with respect to localization* if $\mathcal{C} = \mathcal{C}_M$. Let, further, $\mathcal{E}$ be the set of all $\mathcal{C}^\infty$ functions defined on $\mathbf{R}^n$ called *smooth Euclidean functions*. $\mathcal{C}$ is said to be *closed with respect to superposition with smooth Euclidean functions* if, for any function $\omega \in \mathcal{E}$ and any $n \in \mathbf{N}$, the fact that $f_1, \ldots, f_n \in \mathcal{C}$ implies that $\omega \circ (f_1, \ldots, f_n) \in \mathcal{C}$. The differential space concept is a vast generalization of the manifold concept. *Every* subset of $\mathbf{R}^n$ is a differential space, but there exist many differential spaces which can be embedded in no $\mathbf{R}^n$ (even if $n = \infty$). In particular, each differential manifold is a differential space. A diffeomorphism between two differential spaces can be defined in a natural way. By postulating that a differential space is locally diffeomorphic to a Euclidean space, one recovers the concept of a differential manifold. It turns out that, in spite of its generality, differential space is a workable concept. In particular, one can meaningfully define a tangent space to a differential space at any of its points. Large parts of differential geometry can be developed on differential spaces proper (i.e. on such differential spaces that are not manifolds). It is worthwhile to notice that Walczak and Waliszewski (1981) have published a collection of exercises (in general, difficult), especially adapted to Sikorski's book. Penrose and Rindler (1984) define space–time manifold as a pair $(\mathcal{M}, \mathcal{C})$, without however, generalizing it to space–time differential spaces.

In Gruszczak et al. (1988, 1989) a proposal has been worked out to model space–time by a differential space rather than by a differential manifold. Such a generalized space–time model can find its applications in a quantum gravity regime (when space–time is still a differential space but already not a differential manifold), and seems to be very promising in describing space–time singularities. Unfortunately, Sikorski (1972), and Walczak and Waliszewski (1981) exist only in Polish. For a foreign reader an extensive review by Heller et al. (1989) could be useful of differential space methods and its possible applications to space–time geometry.

There exist many expositions of differential geometry, written especially for physicists. Those by Choquet–Bruhat et al. (1982), Bishop and Goldberg (1980), and O'Neill (1983) are the best for our purposes. Useful nformation concerning the differential structure can be found in the second volume of the book by Richtmyer (1981, chapters 23–28). The reader interested in a short review of modern methods in differential geometry should consult a paper by Schmidt (1972) or introductory chapters of such relativity theory textbooks as: Misner et al. (1973), Schutz (1985), Ryan and Shepley (1975); the last one is on a more advanced level than the first two. Thirring (1978) contains a transparent presentation of elementary concepts of the manifold theory (with applications to classical physics) and a number of instructive exercises.

The manifold model of space–time has been submitted to a detailed analysis by Hawking and Ellis (1973), and Trautman (1964, 1968, 1970a, 1973b, 1976, 1978). See also the book by Raine and Heller (1981), and the works of the present author (Heller 1981a, 1986).

Chapter 2

Relativistic Model of Space–Time

0. Introduction

The space–time model introduced in the previous chapter is very general. To this property it owes its applications in all major theories of macroscopic physics. The manifold structure ensures, among others, the existence of local coordinate systems on a manifold, and the possibility to do calculus on it. Because of that, concepts such as: position, velocity and acceleration are well determined (at least locally) in all macroscopic space–times. However, the manifold model is too poor to allow the full physics to "develop" on it. In particular, such important concepts as space and time measurements cannot be defined within the conceptual framework presupposed by this model. Space and time measurements become meaningful only after introducing a *metric* on a manifold (or on certain of its submanifolds, as it takes place, for instance, in the case of space–time of classical mechanics; see Appendix). The remark in parentheses suggests that various physical theories will differ from each other in the way they introduce a metric structure on a space–time manifold. In the present chapter, we will analyse the metric structure demanded by the relativistic theories of space–time, i.e, by special and general relativity theories, that is to say the *Lorentz metric structure*.

The question "what are the (necessary and sufficient) conditions of the existence of the Lorentz metric on a differential manifold?" is of uppermost importance for theoretical physics. If we agree with Maxwell's saying that in physics one does not know of which one speaks unless one has a recipe *how to measure*, we must admit that there is no meaning in speaking of *space–time physics* as long as there is no metric defined on a corresponding space–time manifold. Therefore, the question of the Lorentz metric existence is, in a sense, the question of the possibility of physics. The metric structure is very rich. It contains several other structures (conformal, projective, affine, ...). These structures can be introduced axiomatically starting from the bare topological and differential structures. In this way, by gradually enriching the construction with new properties, ensured by subsequent axioms, one finally obtains the full Lorentz manifold (see Ehlers et al. 1972). However, in this chapter we will adopt a different strategy: we introduce a "ready–made" Lorentz metric on a manifold which implicitly contains all other afore–mentioned structures; they will be used and, if necessary, further analysed in subsequent parts of the present book.

Mathematical richness of the Lorentz structure is paralleled by its paramount importance in physics. As it is expressed by Sachs and Wu (1977b, p. 12):

"[The Lorentz metric] g somehow remembers the right things and forgets the wrong ones. Concepts of causality, distance, time, velocity, speed, acceleration, rotation, rigidity, simultaneity, orthogonality, gravity, and so on, are derived from g to the extent they are retained at all. g therefore must play many roles; its unifying power is remarkable."

In the present chapter, the organization of the material is the following. Section 1 introduces a metric on a manifold, changing it into a metric manifold, and sec. 2 gives a physical interpretation to this procedure. Section 3 presents, in a more detailed manner, a special case of a metric space–time model, namely the Minkowski space–time which constitutes a geometric arena for the special theory of relativity. Finally, in sec. 4, necessary and sufficient conditions are given for the existence of a Lorentz metric on a differential manifold. Besides the standard treatment of this problem, its formulation in the language of cohomology groups is briefly presented. In *Comments and Remarks* (sec. 5) we elucidate a relationship between the existence of the Lorentz structure on a manifold and the concept of time. Results obtained in this chapter seem to suggest that a certain form of temporality is a necessary condition for physics to happen on a manifold.

1. Metric Manifolds

1.1. Definition. Let X be a differential manifold. A C^k *metric* (or C^k *metric field*) on X is a C^k tensor field of the type $(0,2)$ on X, satisfying the following conditions:
(1) g is symmetric, i.e., for every $x \in X$, the tensor g_x (g at a point x) is a symmetric tensor;
(2) for every $x \in X$, the bilinear form $g_x(.\, ,\, .)$ is non–degenerate, i.e., $g_x(v, w) = 0$, $v, w \in T_x(X)$, for all v, if and only if $w = 0$.

One speaks also of the *metric tensor* g. A differential manifold with a metric, defined in this way, is said to be a *Riemann manifold* (or a *Riemannian manifold*). Sometimes one says that a manifold X is equipped with a *Riemann structure* (or a *Riemannian structure*). A Riemann manifold is said to be a *proper Riemannian manifold* if, for every $0 \neq v \in T_x(X)$, $x \in X$, one has $g_x(v, v) > 0$. If this is not the case, one speaks of a *pseudoriemannian manifold*.

A metric tensor g equips the tangent vector spaces $T_x(X)$, $x \in X$, with scalar products $g_x(v, w) = g_{ij}v^iw^j$, $v, w \in T_x(X)$. As it is well–known, in the Euclidean space, with the help of the scalar product, one can measure lengths of vectors and angles between vectors. These operations are naturally transferred to Riemannian manifolds. In particular, the norm ("length") $\|v\|$ of a vector $v \in T_x(X)$ is defined as $\|v\|^2 = g_x(v, v) = g_{ij}v^iv^j$. In pseudoriemannian manifolds, it can be $\|v\| = 0$, for $v \neq 0$; such a vector v is called *null* (or *lightlike*).

With a suitable choice of a basis (e_i) for $T_x(X)$ the quadratic form $q = g_x(v, v)$ can be written as

$$q = g_{ij}v^iv^j = -\sum_{i=1}^{k}(v^i)^2 + \sum_{i=k+1}^{n}(v^i)^2.$$

The quadratic form q is often denoted by ds^2 and is also called a *line element* of the manifold $\mathcal{M}$. The basis, in which the quadratic form $g_x(v, v)$ assumes the above form, will be called *pseudoorthogonal*. The number k, often denoted by I also, is called the *index* of q; and the number $(n - I) - I$ its *signature*.[1] Because of the metric continuity, the index and signature are the same for all $x \in X$, provided the manifold X is connected. If $I = 0$ (or $I = n$), the manifold is of course a proper Riemannian manifold.

1.2. Definition. A manifold for which $I = 1$ (or $I = n - 1$) is said to be a *Lorentz manifold* (*Lorentzian* or sometimes *hyperbolic*). Vector spaces $T_x(X)$ tangent to Lorentz manifolds are said to be *Minkowski manifolds*. Correspondingly, one speaks of *Lorentz* or *Minkowski metrics*.

All non–zero vectors $v \in T(X)$, where X is a Lorentz manifold, are classified (independently of the choice of a basis) into *timelike, null*, and *spacelike* according to whether $g_x(v, v)$ is less than, equal to, or larger than zero. The set of all null vectors $L_x \in T_x(X)$ is called a *light cone* at x. The light cone L_x separates all timelike vectors in $T_x(X)$ from all spacelike vectors in $T_x(X)$. The set T_x of all timelike vectors in $T_x(X)$ is an open set in $T_x(X)$ and has two connected components, T_x^+ and T_x^-, which are **conventionally** called the *future* and the *past* of x. Analogously, the light cone L_x has two connected components, L_x^+ and L_x^-, which are called the *future light cone* and the *past light cone* of x, respectively.

The set $T \equiv \bigcup_x T_x$, $x \in X$, of all timelike vectors in a Lorentz manifold X can have one or two connected components (see Sachs and Wu 1977b, 1.2). Let X be a connected Lorentz manifold. If the set T has two components, X is said to be *time orientable*; if T has one component, X is said to be *time nonorientable*. If one of the two components of T is defined to be *future oriented*, and the remaining one as *past oriented*, the manifold X is called *time oriented*. One should notice that the geometry itself does not distinguish any of the two possible time orientations. At this stage of our analysis the choice of one of them is a matter of convention.

Let X^n and Y^n be differential manifolds with metrics g_1 and g_2, correspondingly. A mapping $f: X^n \to Y^n$, which is a diffeomorphism and preserves the metric, i.e., $g_1 = f^* g_2$, is said to be an *isometric mapping* or shortly an *isometry*; in such a case, the manifolds X^n and Y^n are said to be *isometric*. A Riemann manifold, isometric with the manifold $(\mathbf{R}^n, \mathrm{id})$ with the metric $g = \sum_{i=1}^{n} \varepsilon_i dx^i dx^i$, where $\varepsilon_i = \pm 1$, is called a *flat space*. If $\varepsilon = +1$, for every i, the manifold is called a *Euclidean space* and is usually denoted by $\mathcal{E}^n$. Minkowski spaces have the metric

$$\eta = -dx^0 dx^0 + \sum_{i=1}^{i-1} dx^i dx^i,$$

therefore, they are *flat spaces*. The bilinear form η is called a *Minkowski metric*.

---

[1] One should notice different definitions of index and signature by various authors, and also different conventions concerning signs of some geometric formulae. A list of these conventions can be found on the inner side of the cover of the textbook by Misner et al. (1973).

2. Metric Model of Space–Time

Differential manifolds equipped with a metric structure model spaces or space–times
of many macroscopic physical theories. For example, instantaneous spaces of classical mechanics are proper Riemann manifolds, more precisely, three–dimensional
Euclidean spaces. One should note, however, that space–time of classical mechanics
is not a metric manifold, since there is no meaningful operation in it allowing one to
measure space–time intervals between pairs of non–simultaneous events (see Raine
and Heller 1981, chapt. 7, and Appendix at the end of the present book). Physical spaces modelled by metric manifolds need not be Euclidean spaces. For instance,
instantaneous spaces of the Friedman–Lemaître cosmological models are three–dimensional, not necessarily flat, Riemann manifolds (see, e.g., Bondi 1965, Weinberg 1972,
Sciama 1975, Raine 1981b).

Successes of special and general relativity have convincingly demonstrated the
efficiency of Lorentz manifolds in modelling the physical space–time. The Lorentz
structure is much more subtle than the proper Riemann structure, and it turns out
that many important physical results are but almost trivial conclusions following
directly from the Lorentz structure.[2] With no exaggeration one can say that four-
dimensional manifolds with a Lorentz structure play the role of a standard model
of the macroscopic space–time in contemporary physics. In the rest of this book
we shall exploit this model. Now, we make a convention that — unless something
to the contrary is directly expressed — by a physical model of space–time we shall
understand the pair $(\mathcal{M}, g)$, where $\mathcal{M}$ is a four–dimensional $\mathcal{C}^k$ manifold with respect
to which all remarks of chapter 1, sec. 6 are valid, and g is a $\mathcal{C}^r$ Lorentz metric,
with the $+2$ signature, defined on $\mathcal{M}$.[3] We shall assume — unless directly expressed
otherwise — that both the manifold $\mathcal{M}$ itself and the Lorentz metric g, on it are
continuously differentiable as many times as it is necessary; in practice, it means
that $\mathcal{M}$ and g are of class $\mathcal{C}^\infty$. The pair $(\mathcal{M}, g)$ will be called a *relativistic model of
space–time*.

Now, some specifics must be introduced. As it can be easily checked, any
isometry is an equivalence relation and, strictly speaking, it is not a single pair $(\mathcal{M}, g)$
but an equivalence class of such pairs that is a model of space–time; two pairs $(\mathcal{M}_1, g_1)$
and $(\mathcal{M}_2, g_2)$ are equivalent, if they are isometric, i.e. if there is a mapping $f: \mathcal{M}_1 \to
\mathcal{M}_2$ such that $f^* g_2 = g_1$. A metric model of space–time does not distinguish between
two manifolds if they are isometric. We should keep this in mind, although one usually
works only with one pair $(\mathcal{M}, g)$ treating it, in a sense, as a representative of the entire
equivalence class (see Hawking and Ellis, p. 56). Traditional physics is interested in
local properties of space–time; cosmology, from its very nature, contemplates global
properties of space–time. And here we are met with a subtle, but rich in consequences,

---

[2]One can see it clearly in the book by Sachs and Wu (1977b) aimed at presenting general relativity
theory in the form of a mathematical text, as far as it is possible. In consequence of this endeavour
significant parts of the theory changed into concise corollaries of mathematical theorems.

[3]From the manifold definition it follows that $r \leq k$.

difficulty: how to guarantee that we are dealing with the "whole of space–time", that no part of if has been "artificially" cut off? First of all, one must make more precise intuitions contained in this question. The following definitions serve this end. A space–time $(\mathcal{M}', g')$ is said to be a C^r *extension* of a space–time $(\mathcal{M}, g)$ if there is an isometric C^r embedding of the manifold $\mathcal{M}$ into the manifold $\mathcal{M}'$, $f: \mathcal{M} \to \mathcal{M}'$. A space–time is said to be C^r *inextensible* if, for every its isometric C^r embedding, $M \to M'$, one has $f(M) = M'$, i.e. space–time is inextensible if it is identical with every its extension (see Hawking and Ellis 1973, pp. 58–59). Geroch (1970) and Clarke (1976) have shown that every Lorentz manifold can be extended as long as it becomes inextensible.

Inextensibility is a mathematical counterpart of the intuitive concept of "totality" of space–time. In the following, we shall assume — unless directly stated otherwise — that all space–times we shall be dealing with are inextensible in a suitable class of smoothness. It is evident that the specification of this class of smoothness is crucial. If we diminish smoothness requirements, an inextensible space–time may become extensible. The question how to suitably choose the smoothness condition has far–reaching consequences in cosmology (see Tipler et al. 1980, pp. 155–158).

Usually, one ascribes some other properties to a metric manifold as a space–time model (besides its inextensibility), for instance: orientability, orientation (see chapt. 1, secs. 6 and 7B), time orientability, etc. (see Sachs and Wu 1977b, sec. 1.3). At the present state of our analysis we shall not assume such additional postulates; they will be discussed and, if necessary assumed, later on along with development of our presentation.

Only after a metric model of space–time has been established, the question becomes meaningful: which are the *largest* and the *smallest* scales for which this models preserves its validity? The question about the upper limit of space–time scales is clearly connected with the cosmological extrapolation (in space and time); this problem will inevitably come back in the following pages of the present book. The question about the lower limit of space–time scales is one of the key problems of the present subquantum physics (see remarks in chapt. 1, sec. 7, E). In the view of Hawking and Ellis (1973, p. 57), there is empirical evidence that space–time has a structure of a smooth manifold down to the scales of 10^{-15} cm. The standard cosmological model suggests that this limit can be further lowered down to the Planck length, i.e., to scales 10^{-33} cm (see, for instance, Hawking and Ellis 1973, pp. 362–364; Misner et al. 1973, pp. 1196–1203; Demiański 1985, pp. 281–283). Doubtlessly, these problems belong to the most fascinating questions of contemporary physics.

3. Space–Time of Special Relativity

The pair $(\mathcal{M}, \eta)$, where $\mathcal{M}$ is a four–dimensional manifold and η the Minkowski metric on $\mathcal{M}$ (see, sec. 1), is a space–time model in special relativity. This theory was created by Einstein in 1905, and it was Herman Minkowski to whom the theory owes its transparent geometric form (Minkowski 1907, 1908a, b; the last work is often quoted).

There is a basis $(e_\alpha) = (e_0, e_k)$ in which the components of the Minkowski metric assume the form $\eta_{\alpha\beta} = \text{diag}(-1, 1, 1, 1)$. This basis will be called *pseudoorthogonal*. The coordinate with the index zero is interpreted as time coordinate, the remaining ones as space coordinates?[4]

The group of linear transformations

$$\mathcal{P}: (e'_\beta) \longmapsto a^\alpha_\beta e_\alpha + b^\beta,$$

where a^α_β and b^β are constants, such that

$$a^\alpha_{\ \mu} \eta_{\alpha\beta} a^\beta_{\ \nu} = \eta_{\mu\nu},$$

that is to say, preserving the bilinear form $\eta_{\alpha\beta} v^\alpha v^\beta$, is called a *Poincaré group* or *non–homogeneous Lorentz group* and is usually denoted by $\mathcal{P}(4)$. If $b^\beta = 0$, the name a *Lorentz* or *homogeneous Lorentz group* is used; it is denoted by $\mathcal{L}(4)$.

It can be shown that $\det[a^\alpha_\beta] = \pm 1$. If $a^0_{\ 0} > 0$, the transformations, being elements of the Lorentz (homogeneous) group, form the so–called *ortochronous Lorentz subgroup*; they preserve the time orientation (i.e., the direction of timelike and null vectors).

If $a^0_{\ 0} > 0$ and $\det[a^\alpha_\beta] = +1$, one obtains α subgroup, called a *proper Lorentz group* which is denoted by $\mathcal{L}_0$.

If $a^0_{\ 0} > 0$ and $\det[a^\alpha_\beta] = -1$, the transformations contain time reflections

$$t: (e_0, e_k) \longmapsto L(-e_0, e_k)$$

and space reflections

$$s: (e_0, e_k) \longmapsto L(e_0, -e_k);$$

they do not form a subgroup (however, they do form a subgroup together with transformations for which $\det[a^\alpha_\beta] = +1$).

The full Lorentz group $\mathcal{L}(4)$ splits into four components: $\mathcal{L}_0$, $t\mathcal{L}_0$, $s\mathcal{L}_0$, and $st\mathcal{L}_0$ (see Choquet–Bruhat et al. 1982, p. 290; Carmeli 1977, chapt. 2).

Special relativity is in fact a theory of invariants of the Lorentz group (or of the Poincaré group) together with a physical interpretation adjoined to it. The classification of vectors into timelike, null and spacelike, and consequently null cones at each point of a space–time manifold (the so–called *cone structure of space–time*) is invariant with respect to transformations forming the Lorentz group. Curves to which the tangent vectors at all their points are timelike, null, or spacelike are called *timelike, null (lightlike)*, or *spacelike*, respectively. In the physical interpretation, timelike curves are treated as representing histories of particles (with a non–zero rest–mass) or of observers, and null curves as representing histories of photons (or more generally of zero rest–mass particles). Spacelike curves, in principle, could represent histories

---

[4]From now on we will stick to the following convention: Greek indices assume 0, 1, 2, 3 values, whereas Latin indices assume 1, 2, 3 values.

of tachions, i.e., of hypothetical particles moving faster than light. Since, however, no empirical results confirm such a hypothesis, in the rest of this book it will be systematically ignored.[5]

In the theory of special relativity, geodesic curves (with respect to the connection defined by the Lorentz metric, see chapt. 4, sec. 6) are also identified as histories of freely falling particles of a non–zero rest–mass. This can be thought of as a version of the principle of inertia postulated by special relativity.

The above interpretation of the cone structure of space–time, from the physical point of view, corresponds to its *causal structure*: it informs us on between which domains of space–time information flow (with the help of light signals or other signals slower than light) can take place (see Hawking and Ellis 1973, chapt. 7; Beem and Ehrich 1982, chapt. 2; the causal structure of the Minkowski space–time is presented in the first of these books on pp. 118–124). In this sense, the causal structure of space–time, as considered from a global point of view, has clear cosmological significance.

With the help of the Minkowski metric the distance between two events in space–time can be defined in a natural way. However, let us note that this distance for two different events can be zero, provided that these two events can be joined by a light signal (see below, sec. 5, A). The existence of the Lorentz metric makes it possible to perform space and time measurements in different reference frames. This, in turn, allows one to construct a relativistic mechanics and, in further perspective, also a relativistic electrodynamics.[6]

Everyday laboratory practice of high energy physicists gives a powerful empirical verification of special relativity.

4. The Existence of the Lorentz Structure

4.1. Theorem. A differential manifold $\mathcal{M}$ with an atlas $\{(\mathcal{U}_\alpha, \phi_\alpha)\}$ (at least of the class $\mathcal{C}^1$) admits a (proper) Riemann structure if an only if $\mathcal{M}$ is paracompact.[7] ∎

Proof. (1) (*proper*) *Riemann structure* $\Rightarrow$ $\mathcal{M}$ *is paracompact* (directly from the Stone theorem, see, for instance, Engelking 1975, pp. 336 and 342–343; even a stronger statement is true: every metrizable space is paracompact).

(2) $\mathcal{M}$ *is paracompact* $\Rightarrow$ (*proper Riemann structure*). First, let us recall an important notion. Let X be a $\mathcal{C}^r$ manifold. A family of $\mathcal{C}^r$ functions Θ_k, defined on X, is said to be a $\mathcal{C}^r$ *partition of unity on* X if (a) the supports of functions Θ_k,

---

[5]In older literature, the term "world line" is used to denote the history of a particle or of an observer; this term was coined by Minkowski himself. For a stricter definition of "history" see chapt. 6, sec. 2.

[6]Let us note that the assumption of a finite velocity of light is contained in the causal structure. The numerical value of this velocity has to be taken from experimental results.

[7]Reminder: a topological Hausdorff space X is *paracompact* if into each its open covering $\{U_i\}$ a locally finite covering $\{V_i\}$ can be *inscribed*. A covering $\{V_i\}$ of a topological space X is *locally finite* if for every $x \in X$ there is a neighbourhood having a non–empty intersection only with a finite number of sets $V_i \in \{V_i\}$. Every compact set is paracompact but not vice versa.

supp Θ_k, are compact, (b) for every $x \in X$, there is a finite number of functions Θ_k such that $\Theta_k \neq 0$, (c) $\sum_k \Theta_k(x) = 1$ for each $x \in X$. Further, let $\{(\mathcal{V}_\alpha, \psi_\alpha)\}$ be an atlas on X. A partition of unity on X is said to be *subordinated* to the atlas $\{(\mathcal{V}_\alpha, \psi_\alpha)\}$ if, for every k, there exists $\mathcal{V}_\alpha$ such that supp $\Theta_k \in \mathcal{V}_\alpha$. Now, let us come back to the proof. Let Θ_i be a partition of unity subordinated to the atlas $\{(\mathcal{U}, \phi_\alpha)\}$ on $\mathcal{M}$. Further, let g_i be a tensor of the type $(0, 2)$ on $\mathcal{U}_\alpha$ having, in a certain basis (called a *natural basis*), the coordinates $(g_i)_{kl} = \delta_{kl}$. As it can be easily seen, a tensor $\sum_i \Theta_i g_i$ of the type $(0, 2)$ defines a (proper) Riemann structure on $\mathcal{M}$.[8] □

Let us note that the argument contained in the above proof does not hold for a Lorentz structure. The sum $\sum_i \Theta_i g_i$, where g_i are quadratic forms with the index 1, need not be quadratic forms with the index 1. Therefore, one must look for different criteria as far as the existence of a Lorentz structure on $\mathcal{M}$ is concerned. Before proving the theorem giving such a criterion, we must introduce a new concept.

A function $\mathcal{D}$, coordinating to every point x of a differential manifold X an h-dimensional subspace $\mathcal{D}(x)$ of the tangent space $\mathcal{T}_x(\mathcal{M})$ is called an *h-dimensional distribution on X*.[9] If, for every point $x \in X$, there exists a neighbourhood $\mathcal{U}$ of x and $\mathcal{C}^r$ vector fields $\mathcal{V}_1, \ldots, \mathcal{V}_h$ defined on $\mathcal{U}$ such that, for every $p \in \mathcal{U}$, the vectors $\mathcal{V}_1(p), \ldots, \mathcal{V}_h(p)$ form a basis of the subspace $\mathcal{D}(x)$, the distribution $\mathcal{D}(x)$ is said to be of the class $\mathcal{C}^r$. One–dimensional distribution on X is called a *direction field* on X. The concept of h-dimensional distribution turns out to be an efficient tool in the theory of partial differential equations of the first order (see Bishop and Goldberg 1980, chapt. 3. 11–12).

4.2. Theorem. A Lorentz structure exists on a paracompact differential manifold X (at least of the class $\mathcal{C}^1$), if and only if a continuous direction field exists on X. ∎

Proof. (1) *Continuous direction field* $\Rightarrow$ *Lorentz structure.* X is paracompact, and consequently (on the strength of the preceding theorem) a (proper) Riemann structure g exists on X. Let u be a unit vector belonging to the direction field, i.e. $u \in \mathcal{D}(x) \subset \mathcal{T}_x(X)$, where $\mathcal{D}(x)$ is a continuous one–dimensional distribution on X. Let us define

$$\mathcal{G}_x(v, v) = g_x(v, v) - 2(g_x(v, u))^2$$

with $v \in \mathcal{T}_x(X)$. It can be easily checked that it is a Lorentz metric (e.g. in the four-dimensional case, by writing down the above expression in coordinates and substituing: $u^0 = u_0 = 1$, $g_{\alpha\beta} = \delta_{\alpha\beta}$)

(2) *Lorentz structure* $\Rightarrow$ *continuous direction field.* Let g — as above — be a (proper) Riemann metric on X. One of the eingenvectors, to be found by solving the

---

[8] From the definition of paracompactness it follows that a paracompact manifold can be covered by a denumerable set of local charts. In fact, the implication in the opposite direction is also true (see Geroch 1971, Appendix).

[9] This concept has nothing in common with distribution as generalized functions (in the sense of Schwartz).

eigenvalue problem

$$\mathcal{G}_{\alpha\beta}u^{\beta} = \lambda g_{\alpha\beta}u^{\beta},$$

is timelike; it defines a continuous direction field on X. (See Choquet–Bruhat et al. 1982, pp. 292–293; Geroch 1971). □

Two comments should be given. Strictly speaking, in order to have a $\mathcal{C}^r$ Lorentz structure on X one should have a $\mathcal{C}^r$ direction field on X: the existence of a *continuous* direction field on X implies only the existence of a *continuous* Lorentz structure on X. However, this is not a serious limitation since, as shown by Steenrod (1951, p. 25 seq.), a continuous direction field on X can be approximated by a smooth direction field (in fact similar approximations are valid as far as continuous cross–sections of any vector bundles are concerned).

Second, the existence of a continuous direction field on X is not only a necessary condition for the existence of a Lorentz structure on X, but also a Lorentz structure on X can always be chosen in such a way that the direction field in question be timelike in this structure.

The existence of a direction field on a manifold is its topological property. As it is known from topology, every non–compact (but paracompact) manifold admits the existence of a continuous direction field; however, a compact manifold admits the existence of a continuous direction field if and only if its Euler–Poincaré characteristic vanishes. It turns out that — on the strength of one of a few possible Euler–Poincaré characteristic definitions — the existence of a Lorentz structure on compact manifolds is strictly connected with the structure of differential forms on these manifolds. Here is an outline of the idea.

Let the space of all p–forms (of the class $\mathcal{C}^k$) on a manifold X be denoted by $\Lambda^p(X)$ (see chapter 1, sec. 4). Since p–forms are antisymmetric, one must have $p \leq n$, where $n = \dim(X)$. The space $\Lambda^p(X)$ is closed with respect to the addition of p–forms and their multiplication by functions (of $\mathcal{C}^k$ class). As it is well–known, an *external product* can be defined

$$\wedge \colon (\Lambda^p(X), \Lambda^q(X)) \longrightarrow \Lambda^{p+q}(X),$$

allowing one to multiply a p–form α by a q–form β to obtain a $(p + q)$–form $\alpha \wedge \beta$ (more about differential forms on a manifold, see: Flanders 1963; Choquet–Bruhat et al. 1982, chapter 4, Bishop and Goldberg 1980, chapt. 4).

The set $\Lambda(X)$ of all forms of any order on X, together with an external product defined in it, is called an *external algebra* or a *Grassman algebra*. A Grassman algebra is a graded algebra, i.e. a set of sets $\{\Lambda^p(X)\}$, indexed by the integers $p = 0, 1, \ldots, n$, together with an associated bilinear mapping $\Lambda(X) \times \Lambda(X) \to \Lambda(X)$; in our case — an external product. An external differentation

$$d \colon \Lambda^p(X) \longrightarrow \Lambda^{p+1}(X)$$

with the property $dd = d^2 = 0$, allows one to treat $\Lambda(X)$ as a *differential* graded algebra. Let us note that the sets $\Lambda^p(X)$ are vector spaces (over $\mathbf{R}$).

Let $\omega \in \Lambda^p(X)$. ω is said to be a *closed form* or a *co-cycle*, if $d\omega = 0$. ω is called an *exact form* or a *co-boundary*, if $\omega = d\Theta$ (of course $\Theta \in \Lambda^{p-1}(X)$).

Evidently, every exact form is closed (since $d^2 = 0$), but not vice versa. The study of closed forms that are not exact brings important information about topology of a given space in the following way.

Let the vector space of closed p–forms on a manifold be denoted by $\mathcal{Z}^p$, and the vector space of exact p–forms on X by $\mathcal{B}^p$. Evidently, one has $\mathcal{B}^p \subset \mathcal{Z}^p$. The quotient space $\mathcal{H}^p = \mathcal{Z}^p/\mathcal{B}^p$ is called a *p–cohomology space* of the manifold X. Since $\mathcal{H}^p$ is an Abelian group with respect to addition it is also called a *cohomology group of pth order* of X. Equivalence classes of closed p–forms are elements of $\mathcal{H}^p$: two closed forms ω_1 and ω_2 are considered to be equivalent if they differ by an exact form, i.e., if $\omega_1 - \omega_2 = d\Theta$.

The dimension of the pth order cohomology group of X is called a pth *Betti number* of the manifold X and is denoted by b^p. The sum $\chi(X) = \sum_{q=0}^{n}(-1)^q b^q$, where $n = \dim(X)$ is called an *Euler–Poincaré characteristic of this manifold*.[10] A compact differential manifold X can be equipped with a Lorentz structure if and only if $\chi(X) = 0$.

Let us note that the vanishing of the Euler–Poincaré characteristic of a compact manifold X is a necessary and sufficient condition of the *global* existence of a Lorentz structure on it. A direction field, and consequently a Lorentz structure, *locally* always exist on X.

5. Comments and Remarks

A. As is well–known, a (proper) Riemann metric on a differential manifold X defines, in a natural way, a (*Riemann*) *distance function* $d_R: X \times X \to [0, \infty)$ on X, satisfying the usual axioms (including the *triangle inequality*). The topology determined by this function, the so–called *metric topology*, is identical with the original manifold topology (see chapt. 1, sec. 2). No Lorentz metric has these nice properties. In terms of the Lorentz structure, one can define a (*Lorentz*) *distance function* $d_L: X \times X \to [0, \infty]$ (which can assume infinite values!), but it satisfies axioms different from those satisfied by d_R (in particular, it satisfies the so–called *reversed triangle inequality*). Any distance measured by d_L along non–timelike curves is, from definition, equal to zero. A topology determined by d_L is essentially weaker than the manifold topology on X. Unfortunately, entering into details would require new concepts going beyond the scope of the present book; the interested reader should consult the book by Beem and Ehrlich (1981, chapters 1 and 3).

---

[10]The Betti numbers and the Euler–Poincaré characteristic of a given manifold can be also defined dually in terms of a triangulation of this manifold. Dual with respect to closed and exact forms are the so–called *cycles* and *boundaries*, respectively. In a similar manner one defines a homology group (or space) $\mathcal{H}_p$ as a quotient space $\mathcal{Z}_p/\mathcal{B}_p$, where $\mathcal{Z}_p$ is a vector space of cycles and $\mathcal{B}_p$ a vector space of boundaries (of the dimension p).

B. As we have seen, the existence of the Lorentz structure on a (paracompact) manifold is equivalent to the existence of a continuous direction field; moreover, the Lorentz structure can always be chosen so as to make the direction field timelike in this structure. This fact seems to have a deep cosmological significance; one can express it by saying that the existence of a Lorentz structure on a manifold is equivalent to the existence of a *local arrowless* time on it. The existence of a continuous timelike direction field on a manifold could be interpreted as a geometric counterpart of the fact that each point of the manifold feels a lapse of a local time. This lapse, however, is determined by a direction and not by a vector, i.e. two possible arrows along the same direction (the past and the future) are equivalent (the present structure does not distinguish between them). It is truly an arrowless time. And it is also a local time: a timelike direction field, besides the rather tolerant requirement of continuity, does not impose any global conditions; in particular, it provides no conceptual tools which could serve to compare "local times" at different points of the space–time manifold.

To sum up: a paracompact differential manifold can carry a Lorentz structure if any only if a continuous direction field exists on it, and this direction field, after the imposing a Lorentz structure on the manifold, can be interpreted as a local arrowless time (see Heller 1986, chapt. 4).

C. If one agrees that without a Lorentz structure physics cannot be done on a manifold (see introduction to the present section), one must admit that the existence of at least a local arrowless time is a condition for the *possibility of physics*. A similar idea was expressed by C. von Weizsäcker (1972). In his opinion, science establishes laws which explain experiments performed in the past, but are verified (or falsified) by predicting outcomes of future experiments, and by comparing them with the results of experiments when they are already present results; in this sense time is a precondition of physics. If one remembers that, from the methodological point of view, retrodictions are as important for science as are predictions (although psychologically predictions play doubtlessly a more important role), one is entitled to assert that the existence of an (at least) arrowless time is a precondition for physics (see Heller 1986, chapt. 4).

D. Although a compact differential manifold with the vanishing Euler–Poincaré characteristic admits the existence of a Lorentz structure, it leads to some pathologies when considered as a model of space–time. Bass and Witten (1957) have shown that a four–dimensional compact manifold with the vanishing Euler–Poincaré characteristic cannot be simply connected.[11] Moreover, it turns out that in every such manifold at least one *closed* timelike curve must exist (see Beem and Ehrlich 1981, p. 23). Physically, this would mean the existence of a closed history of an observer or of a non–zero rest–mass particle (i.e., of a history which would endlessly repeat itself again and again). It can hardly be seen how one could reasonably make this consistent with requirements of the macroscopic causality (for interesting remarks

---

[11]Its first Betti number must be different from zero.

concerning this problem see Hawking 1971, p. 396; Hawking and Ellis 1973, p. 189).
No four-dimensional compact manifold equipped with a Lorentz structure can be
simply connected. However, such a manifold could be treated as a non-compact
Lorentz manifold in which some points have been identified. More precisely, for any
n-dimensional manifold X_0, there is a unique simply connected manifold X covering
the manifold X_0; X is called a *universal covering* of X_0 or a *manifold universally
covering* the manifold X_0 (for definition see Spanier 1966, chapt. 2; Rychtmyer 1981,
chapt. 24). Therefore, from the point of view of local properties, instead of a four-
dimensional compact manifold one could take its universal covering as a model of
the physical space-time. This covering is non-compact and simply connected (see
Hawking and Ellis 1973, p. 190).

   E. Now, we are ready to take up the question posed in chapter 1 (sec. 7, A) concerning the mutual relationship between a coordinate system and a reference frame.
Coordinate system is a mathematical concept which is equivalent to that of a local
chart belonging to an atlas that defines a differential manifold structure. A reference
frame is a physical counterpart of a coordinate system; however, not every transformation of coordinates would be treated by a physicist as a transition to a new
coordinate system. For instance, an affine transformation of the form $x \mapsto ax + b$,
$a, b \in \mathbf{R}$ (changing to new units and new zero point on a scale) or the transition
from Cartesian to polar coordinates would not be considered as going to a new reference frame. On the other hand, "rotating coordinates" with respect to previous
ones would be regarded as being connected with a new frame of reference. Clearly,
the difference between these two classes of coordinate transformations consists in the
fact that transformations of the former class are independent of time whereas the
transformation of the latter class do depend on time. Therefore, the definition of a
reference frame must contain: (a) a definition of rest (with respect to this reference
frame) and simultaneity; (b) a definition of the metric structure of time (t) together
with a unit of time, and (c) a definition of the metric structure of instantaneous
spaces $(t = \text{const})$ together with a distance unit. Owing to these definitions one can
meaningfully speak separately of time and separately of space, but only with respect
to a given reference frame (see Dixon 1978, chapt. 1, sec. 2).

   F. Clarke (1970) has shown that every Lorentz manifold can be smoothly embedded into a flat (pseudo)Euclidean space of suitably many dimensions; this embedding can be of an arbitrarily high differentiability class. In particular, every
compact four-dimensional Lorentz manifold can be smoothly embedded into the 48-
dimensional pseudo-Euclidean space with the index 2 (2 minuses and 46 pluses), and
every non-compact four-dimensional Lorentz manifold can be smoothly embedded
into the 89 dimensional Euclidean space with the index 2 (2 minuses and 87 pluses).

   Bibliographical Note. From among books on differential geometry, enumerated in the
Bibliographical Note after chapter 1, only Choquet-Bruhat et al. (1982), and Bishop and Goldberg
(1980) deal extensively with Lorentz manifolds. The monographs by Hawking and Ellis (1973) and
by Beem and Ehrlich (1981) are entirely devoted to Lorentz manifolds and their applications to

space–time geometry (see also Heller 1981b, 1986 chapt. 3 and 4). The book by O'Neill (1983) should be especially recommended; it develops, step by step, the geometry of spaces with Lorentz metrics, and discusses they use in both special and general relativity. This book can also be treated as a kind of introduction to more advanced texts (like, for example, Hawking and Ellis (1973) or Beem and Ehrlich (1981)).

The reader is urged to make a closer acquaintance with the theory of special relativity. Its basic ideas can be found in every modern textbook of physics. A good introduction, with a penetrating discussion of the conceptual framework of this theory, is Taylor and Wheeler (1966). First chapters of Schutz (1985) are equally good but much shorter. There exist books on special relativity of the monographic character, for example: Aharoni (1959), Synge (1965), Dixon (1978). The history of special relativity can be found in Miller (1981); this book contains also a new English translation of Einstein's original paper (1905) and its extensive analysis.

Chapter 3

Method of Fibre Bundles

0. Introduction

Fibre bundles functioned in physics for quite a time before their strict definition was formulated by mathematicians. For instance, as it is well-known, the Lagrangian of a dynamical system is a real function of coordinates and velocities of particles constituting this system. In other words, if $\mathcal{M}$ is an n–dimensional configuration space of a given system, the Lagrangian is defined on the space of all coordinates and (generalized) velocities, that is, on all tangent vectors to $\mathcal{M}$ at all points x of $\mathcal{M}$. Such a space is called a *tangent bundle* over $\mathcal{M}$, denoted by $T(\mathcal{M})$, and it evidently has $2n$ dimensions. A tangent bundle $T(\mathcal{M})$ is in fact a union of tangent spaces $T_x(\mathcal{M})$ to the manifold $\mathcal{M}$ at all its points $x \in \mathcal{M}$. The space $T(\mathcal{M})$ can be given the structure of a differential manifold; $T(\mathcal{M})$ equipped with this structure is called a *bundle space*, and $\mathcal{M}$ its *base space*. A dynamical system can also be investigated with the help of the Hamilton method. The Hamiltonian of the system is a function of coordinates and (generalized) momenta. Since momenta are linear combinations of (generalized) velocities, one says that the Hamiltonian is defined on a *cotangent bundle* over the configuration space $\mathcal{M}$, where the cotangent bundle over $\mathcal{M}$, denoted by $T_x^*(\mathcal{M})$, is to be understood as a union of dual spaces $T_x^*(\mathcal{M})$ to the spaces $T_x(\mathcal{M})$ for all $x \in \mathcal{M}$. The space $T^*(\mathcal{M})$ can also be equipped with the structure of a differential manifold. (For classical mechanics in the language of bundles, see Abraham and Marsden 1979, Thirring 1978, Arnold 1974.)

The Kaluza–Klein theory can serve as another, perhaps even more sophisticated, example of the functioning of fibre bundles in physics before they were discovered by mathematicians. Kaluza and Klein, yet in 1921, aimed at unifying gravitational and electromagnetic interactions within the framework of a five–dimensional Riemann space on which a one–parameter isometry group was acting (see, for instance, Bergman 1976, chapters 17 and 18). *Ex post* it has turned out that the space, postulated by Kaluza and Klein, is the so–called *circle bundle*, and the electromagnetic potentials, in their approach, are connection forms in it (see, Trautman 1980).

After the Second World War the theory of fibre bundless began gradually to dominate differential geometry and many chapters of topology as well. Today, it constitutes a standard method of the so–called global analysis. After passing through the hands of mathematicians and obtaining a precise, fully formalized form, the fibre bundle method has come back to physics. First, as an efficient tool not only introducing an order into individual physical theories (such as classical mechanics, see above), but also showing how some physical theories can be arranged into evolutionary chains revealing a certain logic of their development (see Appendix). Later on, the theory of fibre bundles was recognized as a suitable mathematical apparatus allowing one to geometrize gauge theories (see, e.g. Drechsler and Mayer 1977, Nash and Sen 1983). It has turned out that the case of the Kaluza–Klein theory was not an exception but rather a signal of a more general rule: gauge potentials, so–called by physicists, are essentially connection forms in principal fibre bundles. If one takes into account the fact that the unification of the electromagnetic and lepton (weak) interactions, achieved by Weinberg and Salam, has been obtained within the framework of the gauge scheme, and that there are strong reasons to believe that the same scheme will work in unifying strong (hadron) interaction with the electroweak one (the so–called grand unification), and perhaps also with gravity (superunification), one can hardly overestimate the role of fibre bundles in contemporary physics (see, for instance, Taylor 1978, Ross 1981).

Today it is also known that grand unification and, even more so, superunification can work only in an environment of extremely high energies. One estimates that the energies needed for grand unification and superunification are of order of 10^{14} GeV and 10^{19} GeV respectively. Such a high energy can be supplied only by the Universe during its very early evolution. According to the standard cosmological model the energy necessary for grand unification was obtainable in the Universe 10^{-33} s after the initial singularity, and the energy necessary for superunification 10^{-44} s after the initial singularity. In this way, the unification of physics seems to be inseparably connected with cosmology.

Einstein's hope for unifying physics in a geometric framework was not in vain. It has only turned out that the traditional space–time geometry is too poor for this end. Physical events cannot be mere points devoid of any structure. The heart of the matter consists in that that such a structure can be given to them by considering the set of suitable geometric objects (tangent spaces, all possible frames, an so on) at a given point of space–time, i.e. by considering a suitable fibre over this space–time point. Collecting together all such fibres (over all points of space–time) one obtains a corresponding fibre bundle with space–time as its base manifold.

The reasons presented above sufficiently justify our aim to present general relativity, with the perspective of its future cosmological applications, from the very beginning in terms of fibre bundles. We shall do that in the next chapter; in the present chapter we focus on the fibre bundle method itself. Section 1 introduces the concept of fibre bundles, and sec. 2 specifies the principal and associated fibre bundles. In sec. 3, the reader will find an outline of the theory of connections in fibre bundles; because of its importance in physics this theory is presented in a greater detail. In sec. 4, the geometry of a principal fibre bundle (connection and its curvature, structure equations, etc.) is constructed, and in sec. 5 this geometry is "projected down" to the base manifold of the bundle. The usual *Comments and Remarks* are

replaced in this chapter by a concise presentation of some applications of the fibre
bundle method in gauge theories in physics.

1. Fibre Bundles

The fibre bundle concept is a generalisation of the Cartesian product. For instance,
a cylinder can be presented as $S^1 \times \mathcal{I}$, where S^1 is a one–sphere and $\mathcal{I} \subset \mathbf{R}^1$, but
something like that cannot be done with respect to a Möbius strip. However, it is not
difficult to see that Möbius strip is "locally" a Cartesian product: if U is a proper
subset of S^1, then $U \times \mathcal{I}$ is "a part" of a Möbius strip. We say that Möbius strip is
a fibre bundle over S^1 with $\mathcal{I}$ as a typical fibre. This situation can be generalised in
the following chain of definitions.

1.1. Definition. The triple $(\mathcal{E}, \mathcal{B}, \pi)$, where $\mathcal{E}$ and $\mathcal{B}$ are topological spaces, and
$\pi \colon \mathcal{E} \to \mathcal{B}$ a continuous surjective mapping called *projection*, is said to be a *bundle*.

1.2. Example. Let $\mathcal{A}$ and $\mathcal{B}$ be topological spaces. The projection $\pi_A \colon \mathcal{A} \times \mathcal{B} \to$
$\mathcal{A}$ is defined by $\pi_A(x_A, x_B) = x_A$, $x_A \in \mathcal{A}$, $x \in \mathcal{B}$. $(\mathcal{A} \times \mathcal{B}, \mathcal{A}, \pi_A)$ is a bundle.

A bundle which is a Cartesian product (like that of example 1.2) is called a
trivial (or *product*) *bundle*.

1.3. Definition. A *fibre bundle* is a set of five elements $\mathcal{E}(\mathcal{B}) = (\mathcal{E}, \mathcal{B}, \pi, \mathcal{G}, \mathcal{F})$,
where $(\mathcal{E}, \mathcal{B}, \pi)$ is a bundle, $\mathcal{E}$ is called a *bundle space* (or a *total space*), $\mathcal{B}$ is a *base*
(or a *base space*). It is assumed that $\mathcal{B}$ has an open covering $\{U_j; j \in \mathcal{J}\}$. The set
$\pi^{-1}(x) \equiv \mathcal{F}_x$, $x \in \mathcal{B}$, is called a *fibre over* x. It is assumed that fibres $\mathcal{F}_x$, for every
$x \in \mathcal{B}$, are homeomorphic to the topological space $\mathcal{F}$, called a *typical fibre*. $\mathcal{G}$ is a
group acting on $\mathcal{F}$ to the left; it is called a *structural group* of the fibre bundle. The
following conditions are supposed to be satisfied:

(1) The fibre bundle is *locally trivial*, i.e., there exists a homeomorphism

$$\varphi_j \colon \ \pi^{-1}(U_j) \longrightarrow U_j \times \mathcal{F}$$

given by $\varphi_j(p) = (\pi(p), \phi_j(p))$, $p \in \pi^{-1}(U_j)$, where

$$\phi_j\big|_{\mathcal{F}_x} \equiv \phi_{j,x} \colon \ \mathcal{F}_x \longrightarrow \mathcal{F}$$

is a surjective homeomorphism, and the diagram

is commutative.

(2) The *consistency condition* holds, i.e., for every $x \in U_j \cap U_k$ and every $j, k \in \mathcal{J}$, the homeomorphic mapping

$$\phi_{k,x} \circ \phi_{j,k}^{-1} \colon \mathcal{F} \longrightarrow \mathcal{F}$$

is an element of the structural group $\mathcal{G}$ (see fig. 3.1).

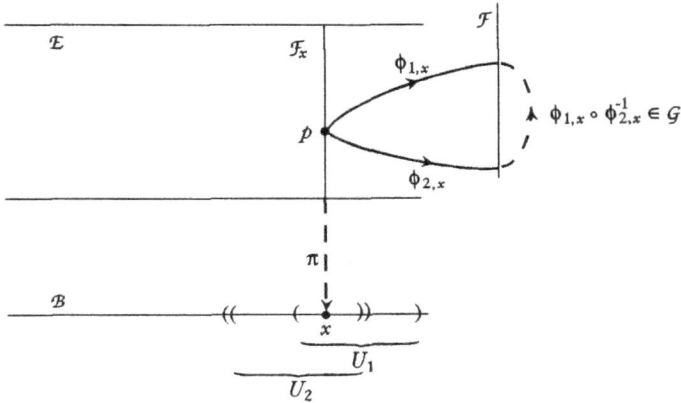

Fig. 3.1. Compatibility condition.

(3) The mappings

$$\gamma_{jk} \colon U_j \cap U_k \longrightarrow \mathcal{G},$$

defined by $\gamma_{jk}(x) = \phi_{j,x} \circ \phi_{k,x}^{-1}$, $x \in U_j \cap U_k$, are continuous; γ_{ij} are called *transition functions*.

1.4. Definition. Two bundles $(\mathcal{E}_i, \mathcal{B}_i, \pi_i)$, $i = 1, 2$, are said to be *equivalent* if there exist mappings $F \colon \mathcal{E}_1 \to \mathcal{E}_2$ and $f \colon \mathcal{B}_1 \to \mathcal{B}_2$ that preserve the bundle structure, i.e., such that the diagram

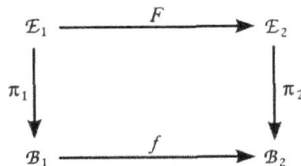

is commutative. The pair (F, f) is said to be a *bundle morphism*. The equivalence of bundles plays a similar role in the fibre bundle theory as the equivalence of atlases does in the theory of manifolds[1] (see chapt. 1, sec. 2): two fibre bundles $(\mathcal{E}_i, \mathcal{B}_i, \pi_i, \mathcal{G}_i, \mathcal{F}_i)$, $i = 1, 2$, are considered to be equivalent if treated locally, as the bundles $(U_i \times \mathcal{F}_i, U_i, \pi_i)$ are equivalent; additionally one assumes that U_i belong to the open covering $\{U_j; j \in \mathcal{J}\}$ of $\mathcal{B}_i$ and $\{U_i\}$ is consistent with the fibre bundle structure, i.e., it satisfies conditions (1) and (2) of the definition of fibre bundle.

The sets $\{U_j, \omega_j\}$ appearing in the definition of a fibre bundle (condition (1)) are called *local trivializations* of the bundle. If the structural group $\mathcal{G}$ has only one element, the bundle is called *trivializable*.

1.5. Definition. A fibre bundle $\mathcal{E}(\mathcal{B}) = (\mathcal{E}, \mathcal{B}, \pi, \mathcal{G}, \mathcal{F})$ is said to be a *differential fibre bundle* (*of the class* C^k) if $\mathcal{E}$, $\mathcal{B}$ and $\mathcal{F}$ are differential manifolds (of the class C^k), the sets U_j covering $\mathcal{B}$ coincide with domains of local charts on $\mathcal{B}$ as a manifold, $\mathcal{G}$ is a Lie group, and mappings π and γ_{ik} are differentiable (of the class C^k). In the following, without saying this directly, we shall consider fibre bundles of the class C^∞

1.6. Example. Let S^2 be a two–sphere with a chosen orientation. A pair of unit orthogonal vectors $(e_1, e_2)_x$, tangent to S^2 at $x \in S^2$, is called a *diad*. The set of all diads having orientation consistent with that of S^2 is denoted by $\mathcal{P}$. Let us define the projection $\pi: \mathcal{P} \to S^2$ by $\pi(e_1, e_2)_x = x$. If $(e_1, e_2)_x$ is a diad at a point $x \in S^2$, then the transformation

$$e_1' = e_1 \cos \varphi + e_2 \sin \varphi,$$
$$e_2' = -e_1 \sin \varphi + e_2 \cos \varphi$$

gives another diad at the same point. Such transformations are elements of the group $\mathcal{SO}(2)$ which turns out to be the structural group of the bundle $\mathcal{P}(S^2) = (\mathcal{P}, S^2, \pi, \mathcal{SO}(2))$. The typical fibre $\mathcal{F}$ is identical with the group $\mathcal{SO}(2)$ treated as a manifold (see below, sec. 2). On the strength of the Brouwer theorem about the impossibility to smoothly comb a vector field on a sphere of an even dimension, the fibre bundle $\mathcal{P}(S^2)$ is not trivial. (See Trautman 1978, pp. 6–7.)

1.7. Example. Möbius strip (see Choquet–Bruhat et al. 1982, pp. 126–127.)

If a fibre bundle $\mathcal{E}(\mathcal{B})$ with a structural group $\mathcal{G}$, after replacing $\mathcal{G}$ by its subgroup $\mathcal{G}_1$, leads to the equivalent bundle structure, the structural group $\mathcal{G}$ is said to be *reducible* to the subgroup $\mathcal{G}$, or the bundle $\mathcal{E}(\mathcal{B})$ with the structural group $\mathcal{G}_1$ is said to be a *reduction* of the bundle $\mathcal{E}(\mathcal{B})$ with the structural group $\mathcal{G}$. If such a reduction occurs, then there exists a family of local trivializatons such that the transition functions γ_{ij} corresponding to them assume values in $\mathcal{G}_1$ (see chapt. 4, sec. 5).

---

[1] Bundles together with bundle morphisms define the *category of bundles*. By suitably generalizing the morphism concept (by taking into account the structure induced by the existence of a typical fibre and of a structural group) one can define the *category of fibre bundles* (see Sulanke and Wintgen 1972).

2. Principal and Associated Fibre Bundles

2.1. Definition. A fibre bundle $\mathcal{E}(\mathcal{B}) = (\mathcal{E}, \mathcal{B}, \pi, \mathcal{G}, \mathcal{F})$, the structural group of which, $\mathcal{G}$, acts to the left on the typical fibre $\mathcal{F}$, i.e., $L_y(f) = gf$, $g \in \mathcal{G}$, $f \in \mathcal{F}$, and such that $\mathcal{F}$ is identical with $\mathcal{G}$ (understood as a manifold), is called a *principal fibre bundle*.

2.2. Example. Let ρ_x be a basis in $T_x(X^n)$, i.e., $\rho_x = (e_i)$, $i = 1, \ldots, n$, where e_i are linearly independent vectors; these vectors, in turn, can be expressed as linear combinations of vectors $\tilde{e}_i$ which form another basis in $T_x(X)$. Of course, one has

$$e_i = a^j{}_i \tilde{e}_j, \quad (a^j{}_i) = a \in \mathcal{GL}(\mathbf{R}^n).$$

Thus, there is a bijection between the set of all bases in $T_x(X^n)$ and the group $\mathcal{GL}(\mathbf{R}^n)$. The group $\mathcal{GL}(\mathbf{R}^n)$ is a Lie group and, as a manifold, it can be identified with an open subset in $\mathbf{R}^{n^2}$. Let us define $\mathcal{F}(X^n)$ to be the set of pairs (x, ρ_x), for every $x \in X^n$, and the projection $\pi(x, \rho_x) = x$. In this way, we have obtained the fibre bundle $\mathcal{F}(X^n) = (\mathcal{F}(X^n), X^n, \pi, \mathcal{GL}(\mathbf{R}^n))$ which is called the *fibre bundle of linear frames*.

2.3. Example. The fibre bundle of diads over a sphere is a principal fibre bundle (see example 1.6).

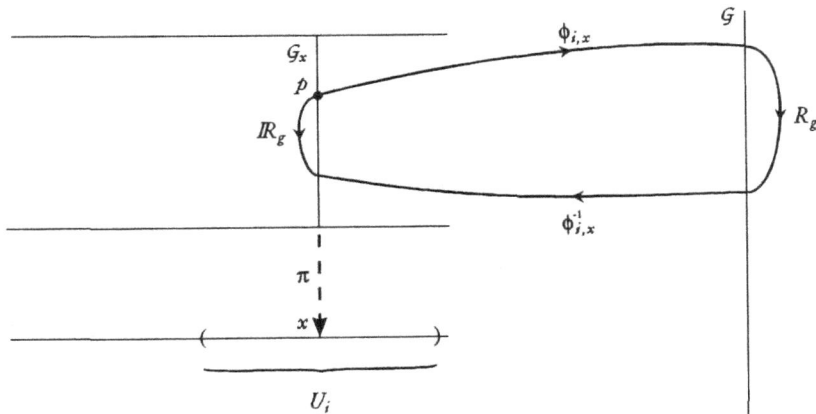

Fig. 3.2. Right action of the structural group $\mathcal{G}$ on a fibre $\mathcal{G}_x$ of a principal fibre bundle.

Let $\mathcal{E}(X) = (\mathcal{E}, X, \pi, \mathcal{G})$ be a principal fibre bundle. Let further $p \in \mathcal{G}_x \equiv \pi^{-1}(x)$, $x \in U_i$, with U_i belonging to an open covering of X which defines the fibre bundle structure. Let us remember that $\phi_{i,x}: \mathcal{G}_x \to \mathcal{G}$, and let $g_i \equiv \phi_{i,x}(p)$. We define the *right action* $\mathbb{R}_g$, $g \in \mathcal{G}$, of the structural group $\mathcal{G}$ on the fibre bundle in the following way (see fig. 3.2)

$$(\mathbb{R}_g p)_i = \phi_{i,x}^{-1} \circ R_g \circ \phi_{i,x} = \phi_{i,x}^{-1}(R_g g_i) = \phi_{i,x}^{-1}(g_i g).$$

The definition is consistent with the fibre bundle structure, i.e., for $p \in \pi^{-1}(U_i \cap U_j)$, one has $(\mathbb{R}_g p)_i = (\mathbb{R}_g p)_j$ (the reader is urged to prove this as an exercise, see Choquet–Bruhat et al. 1982, p. 130). One can easily see that $\mathbb{R}_g$ acts (transitively) on a single fibre, i.e., $(p \in \mathcal{G}_x) \Rightarrow (\mathbb{R}_g p \in \mathcal{G}_x)$, $x = \pi(p)$.

2.4. Definition. A fibre bundle such that its typical fibre is a vector space and its structural group is a linear group is said to be a *vector fibre bundle*. A vector fibre bundle $\mathcal{E}(X) = (\mathcal{E}, X, \pi, \mathcal{F}, \mathcal{G})$ is said to be *associated* with the principal fibre bundle $\mathcal{P}(X) = (\mathcal{P}, X, \pi, \mathcal{G})$ if there is a representation[2] $\rho \colon \mathcal{G} \to \mathbb{L}(\mathcal{F}, \mathcal{F})$ of the structural group $\mathcal{G}$ in the space $\mathcal{F}$ such that the transition functions γ_{ij}^1 of the associated bundle $\mathcal{E}(X)$ are images (under ρ) of the corresponding transition functions of the principal bundle $\mathcal{P}(X)$, i.e., $\gamma_{ij}^1 = \rho(\gamma_{ij})$, or more precisely: Let $\{U_i, \psi_i\}$ be local trivializations of the principal bundle $\mathcal{P}(X)$, and $\{U_i, \varphi_i\}$ local trivializations of the associated bundle $\mathcal{E}(X)$. In such a case the transition functions γ_{ij} of the principal bundle $\mathcal{P}(X)$ are given by

$$\gamma_{ij} \colon U_i \cap U_j \longrightarrow \mathcal{G},$$
$$x \longmapsto \gamma_{ij}(x) = \Psi_{k,x} \circ \Psi_{i,x} \in \mathcal{G},$$

where $\Psi_{k,x} \colon \mathcal{G}_x \to \mathcal{G}$, and the transition functions γ_{ij}^1 of the associated bundle $\mathcal{E}(X)$ are given by

$$\gamma_{ij}^1(x) = \phi_{j,x} \circ \phi_{i,x}^{-1} = \rho(\gamma_{ij}(x)) \in \mathbb{L}(\mathcal{F}, \mathcal{F}),$$

where $\phi_{i,x} \colon \pi^{-1}(x) \to \mathcal{F}$.

Sometimes one stresses directly that a given bundle $\mathcal{E}(X)$ is associated with a principal bundle $\mathcal{P}(X)$ via the representation ρ of the structural group. Having a vector bundle $\mathcal{E}(X)$ it is always possible to construct effectively a fibre bundle $\mathcal{P}(X)$ with which $\mathcal{E}(X)$ is associated (see Choquet–Bruhat et al. 1982, p. 368).

2.5. Example. Let $T(X^n)$ be the space of pairs (x, v_x), $x \in X^n$, $v_x \in T(X^n)$, and let the projection be defined as $\pi(x, v_x) = x$. Let us further suppose that $\{(U_j, \psi_j)\}$ is an atlas on the manifold X^n. Let us consider two local charts (U_1, ψ_1) and (U_2, ψ_2), and $x \in U_1 \cap U_2$, $v \in T(X^n)$. We have

$$\psi_{1,x}' \colon v_x \longmapsto V_1 \in \mathbf{R}^n$$
$$\psi_{2,x}' \colon v_x \longmapsto V_2 \in \mathbf{R}^n$$

The transformation $V_1 \to V_2$ is given by

$$\psi_{1,x} \circ \psi_{2,x}^{-1} \colon \mathbf{R}^n \longrightarrow \mathbf{R}^n \in \mathcal{GL}(\mathbf{R}^n).$$

In this way a fibre bundle $T(X^n) = (T(X^n), X^n, \pi_1, \mathcal{GL}(\mathbf{R}^n), \mathbf{R}^n)$ has been defined; it is called a *tangent bundle* to the manifold X^n. Since there is a representation

---

[2] We remind that a *representation* of a group $\mathcal{G}$ on a space $\mathcal{F}$ is a homeomorphism $\mathcal{G} \to \mathbb{L}(\mathcal{F}, \mathcal{F})$, where $\mathbb{L}(\mathcal{F}, \mathcal{F})$ denotes the set of linear transformations $f \colon \mathcal{F} \to \mathcal{F}$.

$\rho: \mathcal{GL}(\mathbf{R}^n) \to \mathbb{L}(\mathbf{R}^n, \mathbf{R}^n)$ the tangent bundle is associated with the bundle of frames $\mathcal{F}(X^n) = (\mathcal{F}(X^n), X^n, \pi, \mathcal{GL}(\mathbf{R}^n))$.

A $\mathcal{C}^r$ mapping $f: \mathcal{B} \to \mathcal{E}$ such that $\pi \circ f = \text{id}$ is called a $\mathcal{C}^r$ *cross–section* of the bundle $(\mathcal{E}, \mathcal{B}, \pi)$. A $\mathcal{C}^r$ *field of contravariant vectors* on a manifold X^n can be defined as a $\mathcal{C}^r$ cross–section of the tangent bundle $T(X^n)$.

2.6. Example. Let $T^*(X^n)$ denote the space of pairs (x, ω_x), $x \in X^n$, $\omega_x \in T_x^*(X^n)$, and let the projection be defined as $\pi_2(x, \omega_x) = x$. The fibre bundle $T_x^*(X^n) = (T_x^*(X^n), X^n, \pi, \mathcal{GL}(\mathbf{R}^n), \mathbf{R}^n)$ is called the *cotangent bundle* to the manifold X^n. It is also associated with the frame bundle. A cross–section of the cotangent bundle is a *field of covariant vectors* or a *1–form* on X^n.

2.7. Example. By generalising examples 2.5 and 2.6 we define a *tensor fibre bundle* of the type (q, p) over a manifold X^n as $T_p^q(X^n) = (T_q^p(X^n), X^n, \pi, \mathcal{G}, \mathcal{F})$, where

$$T_q^p(X^n) \quad \text{is the space of pairs} \quad (x, t_p^q(x)), x \in X^n,$$
$$t_p^q(x) \in \otimes^q T_x(X^n) \otimes^p T_x^*(X^n),$$
$$\pi(x, t_p^q(x)) = x,$$
$$\mathcal{G} = \otimes^{p+q} \mathcal{GL}(\mathbf{R}^n), \quad \mathcal{F} = \otimes^q \mathbf{R}^n \otimes^p \mathbf{R}^{*n}.$$

It is a fibre associated with the frame bundle. A cross–section of a tensor bundle of the type (q, p) is a tensor field of the type (q, p) on the manifold X^n.

Tensor bundles over a paracompact manifold always admit cross–sections (see Choquet–Bruhat et al. 1982, pp. 385–386).

The concept of reduction of a principal fibre bundle, introduced in the preceding section, can be reformulated in the following way. A principal fibre bundle $\mathcal{E}(X) = (\mathcal{E}, X, \pi, \mathcal{G})$ is said to be *reducible* to a principal fibre bundle $\mathcal{E}_1(X) = (\mathcal{E}_1, X, \pi_1, \mathcal{G}_1)$, where $\mathcal{E}_1 \subset \mathcal{E}$, and $\mathcal{G}_1$ is a subgroup of $\mathcal{G}$, if there is an injective mapping $f: \mathcal{E}_1 \to \mathcal{E}$ which is a morphism of the bundles $\mathcal{E}_1(X)$ and $\mathcal{E}(X)$ and which commutes with the action of the group $\mathcal{G}_1$, i.e., $f(\mathbb{R}_g p) = \mathbb{R}_g f(p)$, $g \in \mathcal{G}$, for every $p \in \mathcal{E}_1$ and $\pi f(p) = \pi_1(p)$, for every $p \in \mathcal{E}_1$.

3. Connections in a Principal Fibre Bundle

Connection in a principal fibre bundle is a structure of crucial importance for our further considerations. Therefore, we will give here three (equivalent) definitions of this structure, beginning from the one which seems to be intuitively most clear and ending with the most operative one. The third definition gives a link between the remaining two.[3]

3.1a. Definition. Let $\mathcal{P}(X) = (\mathcal{P}, X, \pi, \mathcal{G})$ be a principal fibre bundle. A mapping $\sigma_p: T_x(X) \to T_p(\mathcal{P})$, for every $p \in \mathcal{P}$, $\pi(p) = x \in X$, is said to be a *connection* in $\mathcal{P}(x)$, if

---

[3]Here I closely follow Choquet–Bruhat et al. 1982, pp. 357–361.

(1) σ_p is linear,

(2) $\pi' \circ \sigma_p = \text{id}$ on $T_x(X)$,

(3) σ_p depends differentiably on p,

(4) $\sigma_{I\!R_g p} = I\!R' \sigma_p$, for every $g \in \mathcal{G}$.

The above conditions are chosen in such a way that the parallel transport made with the help of this connection is consistent with the structure of the fibre bundle as a differentiable and principal fibre. If $\mathbf{R}^1 \supset \mathcal{I} \to X$ is a curve $t \mapsto C(t)$, $t \in \mathcal{I}$, in X passing through a point $x_0 = \pi(p_0)$, $p_0 \in \mathcal{P}$, the parallel transport along the curve C is given by the curve $\tilde{C}: \mathbf{R}_1 \supset \mathcal{I} \mapsto \mathcal{P}$ which is defined through the following relationships[4],

$$\frac{d\tilde{C}(t)}{dt} = \sigma_p \frac{dC(t)}{dt},$$

$$\tilde{C}(0) = p_0, \quad \tilde{C}(t) = p.$$

The curve $\tilde{C}(t)$ in $\mathcal{P}$ is called the *horizontal lift* of the curve $C(t)$ in X. A horizontal lift $\tilde{C}(t)$ of a curve $C(t)$ determines a correspondence between every two fibres along $\tilde{C}(t)$, that is to say, it connects the bundle into an entirety. Condition (4) guarantees that two different horizontal lifts $\tilde{C}_1$ and $\tilde{C}_2$ of the same curve C in X, to two different points p_1 and p_2 of the same fibre $\pi^{-1}(x)$ over the point $x \in X$, are connected with each other with the help of the right action $I\!R_g$ of the structural group on the fibre $\pi^{-1}(x)$ in such a way that $I\!R_g$ transforms p_1 into p_2.

Let us define a *space of horizontal vectors* (or a *horizontal space*) at a point $p \in \mathcal{P}$ as

$$\mathcal{H}_p \equiv \{\sigma_p(T_x(X))\}, \quad x = \pi(p).$$

A *space of vertical vectors* (or a *vertical space*) can be defined, independently of any connection, as the set of vectors which are tangent to a fibre, i.e.,

$$\mathcal{V}_p \equiv \{v \in T_p(\mathcal{P}): \pi' v = 0\}, \quad p \in \mathcal{P}.$$

It is easily seen that the tangent space $T_p(\mathcal{P})$ to the bundle at a point p can be decomposed into a direct sum of the horizontal and vertical spaces:

$$T_p(\mathcal{P}) = \mathcal{H}_p \oplus \mathcal{V}_p,$$

i.e., for every $v \in T_p(\mathcal{P})$, $v = v_H + v_V$, with $v_H \in \mathcal{H}_p$, $v_V \in \mathcal{V}_p$. Evidently, one has

$$\pi' \mathcal{H}_p = T_x(X) \quad \text{and} \quad \pi' \mathcal{V}_p = 0, \quad x = \pi(p).$$

The first of these relations defines the isomorphism of the horizontal space $\mathcal{H}_p$ and the tangent space $T_x(X)$, and suggests that $\mathcal{H}_p$ can be used to an alternative definition of the connection.

---

[4]In the following, magnitudes defined in a bundle will be denoted by the same symbols as the corresponding magnitudes in the base manifold but equipped with an additional tilde above the corresponding symbol.

3.1b. Definition. A *connection* in a principal fibre bundle $\mathcal{P}(X) = (\mathcal{P}, X, \pi, \mathcal{G})$ is a field of vector spaces $\mathcal{H}_p \subset \mathcal{T}_p(\mathcal{P})$ satisfying the following conditions:
(1) the mapping $\pi': \mathcal{H}_p \to \mathcal{T}_x(X)$, $x = \pi(p)$, is an isomorphism,
(2) the field $\mathcal{H}_p$ depends differentiably on p,
(3) $\mathcal{H}_{\mathbb{R}_g p} = \mathbb{R}'_g \mathcal{H}_p$.

The transition to the third definition is the following. A correspondence between the spaces $\mathcal{H}_p$ and $\mathcal{V}_p$ suggests that the space $\mathcal{V}_p$ can be also used to define connection. This suggestion is correct. It turns out that vertical spaces $\mathcal{V}_p$ are isomorphic with the tangent space $\mathcal{T}_e(\mathcal{G})$ to the structural Lie group (as a manifold) at its neutral element e, i.e., with the Lie algebra G of the structural group $\mathcal{G}$. It is this isomorphism that leads to the third definition.

As is well-known, the space of all left invariant vector fields on a Lie group $\mathcal{G}$ (equipped with the Lie bracket operation) is the Lie algebra G of the Lie group $\mathcal{G}$. The Lie algebra G (as a vector space) of a Lie group $\mathcal{G}$ can be identified with a tangent space to the Lie group $\mathcal{G}$ at its neutral element e through the mapping $G \ni \vartheta \mapsto \vartheta(e) \in \mathcal{T}_e(\mathcal{G})$. On the other hand, to any element ϑ of the Lie algebra G of the structural group $\mathcal{G}$ there corresponds a vector field $\lambda(\vartheta)$ on $\mathcal{P}$ in the following way. If $\vartheta \in G$, then $\exp(\vartheta t)$ is a one-parameter subgroup of $\mathcal{G}$ which acts on $\mathcal{P}$ to the right. This allows us to draw the curve $p_t = \mathbb{R}_{\exp(\vartheta t)}(p)$ through every point $p \in \mathcal{P}$. Now, for any real-valued function f on $\mathcal{P}$ we define

$$\lambda(\vartheta)_p(f) = \frac{d}{dt}\left(f(p_t)\right)\Big|_{t=0}$$

As can be easily seen, the vector $\lambda(\vartheta)_p$ is tangent to the fibre $\pi^{-1}(\pi(p))$ at p; it is, therefore, a vertical vector, $\lambda(\vartheta)_p \in \mathcal{V}_p$. The mapping $\lambda: G \to \mathcal{V}_p$ establishes a *canonical isomorphism* between the Lie algebra G of the structural group $\mathcal{G}$ and a vertical space $\mathcal{V}_p$, for every $p \in \mathcal{P}$. The vector field $\lambda(\vartheta)$ on $\mathcal{P}$ is called a *fundamental vector field*.

Now, a Lie algebra valued 1-form on $\mathcal{P}$,

$$\omega_p: \mathcal{T}_p(\mathcal{P}) \longrightarrow G,$$

can be defined in the following way: if $v \in \mathcal{T}_p(\mathcal{P})$ then $\omega_p(v) = \vartheta \in G$ such that $\lambda(\vartheta) = v_V \in \mathcal{V}_p$. Or, in other words, one demands that $\omega_p(v)$ should be equal to the vertical component of the vector v which, in turn, by the canonical isomorphism is identified with an element of the Lie algebra. The 1-form ω, called a *connection form*, determines a connection in the principal fibre bundle according to the following.

3.1c. Definition. A *connection* in a principal fibre bundle $\mathcal{P}(X) = (\mathcal{P}, X, \pi, \mathcal{G})$ is a 1-form ω on $\mathcal{P}$ satisfying the following conditions:
(1) For every $v \in \mathcal{V}_p$, $\omega_p(v) = \vartheta$ such that between v and ϑ there is a canonical isomorphism,
(2) ω_p depends differentiably on p,
(3) $\omega_{\mathbb{R}_g p}(\mathbb{R}'_g v) = \mathrm{Ad}(g^{-1})\omega_p(v)$.

Let us notice that the set of all vectors $v \in T_p(\mathcal{P})$ for which $\omega_p(p) = 0$ determines the horizontal space $\mathcal{H}_p$. This fact allows one immediately to check the equivalence of the second and third definitions of connections. Condition (3) of the above definition is analogous to conditions (4) and (5) of the first and the second definition, correspondingly, but it is less transparent and requires a few words of comment.

The restriction of the form ω to a single fibre $\mathcal{G}_x = \pi^{-1}(x)$ gives

$$\omega(v_V)\big|_{\mathcal{G}_x} = \vartheta_V, \quad v_V \in T(\mathcal{G}_x) \equiv V_p$$

If we perform the mapping $s: \mathcal{G} \to \mathcal{G}_x$ such that $s(h) = p$ if and only if $p = I\!R_h p_0$, where $h \in \mathcal{G}$ and p_0 is a fixed point in $\mathcal{G}_x$, the form $\omega|_{\mathcal{G}_x}$ can be regarded as the Maurer–Cartan form of the Lie group $\mathcal{G}$.[5] It can be proved that for a Maurer–Cartan form one has

$$I\!R_g^*\omega(v) = \mathrm{Ad}(g^{-1})\omega(v),$$

where $\mathrm{Ad}(g)$; $g \in G$ is the so-called *adjoint representation* of the group $\mathcal{G}$; it is a representation in the Lie algebra G of this group (see Choquet–Bruhat et al. 1982, pp. 168–169).

Keeping in mind that the action $I\!R_g$ preserves the decomposition of a vector $v \in T_p(\mathcal{M})$ into its horizontal component v_H and its vertical component v_V (see condition (3) of the second definition of connection), it is easy to compute the pull back of ω by $I\!R_g$

$$I\!R_g^*(\omega)v = \omega(I\!R_g'v) = \omega(I\!R_g'v_H + I\!R_g'v_V) = \omega(I\!R_g'v_V).$$

By treating ω as a Maurer–Cartan form on $\mathcal{G}$, one immediately obtains condition (3) of the third definition of connection.

To conclude this section, let us define parallel transport in an associated fibre bundle.

3.2. Definition. Let $\mathcal{E}(X) = (\mathcal{E}, X, \pi, \mathcal{G}, \mathcal{F})$ be an associated fibre bundle with a principal fibre bundle $\mathcal{P}(X) = (\mathcal{P}, X, \pi, \mathcal{G})$ by the representation $\rho: \mathcal{G} \to I\!L(\mathcal{F}, \mathcal{F})$. The *parallel transport* of an element $v_1 \in \mathcal{E}$, $\pi_1(v_1) = x_1$, along a curve C from the point $x_1 \in X$ to a point $x_2 \in X$, gives — from the definition — the element $v_2 \in \mathcal{E}$, $\pi_1(v_2) = x_2$ such that

$$\phi_i(v_2) = \rho(\Psi_i(p_2) \circ \Psi_i(p_i)^{-1})\phi_i(v_1),$$

where $p_2 \in \mathcal{P}$ is obtained by the parallel transport of the element $p_1 \in \mathcal{P}$, $\pi(p_1) = x_1$, in the principal fibre bundle $\mathcal{P}(X)$, along the curve C from the point x_1 to the point x_2. All other symbols are the same as in sections 1 and 2 with the exception that in symbols $\phi_{i,x}$ and $\Psi_{i,x}$, the index x was omitted for simplicity.

---

[5]Let us remember that the *Maurer–Cartan form of the Lie group* $\mathcal{G}$ is a form ω, defined on $\mathcal{G}$, with values in the Lie algebra G of the Lie group $\mathcal{G}$, such that $\omega(v_g) = \vartheta$, where $\vartheta = L_g^{-1}v_g \in T_e\mathcal{G}$.

The parallel transport in an associated fibre bundle, defined in the above way, depends on neither the choice of p_1 in a fibre $\pi^{-1}(x_1)$ nor the choice of a trivialization of the fibre bundle.

4. Geometry of Principal Fibre Bundles

As we have seen, connection in a principal fibre bundle $\mathcal{P}(X)$ is a 1–form on $\mathcal{P}$ with values in a vector space, namely in the Lie algebra G of the structural group $\mathcal{G}$. Let us recall that a p–form on a differential manifold X with values in a vector space $\mathcal{V}$ is defined as

$$\varphi = \varphi^\alpha \otimes e_\alpha,$$

where φ^α is an ordinary p–form on X (with scalar values), and e_α are basis vectors in $\mathcal{V}$. It is assumed that $\mathcal{V}$ is a real and finite–dimensional space.

An *external derivative* of a p–form φ is defined to be

$$d\varphi = d\varphi^\alpha \otimes e_\alpha.$$

It can be easily checked that this definition does not depend on the choice of a basis in $\mathcal{V}$. Moreover, if $\mathcal{V}$ is a Lie algebra with the bracket operation $[v_\alpha, v_\beta]$ between every pair of vectors $v_\alpha, v_\beta \in \mathcal{V}$, then the bracket operation can be extended to forms assuming their values in $\mathcal{V}$. Let $\varphi = \varphi^\alpha \otimes e_\alpha$ and $\psi = \psi^\beta \otimes e_\beta$ be p–forms defined on X with values in $\mathcal{V}$. From the definition, we have

$$[\varphi, \psi] = (\varphi^\alpha \wedge \psi^\beta) \otimes [e_\alpha, e_\beta],$$

where $\wedge$ is the external product of forms. This definition is independent of the choice of a basis in $\mathcal{V}$.

Let us consider a principal fibre bundle $\mathcal{P}(X) = (\mathcal{P}, X, \pi, \mathcal{G})$ with a connection in it defined as a 1–form ω. Let

$$h \colon \mathcal{T}_p(\mathcal{P}) \longrightarrow \mathcal{H}_p$$

be a mapping defined by $v \mapsto v_H, v \in \mathcal{T}_p(\mathcal{P})$. Let also $\varphi = \varphi^\alpha \otimes e_\alpha$ be an r–form with values in a vector space $\mathcal{V}$ having a basis e_α. An *external covariant derivative* of an r–form φ is defined to be

$$D\varphi(v_1, \ldots, v_r, v_{r+1}) = d\varphi(hv_1, \ldots, hv_r, hv_{r+1}),$$

where, as usual, $d\varphi = d\varphi^\alpha \otimes e_\alpha$.

An r–form α on $\mathcal{P}(X)$ with values in a vector space $\mathcal{V}$ is called *horizontal* when $\alpha(v_1, \ldots, v_r) = 0$, if at least one of the vectors $v_1, \ldots, v_r$ is vertical.

4.1. Definition. An r–form α on $\mathcal{P}(X)$ with values in a vector space $\mathcal{V}$ is said to be a *form of the type* $(\rho, \mathcal{V})$, where ρ is the representation of the structural group $\mathcal{G}$ in the space $\mathcal{V}$, $\rho: \mathcal{G} \to \mathbb{L}(\mathcal{V}, \mathcal{V})$, if

$$\mathbb{R}_g^* \alpha = \rho(g^{-1}) \alpha.$$

If this relation takes place the form α is said to be *equivariant* under the action of $\mathbb{R}_g$ with respect to the representation ρ. If additionally the r–form α is horizontal, it is called a *tensorial form of the type* $(\rho, \mathcal{V})$. Let us note that the type $(\rho, \mathcal{V})$ of a tensorial form informs us about the character of the tensor field in which, in the general case, the given form assumes its values; namely, the dimension of the space $\mathcal{V}$ is equal to the number of factors in the tensor product of tangent and cotangent spaces defining the given tensor, and the representation $\rho(g^{-1})$ conveys information about the properties of this tensor.

4.2. Definition. An external covariant derivative of a connection one–form ω on $\mathcal{P}(X)$ is said to be a *curvature form of the connection* ω and is denoted by Ω, i.e., $\Omega = D\omega$. Of course, Ω is a 2–form on $\mathcal{P}(X)$ with values in the Lie algebra G.

By using the relationship $\mathbb{R}_g^* \omega = \mathrm{Ad}(g^{-1}) \circ \omega$ (see sec. 3), one shows that the curvature form is a tensorial form of the type (Ad, G).

If $\Omega = D\Omega \equiv 0$ the connection is said to be *flat*.

4.3. Theorem. A connection form can be expressed with the help of the following equation, called the *Cartan* or *structural equation* (or *Cartan structural equation*)[6]

$$D(u,v) = d\omega(u,v) + [\omega(u), \omega(v)]. \quad \blacksquare$$

Before proving this theorem let us recall a useful formula expressing the external derivative of a form Θ,

$$d\Theta(u,v) = L_u \Theta(v) - L_v \Theta(u) - \Theta([u,v]),$$

where $L_u(v)$ is a Lie derivative of the vector v in the direction of the vector u. This formula can be proved by working it out in components.

Proof.

$$\begin{aligned}
\Omega(u,v) &= D\omega(u,v) = d\omega(hu, hv) = d\omega(u - u_V, v - v_V) \\
&= d\omega(u,v) - d\omega(u, v_V) - d\omega(u_V, v) + d\omega(u_V, v_V) \\
&= d\omega(u,v) - d\omega(u_H, v_V) - d\omega(u_V, v_H) - d\omega(u_V, v_V).
\end{aligned}$$

The second term of the middle row, $d\omega(u, v_V)$, has been replaced by the term $-d\omega(u_H + u_V, v_V)$, and analogously, the third term in this row. Two middle terms in the last row are equal to zero. Indeed, let us consider, for example, the second of them,

$$d\omega(u_H, v_H) = L_{u_V}\omega(v_H) - L_{v_H}\omega(v_V) - \omega[u_V, v_H].$$

---

[6]Sometimes the last term on the right hand side of the Cartan equation is multiplied by $\frac{1}{2}$; it depends on conventions assumed in the definition of external derivative.

Let us notice that $\omega(v_V) = \vartheta_V \in G$, and $\vartheta_V = \text{const.}$ Therefore, the Lie derivatives vanish. However, the last term is a horizontal vector since

$$[u_V, v_H] = L_{u_V} v_H = \lim_{t \to 0} \frac{1}{t}(R'_{g(t)} v_H - v_H),$$

that is to say, $d\omega(u_V, v_H) = 0$.

It remains to compute the term $d\omega(u_V, v_V)$. By using the same formula as before, one obtains

$$d\omega(u_V, v_V) = -\omega[u_V, v_V] = -[\omega(u_V), \omega(v_V)] = -[\omega(u), \omega(v)]. \ \square$$

By choosing a basis (e_α) in the Lie algebra G, the Cartan equations can be expressed in coordinates

$$\Omega^\alpha = d\omega^\alpha + \frac{1}{2} c^\alpha{}_{\beta\gamma} \omega^\beta \wedge \omega^\gamma$$

Of course, the coefficients $c^\alpha{}_{\beta\gamma}$ are structure constants of the Lie group $\mathcal{G}$. By differentiating, one obtains

$$d\Omega^\alpha = \frac{1}{2} c^\alpha{}_{\beta\gamma} d\omega^\beta \wedge \omega^\gamma - \frac{1}{2} c^\alpha{}_{\beta\gamma} \omega^\beta \wedge d\omega^\gamma.$$

From the last equation one can see that

$$D\Omega^\alpha(u, v, w) = d\Omega^\alpha(hu, hv, hw) = 0,$$

for any u, v, w. This formula is called *Bianchi identities*; it can be written in the following compact form: $D\Omega = 0$.

5. Geometry of the Frame Bundle

In applications we are often interested in the geometry of a base manifold and individual fibres over its points are treated as an "internal structure" of these points. The bundle manifold is, in such an approach, a kind of a "phase space" of internal states of base points. Such a "phase space of internal states" turns out to be a useful concept in investigating the geometric structure of the base manifold. This is the goal of the present section.

Let $\mathcal{P}(X) = (\mathcal{P}, X, \pi, \mathcal{G})$ be a principal bundle, $\tilde{\omega}$ a connection form in it, and f a smooth (C^r) local cross-section of the fibre bundle, i.e., a cross-section of a region $\pi^{-1}(U)$ of the bundle,

$$f\colon X \supset U \longrightarrow f(U) \subset \mathcal{P},$$
$$\pi \circ f = \text{id}.$$

5.1. Definition. A *connection 1–form* on U with values in G is defined as the pull back of the connection form in $\pi^{-1}(U) \subset \mathcal{P}$ by the mapping f, i.e.,

$$\omega_x = (f^*\tilde{\omega})_x(u) = \omega_{f(x)}(f'u),$$

where $x \in U$, $u \in T_x(X)$.

It can be proved that if a 1–form ω on $U \subset X$ is given, with values in G, and f is a smooth cross–section of the bundle $\pi^{-1}(U)$, then there exists the unique connection $\tilde{\omega}$ in $\pi^{-1}(U)$ such that $\omega = f^*\tilde{\omega}$. The proof consists in, constructing a 1–form $\tilde{\omega}$ at $p \in f(U)$ with the required properties, and extending it to the entire region $\pi^{-1}(U)$. Such a form at a point $p \in f(U)$ is given by

$$\tilde{\omega}_p(v) = \omega_x(\pi'v) + \vartheta_2,$$

where $v \in T_p(\mathcal{P})$, $p = f(x)$, is a vector such that $v = v_1 + v_2$ and, from the definition, $v_1 = (f' \circ \pi')v$, i.e., v_1 is a horizontal vector and, consequently, v_2 is a vertical vector $(\pi'v_2 = 0)$. The vector $\vartheta_2 \in G$ is obtained from the vector v_2 with the help of the canonical isomorphism between the Lie algebra of the structural group and the vertical spaces of the fibre bundle (see sec. 3). Extension of the form $\tilde{\omega}_p$ to all points of $\pi^{-1}(U)$ is made by using the action of the structural group $\mathcal{G}$ on $\pi^{-1}(U)$. (See Choquet–Bruhat et al. 1982, pp. 362–363).

The definition of a local connection on the base manifold X (i.e., of a connection on $U \subset X$) essentially depends on a local cross–section f of the bundle (i.e., on a cross–section of $\pi^{-1}(U)$). Among many such cross–sections one is privileged; let us denote it by s_i and call it a *cross–section canonically associated with the trivialization* (U_i, φ_i) of the fibre bundle. Let us define the mapping $\mathrm{id}_i: U_i \times \mathcal{G}$ by $x \mapsto (x, e)$, where e is the neutral element of the group $\mathcal{G}$; then, from the definition, $s_i = \varphi_i^{-1} \circ \mathrm{id}_i$. The following diagram will help to grasp the meaning of this definition:

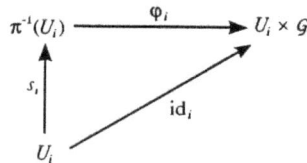

5.2. Definition. The connection form $\omega_i = s_i^*\tilde{\omega}$ will be said to be a *connection form* (or simply a *connection*) *on a local trivialization* (U_i, φ_i). It plays an important role in physical applications.

With the help of a little bit more tedious proof (see Choquet–Bruhat et al. 1982, pp. 364–365) one can show that if (U_i, φ_i) and (U_j, φ_j) are two local trivializations of the fibre bundle $\mathcal{P}(X)$, then two connections ω_i and ω_j on these trivializations, corresponding to the same connection $\tilde{\omega}$ in $\mathcal{P}$, i.e., $\omega_i = s_i^*\tilde{\omega}$ and $\omega_j = s_j^*\tilde{\omega}$ at a point $x \in U_i \cap U_j$, are related by the following formula,

$$\omega_{i,x} = \mathrm{Ad}(\gamma_{ij}^{-1}(x))\omega_{j,x} + (\gamma_{ij}^*\Theta_{\mathrm{MC}})_x,$$

where γ_{ij} are transition functions (see sec. 1), and Θ_{MC} is a Maurer–Cartan 1–form of the structural group $\mathcal{G}$ (see sec. 3, footnote 5).

A local curvature on a base manifold X is defined with the help of a similar procedure as a local connection on X.

5.3. Definition. Let (U_i, φ_i) be a local trivialization of a fibre bundle $\mathcal{P}(X)$ and $\tilde{\Omega}$ a curvature form on $\pi_i^{-1}(U) \subset \mathcal{P}$. A 2–*form of curvature* Ω_i *on a local trivialization* (U_i, φ_i) is defined by

$$\Omega_i = s_i^* \tilde{\Omega}.$$

Two curvature forms Ω_i and Ω_j on two local trivializations (U_i, φ_i) and (U_j, φ_j) respectively corresponding to the same curvature form $\tilde{\Omega}$ on $\mathcal{P}$ at $x \in U_i \cap U_j$ are interrelated with the help of the following transformation,

$$\Omega_i(u, v) = \mathrm{Ad}(\gamma_{ij}(x)^{-1}) \Omega_j(u, v),$$

As it is directly seen, the Cartan equations in a local trivialization (U_i, φ_i) have the following form,

$$\Omega_i = d\omega_i + [\omega_i, \omega_i].$$

6. Applications to Gauge Theories

The number of applications of the fibre bundle geometry to physics quickly increases. The most spectacular, and perhaps the most important, of these applications were mentioned in the introduction to the present chapter. Applications to the theories of space–time, gravity, and cosmology will be studied in the following chapters. As a corollary to this chapter, let us enumerate (after Trautman 1980) some intersecting areas of the theory presented in the above sections and a geometrical formulation of gauge theories.

In the gauge theories one considers fibre bundles (both principal and associated) over space–time that plays the role of a base manifold. To construct a full gauge theory requires the selection of (1) a gauge group, (2) a particle type corresponding to the given gauge field, (3) a form of the field equations. The structure group of the fibre bundle is identified with the gauge group. In the Yang–Mills theories, connection forms on local trivializations (U_i, φ_i) of the principal bundle are what is called by physicists *gauge potentials*, curvature forms Ω_i on (U_i, φ_i) — *gauge fields*, and cross–sections of associated vector bundles — *matter fields*.

To be more precise, let us consider a principal fibre bundle $\mathcal{P}(\mathcal{M}) = (\mathcal{P}, \mathcal{M}, \pi, \mathcal{G})$ over space–time $\mathcal{M}$ and a vector bundle $\mathcal{E}(\mathcal{M}) = (\mathcal{E}, \mathcal{M}, \pi, \mathcal{G}, \mathcal{F})$ associated with $\mathcal{P}(\mathcal{M})$ by a representation $\rho: \mathcal{G} \to \mathbb{L}(\mathcal{F}, \mathcal{F})$. A *Higgs field of the type* $(\rho, \mathcal{F})$ is defined to be a cross–section $\psi: \mathcal{M} \supset \mathcal{N} \to \mathcal{E}$. When reading the literature one should pay attention to the fact that various authors accept different definitions; namely, some authors define a *particle of the type* $(\rho, \mathcal{F})$ as a mapping $\bar{\psi}: \mathcal{P} \to \mathcal{F}$ satisfying

the condition $\bar{\psi}(vg) = \rho(g^{-1})\bar{\psi}(v)$ for every $p \in \mathcal{P}$ and $g \in \mathcal{G}$, and the *Higgs field*
Ψ as the pull back of this mapping by the cross–section s_i of the principal fibre
bundle, i.e., $\Psi = s_i^* \bar{\psi} \colon \mathcal{M} \to \mathcal{F}$ (fig. 3.3) (see Trautman 1980, pp. 296–297). The
relationship between these two definitions of Higgs fields in a local map (U_i, φ_i) is
given by $\psi = \phi_{i,x}^{-1} \circ \Psi$ (definition of $\phi_{i,x}$ in chapt. 3, sec. 1). Particle and Higgs fields
are said to be *matter fields*.

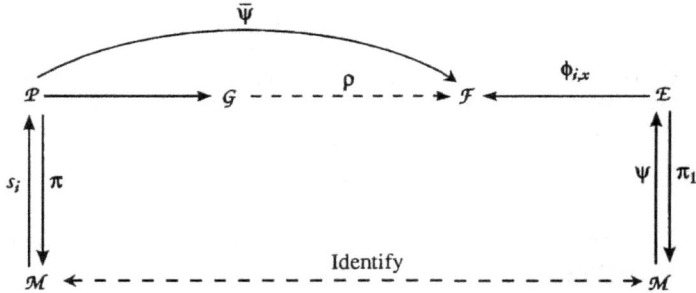

Fig. 3.3. Higgs field (one should remember that $\rho \colon \mathcal{G} \to \mathbb{L}(\mathcal{F}, \mathcal{F})$).

Another important notion is strictly connected with matter fields, namely the
notion of the *spontaneous symmetry breaking*. Let us consider a particle of the type
$(\rho, \mathcal{F})$

$$\bar{\psi} \colon \mathcal{P} \longrightarrow W \subset \mathcal{F},$$

where the subspace W is an orbit of the Group $\mathcal{G}$ in $\mathcal{F}$, i.e., for any two points w_0,
$w \in W$ there is an element $g \in \mathcal{G}$ such that $w = \rho(g)w_0$. Let further $\mathcal{H}$ be an *isotropy
group* (called also a *stability group*) of the element w_0, i.e.,

$$\mathcal{H} = \{g \in \mathcal{G} \colon \rho(g)w_0 = w_0\}.$$

Let us define

$$\mathcal{Q} \equiv \{p \in \mathcal{P} \colon \bar{\psi}(p) = w_0\}.$$

In such a case the fibre bundle $\mathcal{Q}(\mathcal{M})$ with $\mathcal{H}$ as its structural group is a reduction
of the fibre bundle $\mathcal{P}(\mathcal{M})$ having the structural group $\mathcal{G}$ (see sec. 1). This reduction
can be regarded as a geometric counterpart of the breaking of symmetries described
by the group $\mathcal{G}$ to symmetries described by its subgroup $\mathcal{H}$.

It can be shown that the connection form ω_G (corresponding to the structural
group $\mathcal{G}$), after the reduction of the fibre bundle $\mathcal{P}$ to the fibre bundle $\mathcal{Q}$, defines the
connection form ω_H (corresponding to the subgroup $\mathcal{H}$) if and only if ω_H assumes its
values in the Lie algebra H of the group $\mathcal{H}$ which in turn is equivalent to $D\bar{\psi} = 0$
(see Trautman 1980, pp. 297–298; 1981, pp. 409–413).

Let g be a metric of space–time $\mathcal{M}$ (the Minkowski metric is used most often),
and $k_{ij} = k(e_i, e_j)$ be a non–singular metric in the Lie algebra G of the group $\mathcal{G}$ (k_{ij} is

invariant in the representation Ad), where (e_i) is a basis in G. The form $k_{ij} * \Omega^i \wedge \Omega^j$, invariant[7] with respect to the action of the group $\mathcal{G}$, can serve as a the Lagrangian of a gauge theory. In such a case sourceless field equations assume the form: $D * \Omega = 0$.

Bibliographical Note. Classical expositions on the theory of fibre bundles are Steenrod (1951) and Husemoller (1975); one could also use Sulanke and Wintgen (1977). The basic information can be found in every modern textbook of differential geometry, e.g. Kobayashi and Nomizu (1963) or Bishop and Crittenden (1964).

The first gauge theory comes from Weyl (1918); in this form he presented the electromagnetic theory within the framework of its unification with gravitation theory. Only twenty years later, the forgotten Weyl's idea was used by Schwinger (1951, 1953, 1962) in the theory of quantum fields. Great successes of gauge theories begin from the work by Yang and Mills (1954) in which the former ideas were applied to the theory of strong interactions. In the original version of the Yang–Mills theory, bosons transferring strong interactions turned out to be massless particles. This was the reason of the initial lack of interest in this theory. However, the crisis was efficiently overcome by Higgs (1964a, b, 1966) who proposed a mechanism of mass generation, called to–day a *Higgs mechanism*. It consists in a spontaneous symmetry breaking in interactions between the vacuum and the Higgs field. After generalizing, by Utiyama (1956), the Yang–Mills theory to other gauge symmetries, a dramatic expansion of these theories took place, crowned with the unification of electromagnetic and weak interactions by Weinberg (1967) and Salam (1968).

It was Lubkin (1963) who first noticed a connection between gauge theories and geometry of fibre bundles. The gauge theories were afterwards systematically developed in the language of fibre bundles by Trautman (1978, 1979, 1980); see also Dreschsler and Mayer (1977), Daniel and Viallet (1980), Eguchi et al. (1980), Choquet–Bruhat et al. 1982, pp. 401–408, Nash and Sen (1983, chapter 7), Trautman (1984). A popular paper by Berstein and Phillips (1984) presenting applications of fibre bundle methods to physics is recommended.

---

[7]Let us recall that the so–called *star operator* defines the isomorphism

$$*: \Lambda^p(X^n) \longrightarrow \Lambda^{(n-p)}(X^n)$$

by

$$\beta \longmapsto *\beta,$$

where the form $*\beta$, dual with respect to the p–form β, is a form such that $\tau(\alpha|\beta) = \alpha \wedge *\beta$ for every form $\alpha \in \Lambda^p(X^n)$, where τ is a volume element $\tau = \Theta^1 \wedge \cdots \wedge \Theta^n$, and $(\cdot|\cdot)$ denotes an inner product on Λ^p_x (see Choquet–Bruhat et al. 1982, pp. 294–295).

Chapter 4

The Frame Bundle and General Relativity

0. Introduction

One of the main reasons for going from special to general relativity was for Einstein the idea to extend the equivalence of all inertial reference frames to all frames of reference. As is well-known, such an equivalence, in general, can be achieved only locally. If a basis at a given point of space–time is to be considered as the mathematical counterpart of a local reference frame, then it is no wonder that the bundle of bases (*frame bundle*) over a space–time manifold provides a suitable mathematical environment to formulate the general theory of relativity.

The frame bundle is a principal fibre bundle but, being a "concrete" bundle, it has a richer structure than "abstract" bundles, which were studied in the preceding chapter and which are used to geometrize gauge theories (of the Yang–Mills type). The richness of the frame bundle structure and its specific properties are strictly connected with the existence of the soldering form θ in this bundle. This form — as its name picturesquely suggests — solders the frame bundle with its base manifold (space–time) which reflects the fact that the structure of the bundle is, in principle, determined by the structure of the base manifold. From the point of view of general relativity it is a natural property since, in this theory, gravitational field is "reduced" to the space–time geometry.

A fibre $\pi^{-1}(x)$ of the frame bundle over a point x of space–time forms a set of all bases at the point x, i.e., the set of all possible frames of references at x. One could say that in the case of the frame bundle over space–time, an "internal space" of an event x is equipped with a structure which enables one to describe all phenomena occurring in the neighbourhood of x, *with respect* to any frame of reference with its origin at x. In this sense, the idea of relativity is incorporated into the rich concept of the frame bundle over space–time.

It turns out, however, that the structure of the internal space of events of the type "any possible frame at x" is too general (or too vague). As we know from chapter 2, physics can develop on a manifold only if this manifold carries a Lorentz metric. On the other hand, the appearance of a Lorentz metric on the base manifold reduces the frame bundle to the subbundle of orthonormal frames, or, in other words, it corresponds to a spontaneous breaking of symmetries (described by the $\mathcal{GL}(\mathbf{R}^4)$ group) in the frame bundle into symmetries of the Lorentz group. The very possibility of physics enforces such a spontaneous symmetry breaking. This is reflected in the structure of internal spaces in such a way that now this structure admits only pseudoorthogonal local frames. Experimental results tell us that physics distinguishes exactly this realization of the principle of relativity.

In the present chapter the following topics are discussed: geometry of the frame bundle (sec. 2 and 3), the reduction of the frame bundle to the subbundle of orthonormal frames (sec. 4). All these are only a preparation for sec. 5, in which the mathematical structure of general relativity is discussed; we will also focus on the problem of how the frame bundle structure leads to the correct choice of connection.

In *Comments and Remarks* (sec. 6) some of possible generalizations of general relativity (the so–called connection–metric theories) are mentioned. This time, the *Bibliographical Note* at the end of the chapter constitutes its important part; it urges the reader not to be satisfied with a rather compact presentation of general relativity given here but to reach for at least some of available textbooks and monographs.

1. Geometry of the Frame Bundle (continuation)

The subject matter of the present and following sections is the bundle of frames $\mathcal{F}(X^n) = (\mathcal{F}, X^n, \pi, \mathcal{GL}(\mathbf{R}^n))$ (see chapt. 3, example 2.2). Let us notice that a frame at a point $x \in X^n$ can be also defined[1] as a non–singular linear mapping $\rho_x \colon \mathbf{R}^n \to T_x(X^n)$ defined by $\rho_x(u_1, \ldots, u_n) = u \in T_x(X^n)$; the numbers $\rho_x(u_1, \ldots, u_n)$ are components of the vector u in the frame ρ_x. It can be proved that every fibre bundle of the form $\mathcal{P}(X^n) = (\mathcal{P}, X^n, \pi, \mathcal{GL}(\mathbf{R}^n))$ is isomorphic[2] with the frame of bundles if an only if one can define a *soldering form* in the bundle $\mathcal{P}(X^n)$. Let us define it and explore some of its properties.

1.1. Definition. Let the mapping

$$\Phi_p \colon T_x(X^n) \longrightarrow \mathbf{R}^n,$$

where $p = (x, \rho_x) \in \mathcal{F}(X^n)$ is a frame (basis) ρ_x (with the basis vectors (e_i)) at $x \in X^n$, be given by

$$\Phi_p(u) = \theta^i(u), \quad u \in T_x(X^n),$$

where θ^i is a dual basis with respect to (e_i). In other words, the mapping Φ_p coordinates to every $u \in T_x(X^n)$ its components in the frame ρ_x. If one prefers to define a frame as $\rho_x \colon \mathbf{R}^n \to T_x(X^n)$, then $\Phi = \rho_x^{-1}$. A one–form

$$\theta \colon \mathcal{F}(X^n) \longrightarrow \mathbf{R}^n$$

defined by

$$\theta_p(v) = \Phi_p(\pi'v)$$

---

[1] Provided that a base in $\mathbf{R}^n$ has been defined.

[2] In the category of principal fibre bundles.

is called a *soldering form* (or a *canonical form*) on $\mathcal{F}(X^n)$. One speaks also of the *soldering* of the manifold X^n with the bundle of frames. (Since in the present section we are interested principally only in forms defined on frame bundle, we shall not distinguish them from those on the base manifold by the tilde.)

The soldering form can be expressed in coordinates

$$(\theta_p(v))^\gamma = (a^{-1})^\gamma_\beta dx^\beta(\pi'v).$$

Indeed, if $\partial_\alpha = \partial/\partial x^\alpha$, $(\rho_x)_\alpha = a^\beta_\alpha \partial_\beta$, $\theta = \theta^\alpha e_\alpha$, where (e_α) is the natural basis in $\mathbf{R}^n$, then

$$\begin{aligned}
\theta_p(v) &= \Phi_p(\pi'v) = \Phi_p(dx^\beta(\pi'v)\partial_\beta)\\
&= \Phi_p(dx^\beta(\pi'v)(a^{-1})^\gamma_\beta a^\delta_\gamma \partial_\delta)\\
&= dx^\beta(\pi'v)(a^{-1})^\gamma_\beta e_\gamma.
\end{aligned}$$

The soldering form θ on $\mathcal{F}(X^n)$ is a tensorial form of the type $(\mathrm{id}, \mathbf{R}^n)$ (see chapt. 3, definition 4.1). Indeed, from its definition θ is a horizontal form. Moreover, if $g \in \mathcal{GL}(\mathbf{R}^n)$, $\mathrm{id}(g) = g$, $v \in T_p(\mathcal{F}(X^n))$, then

$$\begin{aligned}
(I\!R_g^*\theta)_p(v) &= \theta_p(I\!R'_g v) = \Phi_p(\pi'(I\!R'_g v))\\
&= \Phi_{I\!R_g p}(\pi'v) = g^{-1}\Phi_p(\pi'v) = g^{-1}\theta_p(v).
\end{aligned}$$

The one–before–last equality follows from the fact that if $p = (x, \rho_x) \mapsto I\!R_g p = (x, \rho'_x)$ then $\rho_x \mapsto \rho'_x = g^{-1}\rho_x$.

1.2. Definition. A connection in the bundle of frames is called a *linear connection*. The 2–form $\Theta = D\theta$ is said to be a *torsion form* of the linear connection in $\mathcal{F}(X^n)$.

Now, we shall prove the following.

1.3. Theorem. The torsion form can be expressed with the help of the following equation, called the *Cartan equation* or the *structure equation* (or the *Cartan structure equation*) *for the torsion*,

$$\Theta(u, v) = d\theta(u, v) + [\omega(u), \theta(v)],$$

where $[\omega(u), \theta(v)] \equiv \omega(u)\theta(v) - \omega(v)\theta(u)$, and $\omega(u)\theta(v)$ denotes the action of $\omega(u)$, which is an element of the Lie algebra $GL(\mathbf{R}^n)$ of the group $\mathcal{GL}(\mathbf{R}^n)$, on $\theta(v) \in \mathbf{R}^n$. ∎

Proof.

$$\begin{aligned}
\Theta(u, v) &\equiv D\theta(u, v) = d\theta(u_H, v_H)\\
&= d\theta(u, v) - d\theta(u, v_V) - d\theta(u_V, v) + d\theta(u_V, v_V),\\
d\theta(u_V, v) &= L_{u_V}\theta(v) - \theta([u_V, v]),
\end{aligned}$$

since $L_v\theta(u_V) = 0$ as θ is a horizontal form, but

$$\theta([u_V, v]) = i_{[u_V, v]}\theta = [L_{u_V}, i_v]\theta.$$

We have used the notation $i_v\alpha = \alpha(v)$, where α is a 1-form and v a contravariant vector (i_v can be thought of as an internal product operation), and the following equality,

$$[L_v, i_w]\alpha = i_{[v,w]}\alpha$$

which can be easily checked (by direct computation in coordinates). By joining together the two expressions above after straightforward manipulation, one obtains

$$d\theta(u_V, v) = L_{u_V}\theta(v) - L_{u_V}i_v\theta + i_vL_{u_V}\theta = i_vL_{u_V}\theta.$$

Let us note that

$$\left.\frac{dg(t)}{dt}\right|_{t=0} = \omega(u)$$

(this follows from the definition of connection as a form with its values in the Lie algebra of the structural group). By using the definition of Lie derivative and remembering that $I\!R_g$ acts "vertically", we arrive at

$$i_vL_{u_V}\theta = i_v\frac{d}{dt}(I\!R^*_{g(t)}\theta)\Big|_{t=0} = i_v\frac{d}{dt}(g^{-1}\theta)\Big|_{t=0} = -\omega(u)\theta(v),$$

that is to say,

$$d\theta(u_V, v) = -\omega(u)\theta(v).$$

A similar procedure leads to

$$d\theta(u, v_V) = \omega(u)\theta(v).$$

The term $d\theta(u_V, v_V)$ vanishes since θ is a horizontal form.

By collecting together all the above results we obtain the Cartan structure equation. □

There exists a simple relationship between torsion and curvature forms. By taking external covariant derivative of the torsion form, we find

$$D\Theta = \Omega \wedge \theta.$$

The easiest way to do this is to express the second term of the right-hand side of the Cartan equation as an external product. Let us recall that the external product of a p-form α and a q-form β is defined to be

$$(\alpha \wedge \beta)(v_1, \ldots, v_{p+q}) = \frac{1}{p!\,q!}\sum_\pi (\text{sign } \pi)$$
$$\times \pi(\alpha(v_1, \ldots, v_p)\beta(v_{p+1}, \ldots, v_{p+q})),$$

where π denotes permutation of indices. One should note that sometimes, in place of $1/(p!\,q!)$, the factor $1/(p+q)!$ is used in the above formula; in such a case antisymmetrization makes the factor $1/2$ to appear.

2. Geometry of the Base Manifold

First we will present a method to construct forms in a base manifold X^n (with values in the space of tensor fields of a given type), canonically associated with tensorial forms (of the type $(\rho, \mathbf{R}^{n^q})$) in the bundle of frames $\mathcal{F}(X^n)$. Afterwards, by using this method we shall construct forms on X^n, canonically associated with the soldering form $\tilde{\theta}$, the torsion form $\tilde{\Theta}$, and the connection form $\tilde{\omega}$ in $\mathcal{F}(X)$. In this way, the geometry of the base manifold will be determined in terms of the frame bundle over X^n.

Now, we will generalize to the tensorial case the mapping Φ_p, defined at the beginning of the preceding section. Let us consider a manifold X^n and let Φ_p, $p = (x, \rho_x)$, $x \in X^n$, map any tensor on X^n into its coordinates with respect to the frame ρ_x, i.e.,

$$\Phi_p \colon \otimes^s T_x^*(X^n) \otimes^t T_x(X^n) \longrightarrow \mathbf{R}^{n^q},$$

where $q = s + t$. Let us also consider $v_i \in T_p(\mathcal{F}(X^n))$ and $u_i \in T_x(X^n)$ such that $u_i = \pi'(p)v_i$.

2.1. Definition. An r–form α is said to be *canonically associated* with an r–form $\tilde{\alpha}$ defined on $\mathcal{F}(X^n)$ if

$$\alpha_x(u_1, \ldots, u_r) = \Phi_p^{-1}(\tilde{\alpha}(v_1, \ldots, v_r)),$$

where $x = \pi(p)$, and $\tilde{\alpha}$ is a tensorial form of the type $(\rho, \mathbf{R}^{n^q})$ with values in $\mathbf{R}^{n^q}$.

We shall show that this definition is independent of the choice of v_i and of the choice of the frame ρ_x.

Indeed, (1) it does not depend on the choice of v_i. Let v_i and v_i' be any vectors such that $u_i = \pi'(p)v_i = \pi'(p)v_i'$; then

$$\Phi_p^{-1}(\tilde{\alpha}(v_i)) - \Phi_p^{-1}(\tilde{\alpha}(v_i')) = \Phi_p^{-1}(\tilde{\alpha}(v_i - v_i')) = 0,$$

since $\tilde{\alpha}$ is a tensorial form (and consequently horizontal) and the difference $v_i - v_i'$ is a vertical vector.

(2) The definition does not depend on the choice of the frame ρ_x. Let $p = (x, \rho_x)$ and $p' = (x, \rho_x')$; in such a case there exists $g \in \mathcal{GL}(\mathbf{R}^n)$ such that $p' = I\!\!R_g p$. Let further $v_i' = I\!\!R_g' v_i$. Since $I\!\!R_g'$ acts vertically, we have $\pi' v_i' = \pi v_i = u_i$. Therefore, we obtain

$$\tilde{\alpha}_{p'}(v_i') = (I\!\!R_g^* \tilde{\alpha})_p(v_i) = \rho(g^{-1})\tilde{\alpha}_p(v_i)$$

and $\Phi_{p'}^{-1} = \Phi_p^{-1}\rho(g)$ on the strength of the definition of Φ_p. From the last two formulae it follows the independence of the definition of the choice of p. $\square$

According to the above recipe the soldering 1–form $\tilde{\theta}$ on $\mathcal{F}(X^n)$ defines a 1–form θ on X^n. Since $\tilde{\theta}$ is a tensorial form of the type $(\mathrm{id}, \mathbf{R}^n)$, θ assumes its values in the space of contravariant vector fields. From the definitions of $\tilde{\theta}$ and θ we have

$$\theta_x(u) = \Phi_p^{-1}\tilde{\theta}(v) = \Phi_p^{-1}\Phi_p \pi' v = \pi' v = u,$$

and consequently θ defines the tensor field $\theta^\alpha_\beta = \delta^\alpha_\beta$.

Let us notice that the connection form in the principal bundle, being a non–tensorial form (it is not horizontal), does not define any form on the base manifold in the way discussed above. As a local connection form on the base manifold X^n we must employ the form $\omega_i = s_i^*\tilde\omega$ where s_i is the cross–section canonically associated with the trivialization (U_i, φ_i) (see chapt. 3, sec. 5). Having in mind future applications, let us compute

$$\omega_i(\pi'v) = s_i^*\tilde\omega(\pi'v) = \tilde\omega(s_i'\pi'v) = \omega(v),$$
$$v \in \mathcal{T}_p(\mathcal{F}(X^n)).$$

Remembering that the form ω_i assumes its values in the Lie algebra of the structural group of the bundle, we can also write $\omega(\pi'v) = \omega^\alpha_\beta(\pi'v)\mathcal{E}^\beta_\alpha$, where $(\mathcal{E}^\beta_\alpha)$ is a basis in $\mathcal{GL}(\mathbf{R}^n)$ and $\omega^\alpha_\beta(\pi'v)$ is formally treated as a numerical matrix. The connection form $\tilde\Omega$ on $\mathcal{F}(X^n)$ is a tensorial 2–form of the type $(\mathrm{Ad}, \mathbf{R}^{n^2})$. Therefore, it defines the 2–form Ω on X^n

$$\Omega_x(\pi'v_1, \pi'v_2) = \Phi_p^{-1}(\tilde\Omega(v_1, v_2)).$$

From the type of the form $\tilde\Omega$ one sees that Ω assumes its values in the space of 1–covariant and 1–contravariant tensor fields. By using the structure equation for curvature, we compute

$$\Omega(\pi'v_1, \pi'v_2) = d\omega(\pi'v_1, \pi'v_2) + [\omega(\pi'v_1), \omega(\pi'v_2)],$$

or in coordinates

$$\Omega = d\omega^\alpha_\beta(\pi'v_1, \pi'v_2) + \omega^\alpha_\beta(\pi'v_1)\omega^\gamma_\delta(\pi'v_2)[\mathcal{E}^\beta_\alpha, \mathcal{E}^\delta_\gamma]$$
$$= d\omega^\alpha_\beta(\pi'v_1, \pi'v_2) + (\omega^\alpha_\gamma \wedge \omega^\gamma_\beta)(\pi'v_1, \pi'v_2)\mathcal{E}^\beta_\alpha.$$

In the last step we have used the equality $[\mathcal{E}^\beta_\alpha\mathcal{E}^\delta_\gamma] = \delta^\beta_\gamma\mathcal{E}^\delta_\alpha - \delta^\delta_\alpha\mathcal{E}^\beta_\gamma$. And finally

$$\Omega^\alpha_\beta = d\omega^\alpha_\beta + \omega^\alpha_\gamma \wedge \omega^\gamma_\beta.$$

It is the very well–known form from the standard expositions of differential geometry, *(Cartan) structural equation for the curvature form on the (base) manifold*.

The torsion form $\tilde\Theta$ on $\mathcal{F}(X^n)$ is a tensorial form of the type $(\mathrm{id}, \mathbf{R}^n)$, therefore it defines the 2–form Θ on X^n

$$\Theta_x(\pi'v_1, \pi'v_2) = \Phi_p^{-1}(\tilde\Theta(v_1, v_2)),$$

which assumes its values in the space of contravariant tensor fields. From the structure equation for the torsion form we obtain

$$\Theta^\alpha(\pi'v_1, \pi'v_2) = d\theta^\alpha(\pi'v_1, \pi'v_2) + \omega^\alpha_\beta(\pi'v_1)\theta^\beta(\pi'v_2)$$
$$- \omega^\alpha_\beta(\pi'v_2)\theta^\beta(\pi'v_1),$$

where (θ^β) should be treated as a basis in $T_x^*(X^n)$. This equation can be put into a more condensed form

$$\Theta^\alpha = d\theta^\alpha + \omega^\alpha{}_\beta + \theta^\beta.$$

If is the well–known *structural equation for the torsion form on the (base) manifold.*

3. Geometry of the Base Manifold (continuation)

In order to make the reader's work easier, we now collect the more important formulae of the differential geometry on the base manifold (in coordinates), without proof. A detailed discussion (with proofs) can be found in standard text-books on modern differential geometry (e.g., Choquet–Bruhat et al. 1982, chapt. 5; O'Neill 1983, chapt. 3; Bishop and Goldberg 1980, chapt. 5; Kobayashi and Nomizu 1963, passim) or in the more extensive presentations of general relativity (e.g., Misner et al. 1973, part. III; Weinberg 1972, chapt. 6). All tensors and forms appearing in the present section are defined on the base manifold.

    The structure equations for torsion and curvature, correspondingly, can be written in the form

$$\Theta^i = d\theta^i + \omega^i{}_l \wedge \theta^l = \frac{1}{2}T^i{}_{kl}\theta^k \wedge \theta^l, \tag{3.1}$$

$$\Omega^j{}_i = d\omega^j{}_i + \omega^j{}_m \wedge \omega^m{}_i = \frac{1}{2}R^j{}_{ikl}\theta^k \wedge \theta^l, \tag{3.2}$$

where $T^i{}_{kl}$ and $R^j{}_{ikl}$ are *torsion and curvature tensors* respectively. Let (e_i) be a *moving frame* and (θ^j) a basis dual to (e_i). One has $[e_k, e_l] = c^i{}_{kl}e_i$; $c^i{}_{kl}$ are called *structure coefficients of the moving frame* (e_i). *Connection coefficients* $\gamma^j{}_{kl}$ are defined by the formula $\nabla e_l = \gamma^j{}_{kl}\theta^k \otimes e_j$, where ∇ denotes covariant derivative. The torsion and curvature tensors in local coordinates assume the following forms,

$$T^i{}_{kl} = T(\theta^i, e_k, e_l) = \gamma^i{}_{kl} - \gamma^i{}_{lk} - c^i{}_{kl}, \tag{3.3}$$

$$R_i{}^j{}_{kl} = R(e_i, \theta^j, e_k, e_l) = e_k(\gamma^j{}_{li}) - e_l(\gamma^j{}_{ki}) + \gamma^j{}_{km}\gamma^m{}_{li} - \gamma^j{}_{lm}\gamma^m{}_{ki} - c^m{}_{kl}\gamma^j{}_{mi}. \tag{3.4}$$

    In the natural basis $(\partial/\partial x^i)$, for which $[\partial/\partial x^i, \partial/\partial x^j] = 0$, formulae (3.3) and (3.4) are reduced to

$$T^i{}_{kl} = \Gamma^i{}_{kl} - \Gamma^i{}_{lk}, \tag{3.5}$$

$$R_i{}^j{}_{kl} = \partial_k\Gamma^j{}_{li} - \partial_l\Gamma^j{}_{ki} + \Gamma^j{}_{km}\Gamma^m{}_{li} - \Gamma^j{}_{lm}\Gamma^m{}_{ki}, \tag{3.6}$$

where $\Gamma^i{}_{kl}$ are connection coefficients in the natural basis, called *Christoffel symbols.* The following formulae are valid in any local coordinates,

$$\sum_{(jli)} R_i{}^k{}_{jl} = \sum_{(jli)} (\nabla_j T^k{}_{li} - T^m{}_{ji}T^k{}_{ml}), \tag{3.7}$$

$$\sum_{(jkl)} \nabla_j R_i{}^m{}_{kl} = \sum_{(jkl)} T^h{}_{kj} R_i{}^m{}_{hl}. \qquad (3.8)$$

The summation should be done after cyclic permutation (jli) of the three indices. Formulae (3.8) are called *Bianchi identities*.

4. The Fibre Bundle of Orthonormal Frames

4.1. Definition. Let us consider the following mapping,

$$\mathcal{F}: \mathcal{P}_1(X_1) = (\mathcal{P}_1, X_1, \pi_1, \mathcal{G}_1) \longrightarrow \mathcal{P}_2(X_2) = (\mathcal{P}_2, X_2, \pi_2, \mathcal{G}_2);$$

$\mathcal{F}$ is called a *bundle homomorphism* if the bundles $\mathcal{P}_1(X_1)$ and $\mathcal{P}_2(X_2)$ are equivalent (see chapt. 3, definition 1.4) and if there is a homomorphism $\phi: \mathcal{G}_1 \to \mathcal{G}_2$ such that $\mathcal{F}(I\!\!R_{g_1} p_1) = I\!\!R_{\phi(g_1)}\mathcal{F}(p_1)$, $p_1 \in \mathcal{P}_1$, $g_1 \in \mathcal{G}$. A bundle homomorphism will be denoted by $(\mathcal{F}, \phi)$.

4.2. Theorem. If between $\mathcal{P}_1(X_1)$ and $\mathcal{P}_2(X_2)$ there is a bundle homomorphism, then a connection on $\mathcal{P}_1(X_1)$ uniquely defines the connection on $\mathcal{P}_2(X_2)$. ■

Proof. Let us assume that $\mathcal{H}_{p_1} \subset T_{p_1}(\mathcal{P}_1)$ defines a connection in $\mathcal{P}_1(X_1)$ (see definition 3.1b, chapter 3). Then $\mathcal{H}_{p_2} \equiv I\!\!R_{g_2}(\mathcal{F}'(p_1))$, $\mathcal{H}_{p_2} \subset T_{p_2}(\mathcal{P}_2)$ defines a connection in $\mathcal{P}_2(X_2)$. From the properties of $I\!\!R_g$ and of homomorphism ϕ it follows that the definition of $\mathcal{H}_p$ does not depend on the choice of p_1 and p_2. (See Choquet–Bruhat et al. 1982, p. 380). □

Let $\mathcal{P}(X) = (\mathcal{P}, X, \pi, \mathcal{G})$ be a reduction of the frame bundle $\mathcal{F}(X)$ (see chapt. 3, sec. 1). The inclusions $\mathcal{P}(X) \subset \mathcal{F}(X)$ and $\mathcal{G} \subset \mathcal{GL}(\mathbf{R}^n)$ define a bundle homomorphism. From this and from theorem 4.2 it follows that a connection in $\mathcal{P}(X)$ uniquely defines the (linear) connection in $\mathcal{F}(X)$.

Let X^n be a Riemann manifold. A basis (e_i) on X^n is said to be an *orthonormal frame* if $(e_i|e_j) = \delta_{ij}$. Having an orthonormal frame (e_j) on X^n, we can obtain from it any other orthonormal frame $(\bar{e}_{j'})$ with the help of the transformation $\bar{e}_{i'} = a^j{}_{i'} e_j$, where $a^j{}_{i'}$ is an element of the group $\mathcal{O}(n)$ of orthogonal matrices $n \times n$. There is, therefore, a natural bijection between the set of orthonormal frames on X^n and the group $\mathcal{O}(n)$.

Let us denote the set of pairs (x, τ_x), where $x \in X^n$ and τ_x is an orthonormal frame at x, by $\mathcal{O}(X^n)$. The bundle of orthonormal frames $\mathcal{O}(X^n) = (\mathcal{O}(X^n), X^n, \pi_1, \mathcal{O}(n))$ can be defined in a natural way; evidently, it is a reduction of the principal fibre bundle of (linear) frames $\mathcal{P}(X^n) = (\mathcal{P}, X^n, \pi, \mathcal{GL}(\mathbf{R}^n))$. On the strength of theorem 4.2 a connection in $\mathcal{O}(X^n)$ uniquely determines a connection on $\mathcal{P}(X^n)$ which is called the *metric connection*. Since an orthonormal basis, when parallelly transported with the help of the metric connection, remains an orthonormal frame, the covariant derivative of the metric tensor on X^n vanishes. Direct computation from (3.1) shows that the torsion of the metric connection is also equal to zero. By using elementary tools of the Riemann geometry one can prove that, in any Riemann space, there is only one connection with these properties; it is also called the *Riemann connection*.

Let (F, ϕ) be a reduction of the bundle $\mathcal{F}(X)$ with the structural group $\mathcal{G}$ to a bundle $\mathcal{P}(X^n)$ with a structural group $\mathcal{G}_1 \subset \mathcal{G}$. The opposite question arises: does a connection in $\mathcal{F}(X^n)$ uniquely determine a connection in $\mathcal{P}(X^n)$? In general, since the mapping F need not be a surjection, the answer is negative. However, if X^n is a Lorentz (pseudoriemannian) manifold and (F, ϕ) reduction of $\mathcal{F}(X^n)$ to the bundle of orthonormal frames $\mathcal{O}_k(X^n)$, where k is an index of the manifold X^n (see chapt. 2, sec. 1 and 3), then the above question chooses one and only one connection in $\mathcal{O}_k(X^n)$ such that the torsion of this connection vanishes. To be more precise we formulate the following.

4.3. Definition. Let $F: \mathcal{P}_1 \to \mathcal{P}$, $\phi: \mathcal{G}_1 \to \mathcal{G}$ be inclusions defining a reduction of a principal fibre bundle $\mathcal{P}(X)$ with a structural group $\mathcal{G}$ to a fibre bundle $\mathcal{P}_1(X)$ with a structural group $\mathcal{G}_1$. One says that a connection ω in $\mathcal{P}(X)$ *can be reduced* to $\mathcal{P}_1(X)$ if there exists a connection ω_1 in $\mathcal{P}_1(X)$ such that $\omega_1 = F^*\omega$.

If a connection ω in $\mathcal{P}(X)$ can be reduced to $\mathcal{P}_1(X)$, then values of the connection form ω_1 belong to a Lie subalgebra G_1 of the Lie algebra G of the structural group $\mathcal{G}$, and the connection in $\mathcal{P}(X)$ uniquely determines a connection in $\mathcal{P}_1(X)$ (see Sulanke and Wintgen 1977, p. 171). As we have mentioned, in general this is not true. However, we have

4.4. Theorem. There exists one and only one linear connection ω in the frame bundle $\mathcal{F}(X^n)$ which can be reduced to the bundle of pseudoorthonormal frames $\mathcal{O}_k(X^n)$, being the reduction of $\mathcal{F}(X^n)$ to the pseudoorthogonal group $\mathcal{O}(n, k)$, such that the torsion Θ of ω is equal to zero. ∎

Proof consists in, first, showing that the linear connection in $\mathcal{F}(X^n)$ can be reduced to the bundle $\mathcal{O}_k(X^n)$ if and only if $Dg = 0$, where g is a Lorentz metric (with index k on X^n). Then, by using the relationship $Dg = 0$, one constructs the unique connection in $\mathcal{O}_k(X^n)$. If turns out that the coefficients corresponding to this connection can be transformed to the well-known form of the Christoffel symbols. The detailed proof would require several additional theorems with their corresponding proofs which, in turn, would mean reproducing a large part of a differential geometry textbook. The interested reader should be referred, for instance, to the book by Sulanke and Wintgen (1972, chapt. 2, sec. 11). □

Let us stress that theorem 4.4 distinguishes exactly one connection in $\mathcal{O}_k(X^n)$; it will be called a *Lorentz connection* (the same name will also be used for the corresponding connection in X^n). Let us notice that, on the strength of theorem 4.2, the Lorentz connection in $\mathcal{O}_k(X^n)$ uniquely determines the linear connection in $\mathcal{F}(X^n)$.

In the light of the above considerations it can be said that the existence of a Lorentz structure in X^n breaks down the full $\mathcal{GL}(\mathbf{R}^n)$ symmetry of the frame bundle to the symmetry $\mathcal{O}(n, k)$ which will be also called the *Lorentz symmetry*.[3] In this

---

[3]It can be proved that the Lorentz structure (with index k) on a manifold X^n is equivalent to

sense, one can think that the Lorentz structure (or Lorentz metric) behaves similarly as a Higgs field in gauge theories (see chapt. 3, sec. 6).[4]

5. General Relativity

The mathematical theory developed in chapters 3 and 4 essentially allows one to enrich the metric model of the space–time presented in chapt. 2 (especially sec. 2). We have seen how the important role of enriching the manifold structure is played by connection. From a logical point of view, one should determine directly which connection one has in mind when speaking of the pair $(\mathcal{M}, g)$, and consequently the pair $(\mathcal{M}, g)$, as a model of space–time, should be replaced by the triple $(\mathcal{M}, \omega, g)$, where ω is a connection in the manifold $\mathcal{M}$. Every choice of connection leads to a different theory of space–time (see below sec. 6, B). The ultimate criterion of the choice of a physical theory is the agreement of theoretical predictions deduced from this theory with empirical data. However, the history of science demonstrates that this criterion usually follows the criterion of mathematical elegance. Very often, a mathematical formalism of a theory distinguishes, in some way (sometimes uniquely), the structure which leads to a good agreement with experimental results.[5]

Let us note that the analyses of the preceding section clearly suggest a correct choice of connection for the space–time model. If the new theory is to be formulated within the framework of the "frame bundle over space–time" scheme, and if it is expected to have special relativity as its limiting case, the pseudoorthogonal group $\mathcal{O}(4,1)$, called also the *Lorentz group* $\mathcal{L}(4)$, ought to be its structural group. However, according to theorem 4.4, there exist one and only one linear connection in the frame bundle which is reduced to the bundle of orthonormal frames, namely the Lorentz connection for which $\mathcal{D}g = 0$ and $\Theta = 0$. Since such a choice of connection is uniquely determined by the metric g, one can continue to denote the space–time model by the pair $(\mathcal{M}, g)$. (One should note that, according to theorem 4.4, a connection in the bundle of orthonormal frames $\mathcal{O}_1(\mathcal{M})$ uniquely determines the linear connection in the frame bundle $\mathcal{F}(\mathcal{M})$, but the reduction of the connection in $\mathcal{F}(\mathcal{M})$ to $\mathcal{O}_1(\mathcal{M})$ uniquely determines the space–time model.)

It turns out that the above criterion leads to a physical theory which remains in very good agreement with experimental data. The choice of the Lorentz connection is characteristic of general relativity, and, as is well-known, this theory has successfully passed all empirical tests.

---

the reduction of the frame bundle $\mathcal{F}(X^n)$ to the bundle of orthonormal frames $\mathcal{O}_k(X^n)$ (see Sulanke and Wintgen, 1972, chapt. 2, sec. 11).

[4]There are more analogies between Lorentz metrics and Higgs fields (see Trautman 1981).

[5]Collins (1977) calls it a "subliminal role of mathematics" in contradistinction to its "sublime role". The latter consists in that "mathematics in physics acts as a language and a logical framework for discussion" Collins notices a strong increase of the "subliminal role" of mathematics in the contemporary physics of gravity.

The only degree of freedom, which is left after the above analysis, is the choice of field equations for the new theory. Einstein, motivated by reasons of a philosophical character, believed that, on the one hand, geometry of space–time should be determined by the mass–energy distribution and, on the other, motions of mass–energy should be determined by the space–time geometry. These intuitions — together with some physical arguments, such as the existence of special-relativistic and Newtonian limits — led Einstein to field equations which will now be, almost axiomatically, introduced.

Coefficients of the Lorentz connection assume the form of Christoffel symbols, and components of the curvature tensor of this connection have the following symmetries,

$$R_\alpha{}^\beta{}_{\mu\nu} = -R_\alpha{}^\beta{}_{\nu\mu}, \qquad\qquad R_{\alpha\beta\mu\nu} = -R_{\beta\alpha\mu\nu},$$
$$R_{\alpha\beta\mu\nu} = R_{\mu\nu\alpha\beta}, \qquad\qquad \sum_{(\alpha\mu\nu)} R_\alpha{}^\beta{}_{\mu\nu} = 0. \tag{5.1}$$

The so-called *Bianchi identities* are also satisfied:

$$\sum_{(\gamma\mu\nu)} \nabla_\gamma R_\alpha{}^\beta{}_{\mu\nu} = 0.$$

Contraction of the curvature tensor $R_\alpha{}^\beta{}_{\mu\beta} = R_{\alpha\mu}$ is called the *Ricci tensor*. Its subsequent (metric) contraction, $g^{\alpha\beta}R_{\alpha\beta} = R$, gives the so-called *scalar of Riemann curvature*. The Bianchi identities, after their contraction (in β and γ) and multiplication by $g^{\alpha\gamma}$, can be reduced to the form

$$\nabla_\beta(R^\beta{}_\mu - \frac{1}{2}R\delta^\beta_\mu) = 0. \tag{5.3}$$

Terms in parenthesis are called the *Einstein tensor* and is denoted by G. By playing with indices, one can also write

$$G_{\mu\nu} = R_{\mu\nu} - \frac{1}{2}Rg_{\mu\nu}. \tag{5.4}$$

It can be demonstrated that the tensor

$$G_{\mu\nu} + \Lambda g_{\mu\nu} = R_{\mu\nu} - \frac{1}{2}Rg_{\mu\nu} + \Lambda g_{\mu\nu}, \tag{5.5}$$

where $\Lambda = $ const, is the most general two–index tensor, constructed out of the metric tensor and its first and second derivatives and having the property (5.3) (vanishing divergence). Tensor (5.5) is a purely geometric quantity. According to Einstein's idea, it should be connected with a term describing the "mass–energy distribution". A natural candidate seems to be the *energy–momentum tensor* $T_{\mu\nu}$ appearing already in special relativity. Every particle in space–time carries an energy and momentum or, more geometrically, to every non–spacelike curve in space–time at each of its points there is a tangent four–vector P^μ, called the *energy–momentum four–vector*. For a particle moving with a four–velocity V^μ (with respect to a given reference frame), the energy–momentum four–vector can be obtained from the energy–momentum tensor with the help of the formula

$$P^\mu = T^\mu_{\;\nu} V^\nu. \tag{5.6}$$

The local conservation law can be written in the form of the vanishing divergence of the energy–momentum tensor, $\nabla_\mu T^\mu_{\;\nu} = 0$, which suggests the following form for the field equations

$$G_{\mu\nu} = -\kappa T_{\mu\nu}. \tag{5.7}$$

or

$$G_{\mu\nu} + \Lambda g_{\mu\nu} = -\kappa T_{\mu\nu}, \tag{5.8}$$

where $\kappa = 8\pi G/c^4$, G being the Newtonian constant of gravity, is called *Einstein's constant of gravity*, and Λ, the *cosmological constant*. Both equations (5.7) and (5.8) were considered by Einstein as the field equations of his theory; the generalization (5.8) appeared in the cosmological context. In traditional non–cosmological applications, effects of the constant Λ are negligible and one can successfully use equations (5.7). Since Einstein the significance of the constant Λ for cosmology was widely discussed. For the sake of generality, and also because of the role played by the cosmological constant in contemporary theories of grand unification, we will usually consider the field equation in the form (5.8).

In the "empty" space–time, where $T_{\mu\nu} = 0$, equations (5.7) and (5.8) assume the form

$$R_{\mu\nu} = 0, \tag{5.9}$$

$$R_{\mu\nu} = \Lambda g_{\mu\nu}. \tag{5.10}$$

For dimensions $n = 2$ and $n = 3$, equation (5.9) implies vanishing of the curvature tensor; for dimension $n = 4$, the Ricci tensor has 10 independent components whereas the curvature tensor has 20 independent components, and consequently equations (5.9) (and of course equations (5.10) as well) have non–trivial solutions.

The field equations can be also obtained from the last action principle by assuming a suitable Lagrangian for gravity and source-fields (see, for example, Landau and Lifszic 1980). However, one should remember that, irrespectively of any theoretical reasons, the only lawful argument on behalf of any field equations is their agreement with actual empirical results. So far equations (5.7) (or (5.8)) successfully satisfy this criterion.

6. Comments and Remarks

A. A geometric space–time structure, indispensable for building up the general theory of relativity, has been introduced in terms of fibre bundle of linear frames over space–time reducible to the bundle of pseudoorthogonal frames with the (homogeneous)

Lorentz group as its structural group. This is not the only possible choice. Equally natural structure would be the fibre bundle of affine frames $\mathcal{A}(\mathcal{M})$ over space–time $\mathcal{M}$ with the affine group $\mathcal{GA}(\mathbf{R}^4)$ as its structural group, reducible to the Poincaré group (see chapt. 2, sec. 3).

The fibre bundle of affine frames $\mathcal{A}(\mathcal{M})$ can be thought of as a Cartesian product of the fibre bundle of linear frames $\mathcal{L}(\mathcal{M})$ and the tangent bundle $\mathcal{T}(\mathcal{M})$, $\mathcal{A}(\mathcal{M}) = \mathcal{L}(\mathcal{M}) \times \mathcal{T}(\mathcal{M})$, with the affine group $\mathcal{GA}(\mathbf{R}^4) = \mathcal{GL}(\mathbf{R}^4) \times \mathbf{R}^4$ as its structural group, where the group $\mathcal{GA}(\mathbf{R}^4)$ acts in the following way: if $e: \mathbf{R}^4 \to \mathcal{T}_x(\mathcal{M})$ is a linear frame at $x \in \mathcal{M}$, and $u \in \mathcal{T}_x(\mathcal{M})$, $g \in \mathcal{GL}(\mathbf{R}^4)$, $b \in \mathbf{R}^4$, then

$$\mathcal{A}(\mathcal{M}) \ni (e, u)(g, b) = (eg, u + e(b)).$$

The construction of general relativity in terms of the fibre bundle of affine frames does not differ essentially from that in terms of the bundle of linear frames (see Trautman 1980, 1981). One could also choose other principal fibre bundles over space–time which would equivalently lead to the Einstein theory of gravity (see Trautman 1976).

As the analyses of the present chapter have shown, general relativity can be treated as a gauge theory with a Lorentz metric on space–time playing a role similar to that of Higgs fields, namely, spontaneously breaking symmetries to a subgroup of the structural group (see chapter 3, sec. 6). Essential differences between Einstein's theory of gravity and other gauge theories come from the existence of the soldering form in the frame bundle. This causes the fact that the geometry of the frame bundle is entirely determined by that of the base manifold, namely of space–time. (For more detail on general relativity as a gauge theory see Trautman 1978, 1980, 1981, 1984).

In sec. 5 of the present chapter, the field equations of general relativity were introduced directly as certain postulated relationships of tensorial type on space–time. By consequently using the theory of frame bundles over space–time, the fields equations can be introduced in the following way.

Let us consider the bundle of orthonormal frames $\mathcal{O}_1(\mathcal{M})$ over space–time $\mathcal{M}$, with the homogeneous Lorentz group $\mathcal{L}$ as a structural group. Let ω and Ω be connection and curvature forms on $\mathcal{O}_1(\mathcal{M})$, correspondingly. Let further g be a metric on $\mathcal{M}$, and h a biinvariant metric on the group $\mathcal{L}$ (i.e., both right– and left–invariant metrics, on $\mathcal{L}$). A metric on $\mathcal{O}_1(\mathcal{M})$ can be constructed in the following way,

$$\gamma(X, X) = g(d\pi(X), d\pi(X)) + h(\omega(X), \omega(X)),$$

where X is a tangent vector to $\mathcal{O}_1(\mathcal{M})$. This metric is invariant with respect to the action of the group $\mathcal{L}$. Let us assume $*\Omega^i_j \wedge \Omega^i_j$ to be a Lagrangian for gravity, compute the Ricci scalar R of the metric γ, and perform the variation of the action $[\int R \times$ volume element on $\mathcal{O}_1(\mathcal{M})]$ with respect to g and ω. We shall obtain the vacuum field equations with the cosmological constant together with suitable gauge equations, $D * \Omega = 0$ (see chapt. 3, sec. 6, and more extensively Trautman 1984, in particular pp. 105–107).

B. Analyses of the present chapter inevitably lead to the question: can we obtain a theory of gravity different from general relativity by choosing different connections in a principal bundle or by performing a reduction of the frame bundle to a different subbundle than that of pseudoorthogonal frames? The answer is positive. The problem has been extensively studied by Yasskin (1979). In such theories, called *metric connection theories of gravity*, the metric g, the connection ω, and the soldering form θ are dynamic variables. Dynamical equations of a given theory are obtained from the stationarity of the action $S(g, \omega, \theta, \psi)$, where ψ is a cross–section of an associated bundle which represents matter fields (see, chapt. 3, sec. 6).

The so–called *theories with a Cartan connection*, i.e., with a connection consistent with the metric but having a non–vanishing torsion, constitute an important subclass of metric–connection theories of gravity. The *Einstein–Cartan theory*, belonging to this subclass, was especially thoroughfully investigated. The torsion in this theory is interpreted as macroscopic spin of particles. Studies concerning this theory go back to Weyl (1919) and Cartan (1922, 1923, 1924, 1925); afterwards the theory was developed by Kibble (1961); Sciama (1962); Hehl (1973, 1974); Trautman (1972 a–c, 1973 a–c, 1975, 1978, 1980, 1981).

Bibliographical Note. More about the structure of the frame bundle can be found in the literature dealing with bundles in general (see *bibliographical note* to chapter 3), in the works by Trautman (1970b, 1976), and especially in the review paper by Dodson (1979).

The presentation of general relativity in sec. 5 of the present chapter was very concise. The reader should consult at least some expositions of the huge literature concerning this beautiful theory. I shall quote only some examples.

The following textbooks are often recommended: Weinberg (1972), Misner et al. (1977), Landau and Liphshitz (1980), Rindler (1977), Schutz (1983). To classical textbooks, which always deserve attention, belong Weyl (1922), Eddington (1923), Tolman (1934), Robertson and Noonan (1968).

Mathematical structure of the theory was elaborated in the following monographs: Petrov (1966), Hawking and Ellis (1973), Sachs and Wu (1977), Carmeli (1977), Beem and Ehrlich (1981). O'Neill (1983) can serve as a good introduction to the mathematical structure of general relativity. No serious work concerning solutions to Einstein's field equations can be done without consulting the fundamental monograph by Kramer et al. (1980). A useful collection of exercises is in Lightman et al. (1975).

Philosophical aspects of general relativity are explored by Tonnelat (1971), Raine and Heller (1981), Friedman (1983), and Toretti (1983).

Chapter 5

The Cosmological Problem

0. Introduction

Relativistic cosmology, or more precisely its theoretical foundation, is a main topic of the present book. The foundations of cosmology are deeply rooted in the contemporary theory of space–time and gravitation. Therefore, the preceding chapters constitute a preparation for the main topic, or perhaps, even more than a preparation: as I shall try to demonstrate in the present chapter, global aspects of space–time and gravitation belong, in a sense, to the field of cosmology. Now, the cosmological problem will be addressed explicitly.

Cosmology aims at investigating the structure of space–time on its largest scale. One often speaks of the Universe, having in mind not only the space–time itself, but also everything that fills it in, i.e., all physical fields defined on space–time. All these concepts are understood more or less in an intuitive way. However, one must admit that it is not a rough intuition, but an intuition shaped by long tradition, especially by the evolution of concepts, which has acquired a great acceleration in our century.

There exist two essentially different "directions" of studying cosmology. The "down–direction" consists in assuming some *a priori* postulates concerning the world structure at large and trying to deduce the "local physics" from them. The "up–direction" accepts the local physics as given and tries to extrapolate from it as far as possible to guess at the global structure of space–time. The first way of doing cosmology was represented by Milne (1935, 1948), and Bondi and Gold (1948); but it was the second approach that has acquired the paradigmatic citizenship in the cosmology of the XXth century. The extrapolating cosmology also cannot avoid doing some more or less aprioristic assumptions concerning either the Universe itself or the nature of cosmology as a science of the Universe. Extrapolation, from its very essence, goes beyond local collecting of information, and if it does not explicitly make some assumptions supervising this process, this means it has made such assumptions implicitly. Unfortunately, the majority of publications in cosmology (not to say all with very few exceptions) choose the second possibility. This is an additional motive to write a chapter in a book on foundations of cosmology discussing such problems, even at the risk of loosing something of the formal consistency of the book. At the present state of its development the analysis of the cosmological problem cannot be fully mathematical. However, an analysis seems indispensable for a mathematical formalism to be able to correctly fill in with a cosmological content. This justifies a different, rather unformal, character of the present chapter.

In sec. 1, I try to determine what it means that cosmology is a global geometry of space–time or a non–local physics. Quite unexpectedly, it turns out that non–local assumptions (with a natural understanding of "non–local") are involved in every empirical prediction of standard physics. Global character of cosmology can be understood in a strong sense as a study of the set of all solutions to cosmological equations (Einstein's field equations), i.e., of the set of all cosmological models (such a set is called an *ensemble of universes*). Understanding of cosmology as a theory of such a set is not only a useful methodological trick but it has been implicitly assumed in the scientific practice for a long time. These problems are considered in sec. 2. One cannot forget that cosmology has the ambition to belong to the family of empirical sciences. Consequently, the cosmological problem cannot be responsibly dealt with without touching its observational aspects. These aspects certainly deserve a separate monograph. One can hope that the works of Ellis' group (quoted in the present chapter) will lead to such a monographic study. Therefore, in sec. 3, I only summarize the most important results of the Ellis programme. These results strengthen the above formulated thesis that the discussion of empirically unverifiable assumptions, which enable cosmology to be a science of non–local structure of space–time, must enter the core of the cosmological problem.

1. Cosmology as a Global Geometry of Space–Time

As we know from the preceding chapters, all dynamical theories of hitherto physics presuppose the manifold structure of space–time. However, the traditional approach to physics makes use almost exclusively of local properties of the manifold model. What matters is only a "correct behaviour" in the neighbourhood of the point at which experiments or observations have been made. In practice, one is interested in how "to translate statements about geometrical entities in space–time into statement about real numbers" (Friedman 1983, p. 33), and the work is smoothly done by local maps which define the manifold structure of space–time. On the other hand, the very concept of a manifold is a non–local notion, and the assumption that physics can develop by using only local methods, i.e., that a "local neighbourhood" could be isolated from global influences, turns out to be a strong assumption of a non–local character.

From the field–theoretical point of view the situation is the following. Local empirical predictions are trustworthy only if unexpected perturbations are excluded in the future comprised by the prediction. In order to predict the state of a field at a certain instant one must know "initial conditions" at an "earlier" spacelike hypersurface (or at a null cone, see Penrose 1980). The assumption that there are no signals, generated by a distant event, which at the next moment could affect the investigated system ("no–news" assumption, see Ellis 1983), implies a knowledge about the field state on a distant (towards the past) hypersurface. In this sense, cosmological assumptions are contained in every empirical prediction (see Heckmann and Schücking 1959).

The above considerations justify the following assertion: *The Universe possesses a certain structure (or it is a structure); this structure can be investigated by using either local or non–local (global) methods; the first is a task of physics, the second — of cosmology.* For the time being, terms "non–local" and "global" are used interchangeably. In the following, I shall explain the meaning of the above thesis and argue on behalf of it. Let us now pause for a while to reflect upon a difference between local and global methods. Even in contemporary mathematics, a substantial part of which consist of "global analysis", these terms are fuzzy and non–precise. In Smale's opinion, "global analysis is simply the study of differential equations, both ordinary and partial, on manifolds and vector space bundles" (Smale 1980, p. 84). As we can see, more or less, such a definition of "globality" has been accepted in the preceding chapters. The manifold model of space–time and fibre bundles over space–time were for us fundamental research tools from the very beginning.[1]

Modern observational astronomy deals with such distant objects (galaxies, radiogalaxies, quasars, ...) that no reasonable interpretation of what is observed can be made without a theory of the space–time over which received signals are propagated. This means that some features of the Universe, investigated by contemporary astronomy, essentially depend on integration of some differential equations over a "large scale" domain of the space–time manifold. In this sense cosmological methods permeate the domain of the extragalactical astronomy. Very often this is obscured by the fact that astronomers usually employ the simplest cosmological models ascribing to the Universe strong symmetries (such as Robertson–Walker symmetries). Such symmetries effectively eliminate the necessity to compute troublesome integrals over space–time. Of course, this fact substantially simplifies the situation but, on the other hand, it "screens out" the role of cosmology in standard astronomical investigations (see Ellis and Sciama 1972).

General relativity suffers a certain dualism. Space–time manifold, equipped with different geometric structures, is a "primitive element" of the theory, but on this space–time manifold various physical fields are defined. It is true that these physical fields are also represented by certain geometric structures on space–time (such as scalar or tensor fields); however, they do not appear naturally in the process of the development of the theory, but are introduced into it by decree. Einstein's field

---

[1] A tradition of global analyses in mathematics goes back to Poincaré. To denote this kind of research, Morse in 1934 used the term "macroanalysis". He noticed that "any problem which is nonlinear in character, which involves more than one coordinate system or more than one variable, or whose structure is initially defined in the large, is likely to require considerations of topology and group theory in order to arrive at its meaning and solution" (quoted after Smale 1980, p. 118). The famous Morse theory can serve as an example of non–local methods in mathematics. This theory, to a large extent, contributed to developments of the global analysis. (See, Smale's essay *What Is Global Analysis?* in Smale 1980, pp. 84–89; see also ibid. pp. 117–127).

equations (see chapt. 4, sec. 5) relate geometry and physical fields with each other. In this way, a non–local character of geometrical methods in cosmology implies a non–local character of physical fields (with the exception of gravitational field which is already incorporated into geometry). One could simply say that cosmology is a non–local physics.

It is interesting to notice that in such exotic (from our present point of view) states at the first fractions of a second after the initial singularity, the very distinction between local and non–local ceases to be clear–cut. The closer we are to the initial singularity, the more strongly will any physical property imply non–local consequences.[2]

The initial singularity itself turns out to be a non–local property of the model. From the geometric point of view singularities are understood as geodesic incompleteness of space–time. A space–time is geodesically incomplete if there is at least one non–spacelike geodesic in it along which the affine parameter cannot assume arbitrarily large values (see Hawking and Ellis 1973). The well–known Penrose–Hawking–Geroch singularity theorems (see ibid.) show that if a space–time possesses certain properties, they cannot be "fitted together" with the requirement of the geodesic completeness. Among those properties there are always some of a manifestly non–local character, for instance: spatial compactness, non–existence of closed timelike curves, existence of a Cauchy surface, etc. It is worthwhile to note that in the initial singularity of the standard cosmological model (in the so–called *Big Bang*) all non–spacelike geodesics break down; this fact additionally strengthens a global character of the singularity.

Even in space–times in which not all non–spacelike geodesics break down at a singularity, the appearance of a singularity causes the appearance of horizons and boundaries which are typically global properties of space–time (see Penrose 1968, Hawking and Ellis 1973, Geroch and Horowitz 1979, Tipler et al. 1980).

There are different kinds of singularity (see Clarke 1975, Clarke and Schmidt 1977, Ellis and Schmidt 1977, Tipler et al. 1980), and some of them are "more global than others". A singularity is said to be a *curvature singularity* if in a given space–time there exists at least one non–spacelike curve such that the components of the curvature tensor, in a frame parallelly propagated along this curve, diverge as one approaches the singularity. A singularity is said to be *quasi–regular* if in such a situation components of the curvature tensor, in any such frame, behave correctly. Suppose that an observer approaches a quasi–regular singularity. He is in no way locally warned of the imminent danger. Indeed, on the strength of the Clarke theorem (1973), his history (a timelike curve) has a neighbourhood isometric to a one in a complete space–time manifold. "This means that the 'bad behaviour' that causes the incompleteness is of a global nature: near a single curve, one does not realize that anything is wrong" (Tipler et al. 1980, p. 159). Ellis and Schmidt (1977) have shown

---

[2]This idea is contained in the primeval atom hypothesis of Georges Lemaître (see Godart and Heller 1985).

that space–time in a neighbourhood of a quasi–regular singularity can be thought of as being glued together out of several copies which are perfectly regular. As we can see, in such a situation everything seems to be in order, and the geodesic incompleteness has clearly a global character.

2. Cosmology as a Theory of the Ensemble of Universes

In mathematics, there is also another understanding of what global could mean. We face it when, for example, we are dealing, not with a single differential equation on a manifold $\mathcal{M}$, but with a space of all possible equations (of a given type) on a manifold $\mathcal{M}$. Steve Smale, in his short essay entitled *What is global analysis?*, after having quoted Peixoto's theorem asserting that the structurally stable differential equations on a compact two–dimensional manifold $\mathcal{M}$ form an open and dense set, writes: "This theorem is an excellent theorem in global analysis. One sees in two ways how it is global. First, the differential equation is defined over a whole manifold, and structural stability depends on its behaviour everywhere. Second, the theorem makes a conclusion about the space of all differential equations on $\mathcal{M}$" (Smale 1980, p. 88). In the preceding section we have explored the first meaning of the globality as applied to cosmology, now we shall comment upon its second understanding (which could be termed "superglobality"). In general relativity things are much more difficult than in any "classical" theory of differential equations. Einstein's equations themselves determine the manifold on which they act, and Smale's second understanding of what global means leads to much more difficult — and, one must admit, philosophically much more fascinating — problems. There is no other way; such an understanding must be reduced to the study of all admissible solutions to Einstein's field equations. A mere look upon the history of the twentieth century cosmology reveals that it has developed, as far as its theoretical aspects are concerned, by constructing newer and newer cosmological models, that is to say, by exploring the space of all possible solutions of Einstein's equations.

In agreement with a commonly accepted custom among relativists, the space of all solutions of Einstein's field equations will be called the *ensemble of universes* or shortly the *ensemble*. We shall try to look at cosmology as at a theory of the ensemble. We will argue that such an approach to cosmology is unavoidable.

Here we meet a crucial point. It is commonplace, both in scientific and in popular literature, to stress that all peculiarities of cosmology, as a science, follow from the fact that "the Universe is given to us in one copy". I know of no research work that would go more deeply into this methodological peculiarity of the science of the Universe. I will dwell on the problem for a while.

For the empirical method it seems essential to have many instances of a similar type. Laws of physics are formulated in the form of differential equations which express a structure built up from relations among quantitative properties common to many phenomena. Individual characteristic of every phenomenon are accounted for by choosing suitable initial or boundary conditions which select a particular solution from the class of all admissible solutions. The uniqueness of the Universe makes it impossible to transfer this strategy directly to cosmology. However, one could hardly see how to develop a science otherwise. Cosmology, therefore, has chosen the following tactics. One assumes that a structure called the Universe is described by a solution of a certain equation (usually, Einstein's field equations); one attempts to find all its possible solutions, and then to identify those initial or boundary conditions which correspond to the Universe we actually observe. All other solutions describe other possible universes. Owing to this strategy the actual Universe is placed in the environment of many universes. Now, one has many "individuals", and one can proceed in the same manner as in other departments of physics. In this sense, as Bondi (1960, p. 10) remarks, "we select the important (as opposed to the 'accidental') features of the actual universe by their relation to the theory chosen rather than by any independent criterion". In the following, the set of all possible solutions to Einstein's field equations will be called the *ensemble of universes*, or the *ensemble*, for brevity.

The richness of all possible solutions of Einstein's field equations is so great that usually we explore only "small" subsets of the set of all possible solutions (for instance, spatially homogeneous or homogeneous and isotropic solutions). However, the fact that we can explore subsets of the ensemble is essential for cosmology. Recently, one can notice quite a pronounced tendency: it seems that cosmologists are more interested in, say, how a given property (e.g., the existence of singularities, the appearance of horizons, etc.) is distributed within the ensemble rather than in the structure of individual world models (of single points of the ensemble). This tendency could be explained by methodological reasons. Because of the measurement error inherent in every cosmological observation, a solution which pretends to describe the real Universe can do it only up to a finite degree of accuracy. If we improve our measurements, we must be ready to change to some "nearby" solution which could describe the Universe to a better approximation. In this sense, a *structural stability* of solutions must be postulated (see Thom 1977, Arnold 1980). The problem is by no means a trivial one. It must be decided which solutions are to be defined as identical (or equivalent), and when the distance between two different (non–equivalent) solutions is to be said "small". The decision must be both mathematically and operationally realistic. From the mathematical point of view, the space of all solutions must be equipped with a metric, or at least with a suitable topology. From the operational point of view, the adopted definitions should correspond to what, and with which accuracy, is measured in observational cosmology.

The stability property is also interesting with respect to other features, not necessarily directly observable, which we would like to ascribe to the actual Universe (for instance with respect to such properties as the existence of singularities or horizons, different degrees of causality, and so on). If a property of a certain model is shared by "neighbouring" models it is said to be a *typical property* (or a *generic property*).[3]

---

[3]More precisely: a subset of a topological space is said to be a *Bair set* if it is an intersection

This presupposes a metric or at least a topology on the ensemble which have to be both mathematically workable and leading to physically reasonable results. So far a "natural" topology on the ensemble is not known, and to investigate the stability of different properties one must use different topologies (see below). Theoretical aspects of the structural stability of certain classes of cosmological models were investigated, among others, by Collins (1974, 1977), Golda et al. (1983), Szydłowski et al. (1984); observational aspects of structural stability still await elaboration (see remarks at the end of sec. 3).

The notion of the ensemble of universes was implicitly used in cosmology for a long time. For instance, the Cauchy problem, when considered within the cosmological context, leads to this notion almost immediately: the set of all admissible Cauchy data (on a certain spacelike hypersurface) may, quite naturally, be identified with the set of all universes which can be developed out of these data. Explicitly, the concept of the ensemble began to emerge in the work by Dicke (1961) who, for rather philosophical reasons, postulated that there exists not only the Universe but a whole class of universes with all possible values of physical constants. These ideas found a more elaborate form in Carter's talk during the Symposium of the International Astronomical Union in Cracow, 1973. Carter used the concept of the ensemble of universes to formulate a doctrine called by him the anthropic principle. He used also the name "world–ensemble". "By this I mean — he wrote — an ensemble of universes characterized by all conceivable combinations of initial conditions and fundamental constants (the distinction between these concepts, which is not clear–cut, being that the former refers essentially to local and the latter to global features)" (Carter 1974). According to the author, "the existence of any organism describable as an observer will only be possible for certain restricted combinations of the parameters, which distinguish within the world–ensemble an exceptional cognizable subset". Since we are certainly "describable as organisms", the Universe we are living in must belong to this distinguished subset.[4]

---

of countably many open and dense subsets of this space. A property is called *typical* if it is shared by all elements of a Bair set. Property of the complement of a Bair set is called *non–typical*. The difference between typical and non–typical properties cannot be reduced to whether a corresponding set is dense or not (complements of Bair sets can be dense) or to the measure criterion (in the case of the space of a finite dimension, a Bair set can have zero Lebesgue measure). However if, between two contradictory properties that can be possessed by a model, one is typical, the other must be non–typical, and vice versa. An important notion is that of a strongly typical property. A property is *strongly typical* if it is possessed by elements of an open and dense subset of a given topological space (some authors, when writing of typical properties, have in fact in mind strongly typical ones). Correspondingly, one speaks about *strongly non–typical properties*. If a model has a strongly typical property, it preserves it after being slightly perturbed. Moreover, a property is preserved if a perturbed model is perturbed again provided the second perturbation is "smaller" than the first one. On the contrary, if a model has a strongly non–typical property, a slight perturbation can always be found which liquidates this property. (See Richtmyer 1981, pp. 304–311).

[4]Later on, the anthropic principle gained considerable popularity among cosmologists (see, for instance, Carr and Rees 1979). The possibility of reintroducing the man back into the post–Copernican

In the last decade — independently of any philosophical inspirations — the space of solutions of Einstein's field equations was a subject of intensive mathematical research. Usually, one first considers the space $\text{Lor}(\mathcal{M})$ of all Lorentz metrics on a manifold $\mathcal{M}$. Serious difficulties are encountered already at this stage. It turns out that one can introduce infinitely many topologies on $\text{Lor}(\mathcal{M})$, and neither of them is natural. To show the nature of the problem let us consider it in a more detailed way.

The neighbourhood of a metric $g \in \text{Lor}(\mathcal{M})$ can be understood as a set of Lorentz metrics on $\mathcal{M}$, whose derivatives up to an order r only slightly differ from the corresponding derivative of the metric g on a compact set $U \subset \mathcal{M}$. The topology defined with the help of such neighbourhoods is called C^r-compact-open topology on $\text{Lor}(\mathcal{M})$. If one assumes $U = \mathcal{M}$, one obtains the so-called F^r-open topology on $\text{Lor}(\mathcal{M})$.

It turns out that every C^{r+1}-compact-open topology is richer than C^r-compact-open topology. Similarly, every F^{r+1}-open topology is richer than F^r-open topology. And every F^r-open topology is richer than C^r- compact-open topology.

Moreover, one can define the so-called W^r Sobolev topology on $\text{Lor}(\mathcal{M})$, by replacing the postulate of the closeness of metric derivatives with the postulate of the closeness of integrals of squares of these derivatives.[5] It can be shown that every W^{r+3} (compact-open) Sobolev topology is richer than C^r-compact-open topology which, in turn, is richer than W^r Sobolev topology.

This hierarchy of topologies on $\text{Lor}(\mathcal{M})$ can be further enriched in many ways (for more details see Hawking 1971; Hawking and Ellis 1973, pp. 198, 252–253; Beem and Ehrlich 1981, pp. 28–29). Neither of these topologies is "natural" on $\text{Lor}(\mathcal{M})$, and one must suitably choose different topologies for different problems. For instance, Sobolev topologies turn out to be suitable for discussing the stability of initial conditions for Einstein's equations[6], but to define the so-called stable causality[7] compact-open topologies are preferred. However, when considering the ensemble problem, we are not interested in all Lorentz metrics on $\mathcal{M}$, but in all those Lorentz metrics on $\mathcal{M}$ which are solutions of Einstein's equations; they constitute the subset $\mathcal{E}$ of the space $\text{Lor}(\mathcal{M})$. The space $\mathcal{E}$ usually has the structure of a smooth manifold. Moreover, there exists its local representation in the space of four functions of three variables. It turns out that Einstein's equations act in $\mathcal{E}$ as a Hamiltonian system.

One–parameter families of solutions for Einstein's equations are curves in $\mathcal{E}$,

---

universe caused far too much of metaphysical emotions; see a responsible review by Demaret et Barbier (1981) and a balanced evaluation by Barrow (1983). However, the main reference to the anthropic principle, as far as its both physical and metaphysical aspects are concerned, is a book by Barrow and Tipler (1983).

[5]These integrals and squares are defined in terms of a certain positively defined metric on $\mathcal{M}$ which always exists.

[6]The problem is how to formalize the postulate that a slight change of the initial conditions leads to evolutions only slightly different from each other.

[7]Space–time $(\mathcal{M}, g)$ is said to be *stably causal* if a slight perturbation of the metric g does not produce closed timelike curves.

and spaces of solutions of linearised Einstein's equations form tangent spaces to $\mathcal{E}$
(fig. 5.1). Of course, one can meaningfully speak of a tangent space to $\mathcal{E}$ only if $\mathcal{E}$ has
the structure of a smooth manifold in the neighbourhood of a given solution. In such
a region, Einstein's equations are said to be *stable with respect to linearization*. This
is not always the case. One of the most interesting results, as far as investigations of
the ensemble are concerned, is the theorem stating that Einstein's equations (with the
vanishing energy–momentum tensor) in the neighbourhood of a solution $g_0 \in \mathcal{E}$, such
that the space–time $(\mathcal{M}, g_0)$ has a compact Cauchy surface, are stable with respect
to linearization if and only if the space–time $(\mathcal{M}, g_0)$ admits no Killing vector field
(see Fisher and Marsden 1979, p. 206). This result points to the fact that space–times
with symmetries (with Killing vector fields) violate the smooth manifold structure of
the solution space $\mathcal{E}$. Indeed, it turns out that the space $\mathcal{E}$ has a cone singularity in
the naighbourhood of a solution with symmetries (see Fisher et al. 1980; Brill 1982).[8]
This fact has serious consequences in the theory of "small perturbations" of Einstein
equations (e.g., in the problem of the origin of galaxies and clusters of galaxies). Since
in the neighbourhood of a solution with symmetries the space $\mathcal{E}$ is not stable with
respect to linearization, small perturbations of such solutions produce, in principle,
spurious perturbations (unless they satisfy the so–called Taub condition, see Fisher
et al. 1980).

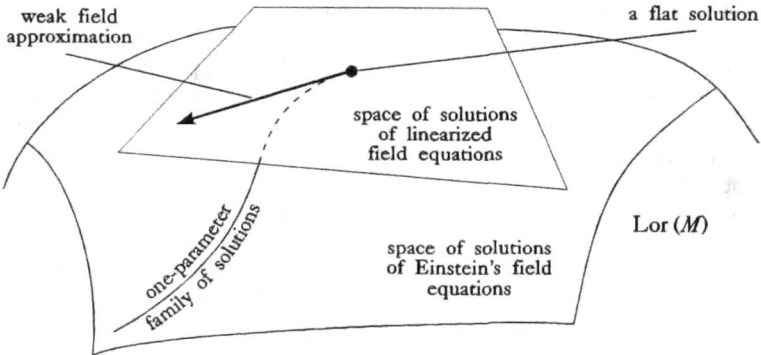

Fig. 5.1. The space of solutions of Einstein's equations.

Let these few results suffice as illustration of the subtle nature of the ensemble.
Our aim is not to review this field of research (for more see Montcrief 1975; Marsden
1981, chapters 9 and 10, and already quoted works), but only to point out that the
ensemble, from its very nature, is a natural environment of cosmology.

---

[8]Analogously to the "cone" $F(X, Y, Z) = X^2 + Y^2 - Z^2$ at the point $(0, 0, 0)$. Everywhere
besides this point, the linearized equation $dF/ds|_{s=0} = 2XX' + 2YY' - 2ZZ' = 0$ uniquely defines
the tangent space.

It might turn out that the space of all solutions approach to cosmology be indispensable in formulating quantum theory of gravity and its applications to understanding the evolution of the Universe. For instance, Hawking (1979, 1982) argues that a suitable method to quantize gravitational field should consist in adapting to this goal Feynman's path–integral method. The method, as applied to gravitational field, consists in specifying a three–metric h_1 and the values of matter fields φ_1 on a spacelike surface S_1, and going to a three–metric h_2 and the values of matter fields φ_2 on a spacelike surface S_2, as a sum over all four–geometries having S_1 and S_2 as their boundaries, with matter fields suitably matching φ_1 and φ_2 correspondingly. The method was elaborated and applied to cosmology by Hartle and Hawking (1983) and by Hawking himself (1984). These authors assumed that spacelike three–geometries are compact (the closed universe). Quantum states of such a universe are described by a wave function defined as a functional on the compact three–geometries and the values of matter fields on them, and satisfying a second order functional differential equation, called the *Wheeler–DeWitt equation*. Specifying the states of the universe is equivalent to specifying the boundary conditions for this equation.

By a rotation to Euclidean time τ ($t \mapsto -i\tau$) in the action, one changes to the integration over all positive–definite four–geometries having suitable three–geometries as their boundaries (the so–called Euclidean approach, see Hawking 1979). Hartle and Hawking put forward a proposal that the ground–state amplitude for a given three–geometry is that given by an integral over all *compact positive–definite* four–geometries having this three–geometry as a boundary.[9] "This means that the Universe does not have any boundaries in space or time (at least in the Euclidean regime). One can interpret the functional integral over all compact four–geometries bounded by a given three–geometry as giving the amplitude for that three–geometry to arise from a zero three–geometry, i.e., a single point. In other words, the ground state is the amplitude of the Universe to appear from nothing" (Hartle and Hawking 1983, p. 2961).

As it can be seen, the Hartle–Hawking method presupposes the space of all possible states of the Universe (of all compact three–geometries). This set is called *superspace* (having nothing in common with the superspace of supergravity theory), and in quantum cosmology it plays a similar role to that played by the ensemble in classical cosmology.

Finally, let us note that historical developments of the relativistic cosmology can be naturally described in terms of the ensemble. Theoretical investigations in cosmology had penetrated larger and larger domains of the ensemble by discovering more and more new solutions. Later on theoreticians were interested in how certain properties of world models were distributed in the ensemble. It is interesting that world models first discovered historically (the static Einstein model, the empty de Sitter model ...) or classes of world models (spatially homogeneous but isotropic

---

[9] Matter fields should be regular on these geometries, and only space–times with the cosmological constant $\Lambda > 0$ are considered, $\Lambda = 0$ is regarded as a limiting case.

models, spatially homogeneous but anisotropic models ...) usually turned out to be non–typical in the space of solutions. The direction of research in observational cosmology is just the opposite: investigations in this field always tend to narrow a certain domain in the solution space, i.e., to distinguishing this class of solutions (or, in the highly idealized case, this single solution) which best fits the actual observational data. Without its observational image modern cosmology would loose its identity. Only if we remember about its observational side can we claim that cosmology is a theory of the ensemble.

3. Observational Problems of Cosmology

Since Lemaître's work of 1927, in which this author for the first time demonstrated that actual red shift measurements in galactic spectra do not contradict predictions of a relativistic world model, comparisons of theoretical models with observational data became an important chapter of cosmology. Hubble's (1936, 1937) ambition was to treat cosmology as an observational science in the following sense: the task of astronomical research is to describe the "observable universe"; if one assumes that it provides a "fair sample of the Universe", a simple extrapolation will allow one to arrive at a satisfactory cosmological model. However, soon it became evident that Hubble's approach was unrealistic. Modern astronomical (and radioastronomical) techniques reach such faraway domains of space–time that it is impossible to interpret the registered "data" without assuming a space–time model (at least as a working hypothesis). In practice, the following procedure, called the *empirical testing of cosmological models*, has been established. Basing on certain postulates, usually inspired by philosophical views (such as cosmological principle, postulate of the spatial closure of the Universe, and so on) or by empirical results believed to be well–founded (e.g., symmetrical matter distribution in the Universe), a certain cosmological model or a class of models is assumed. In the second step, one deduces relationships between observable magnitudes (the so–called *observables*) from this world model or from this class of world models, and then one compares theoretically deduced relationships with those established observationally. Results of this procedure were initially only able to show that there were no contradictions between a given model (or a given class of models) and the observational data.

The first major success of the observational cosmology consisted in making the steady–state cosmology (Bondi and Gold 1948, Hoyle 1948) "non–satisfactory" and "less attractive" than its rival evolutionary world models. However, this became possible, not owing to the traditional method of observational testing, but rather to the discovery of the microwave background radiation (Penzias and Wilson 1965, Dicke et al. 1965). The steady–state cosmology was not able to explain the genesis of this radiation.

In the seventies the so–called *standard cosmological model* has been established as a result of several factors. Accumulation of new empirical data indeed seemed to corroborate postulates assumed so far on purely theoretical basis (see, for instance, Symposium of the International Astronomical Union in 1973 on comparing observational data with theoretical models (Longair 1974)). An important role was also played by stronger and stronger connections between the evolutionary world model and standard physics: high energy physics has found its "natural laboratory" in the very early Universe. On the other hand, typically relativistic methods began to infiltrate the field of astrophysics. People get gradually accustomed to the fact that the Universe is a relativistic object. However, during this entire period the method of testing of cosmological models was based on intuitions and many tacit assumptions, believed to be self–evident, rather than on a solid mathematical analysis of the problem. It is symptomatic that a paper by Sandage (1961) was long treated as a classical work in this field in spite of the fact that it dealt with observational possibilities of the 200–inch telescope rather than with observational foundations of cosmology.

A paper by Kristian and Sachs (1966) was the first in which the authors began systematically to analyse the problem. They treated it in a general way, not limiting themselves to a chosen class of models. They assumed that (1) space–time is a four-dimensional pseudo–Riemannian manifold with a structure enabling one to express all the investigated observational relationships as a series around the here–now of the earthly observer; (2) null geodesics are histories of light beams; (3) the dynamics of the world is ruled by Einstein's field equations with the dust energy–momentum tensor. Assumption (1) is the most limiting one. It implies that all investigated relationships remain correct for space–time distances which are small when compared with both the scale factor of a given world model and the scale of local inhomogeneities. In the case of relativistic cosmology, assumption (2) can be deduced from the postulates of general relativity as the limit to geometric optics. In the work of Kristian and Sachs, assumption (3) interferes only with calculations leading to numerical estimates of some observational magnitudes; the authors succeeded in making no use of dynamical equations in all other considerations.

However, cosmology owes systematical analysis of its observational possibilities to G.F.R. Ellis and his group. In a series of works (Ellis 1975, 1980, 1984; Ellis and Perry 1979; Ellis et al. 1978) the problem has been clearly stated, and some of its aspects discussed in a detailed way. An extensive paper of Ellis et al. (1985) marks the close of a stage of program implementation. It certainly deserves a closer look.

In their first work, the authors attempted at decoding information of cosmological significance contained in the ideal astronomical observations, i.e., with the assumption that all measurements can be performed with any desired precision and without making use of any dynamical equations. The assumption remains in power that the pair $(\mathcal{M}, g)$, where $\mathcal{M}$ is a four–dimensional smooth manifold, and g a Lorentz metric defined on it, is a model of space–time. As usual, null geodesics are histories of photons, and timelike geodesics histories of freely falling particles with non–zero rest–mass. The program consists in expressing the metric g with the help of the so–called *observational coordinates*, i.e., such coordinates as can be directly translated into observational terms. If derivatives of the components of the metric tensor (in observational coordinates) can also be determined by measurements, then, by computing the coordinates of the connection and of the curvature tensor in the usual way, one could fully reconstruct the space–time model from astronomical observations. In this method one need not expand any observable into a series, and consequently one need no assumption about the smallness of the region controlled by actual observations.

The observational coordinates $\{x^i\} = \{w, y, \theta, \varphi\}$ are defined with respect to the history C of our Galaxy or of our Cluster of Galaxies; one assumes that this history is a suitably smooth timelike geodesic (at least in a neighbourhood of the instant in which observations are made) (see fig. 5.2). The surfaces $w \equiv w_0 = $ const are past light cones of events situated on the curve C. The parametrization of the curve C is normalised in such a way that w measures the proper time along C. The instant $w = w_0$ corresponds to the event q which is our "here and now".

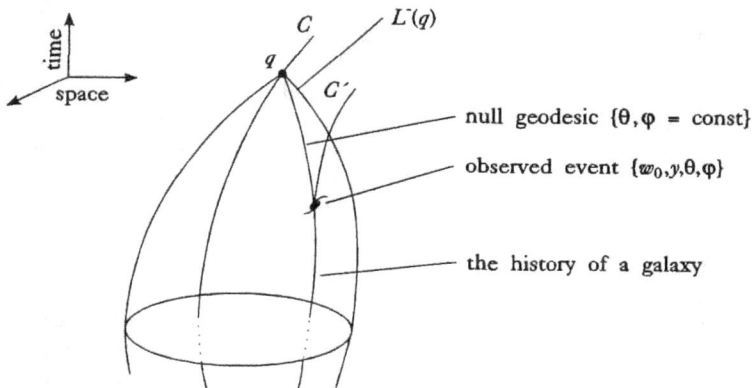

Fig. 5.2. Observational coordinates.

Let us denote the light cone of the event q by $L^-(q)$. Null geodesics generating $L^-(q)$ are given by $\{x^2 \equiv \theta = $ const, $x^3 \equiv \varphi = $ const$\}$ on the surface $w = $ const. The coordinates θ and φ can be normalised in such a way that they become the usual spherical coordinates on the celestial sphere with respect to a non–rotating reference frame.

The coordinate $x^1 \equiv y$ measures space–time distance along null geodesics. One should notice that the concept of a space–time distance contains both a space distance to the curve C and a temporal distance to the event q. There are a few possibilities as to how practically identify the coordinate y with the actually measured "distances".

It may happen that the coordinates $\{w, y, \theta, \varphi\}$ do not cover the entire space–time manifold (even if one does not take into account the coordinate singularity

$\theta = \varphi = 0$). However, they do cover the region $\mathcal{N} \subset \mathcal{M}$ which can be observed from the curve C (these coordinates can also give a "one–many" covering, i.e., the same coordinates can designate distinct events; in such a case, ghost images of galaxies may appear).

Usually, one assumes that the length of the period during which the humanity performs its astronomical observations is negligibly small as compared with the age of the Universe, and one can treat it as an event q on the curve C. If — with a bit of optimism — one replaces the event q by a finite proper time interval $\mathcal{I}$, the observational situation is slightly improved (fig. 5.3).

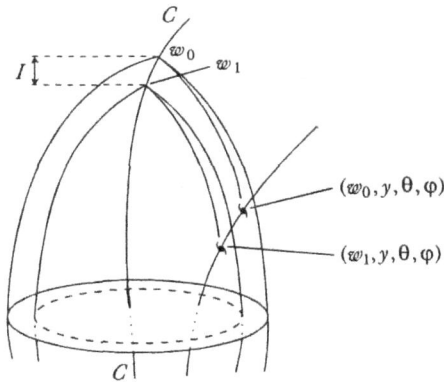

Fig. 5.3. Observational situation with the assumption of a finite period of performing observations.

The general conclusion from the detailed analysis — which, unfortunately, cannot be carried out here — is rather pessimistic (as it should be expected): ideal astronomical observations are not enough to uniquely reconstruct the geometry of space–time on our past light cone $L^-(q)$. It turns out, in particular, that we are not able to observationally determine (with no help of dynamical equations) how the three–dimensional hypersurface forming our past light cone is embedded in the four–dimensional space–time manifold. Moreover, it is impossible to demonstrate observationally that space–time is spherically symmetric around us; neither the opposite statement can be demonstrated. These pessimistic conclusions remain in force even if one assumes that our observations cover a finite span of time. Therefore, observational data collected from $L^-(q)$ leave a considerable cosmological indeterminacy: without help from dynamical equations (such as Einstein's field equations) many different world models can still show consistency with observational data.

How the situation is improved when one takes into account dynamical equations? One should be aware of the fact that after such equations have been accepted, observational data become theory–impregnated to a much higher degree than before; and one cannot hope that one would still be able to discern in a cosmological model what comes from observation and what is theoretically introduced into it. The problem was addressed in the second part of the work by Ellis et al. (1985). Einstein's field equations have been assumed as dynamical equations in cosmology; therefore, in this work, observational implications of the relativistic cosmology are investigated. No space–time symmetries are *a priori* assumed. On the contrary, the aim of the work is to demonstrate the existence or non–existence of space–time symmetries (or of other properties of a model) from observations helped by the theory of general relativity.

The observational coordinates are also used in this approach, but they have to be adapted to the field equations (afterwards they were analysed with the help of the Newman–Penrose formalism). The problem consists in determining (in terms of ideal astronomical observations) the Cauchy data for the field equations on the past light cone $L^-(q)$, and then in integrating these equations on and outside the past light cone $L^-(q)$. However, one must formulate the *final value problem*, characteristic for cosmology: one integrates the equations from the Cauchy data on $L^-(q)$ backward to a certain value $y = y^*$. This limitation for the value of y is a consequence of the fact that the very early Universe was not transparent to electromagnetic radiation. The backward integration on the light cone $L^-(q)$ can meet obstacles in a caustic form which may have origin either in the topology of space–time or in the phenomenon of *gravitational lensing*. This, of course, creates serious difficulties for the program of observational cosmology (see Friedrich and Steward 1982).

By using the field equations, the coordinate y can be (uniquely) replaced by the red shift z. To the limiting value y^* now corresponds the value z^*.

The Cauchy data, consisting of the most detailed cosmological information we could hope to obtain, form the *maximal data set*; let it be denoted by $\mathcal{D}(w_0, z)$. It consists in the following ideal observations: red shifts, proper motions, distances to galaxies (the so–called area distances)[10], distortion of images of celestial objects and their groups (provided that "proper shapes" are known), results of galaxy and quasar counts up to a limiting value $z = z^*$. It is worthwhile to notice that the background radiation is not an element of $\mathcal{D}(w_0, z^*)$. This radiation does not carry any information about particular regions of space–time, but rather imposes certain global constraints on the space–time structure which have to be satisfied by every cosmological model purporting to describe the real Universe.

Here an interesting problem arises, namely the problem of the observational space–time. A space–time $(\mathcal{M}, g)$, being a solution of Einstein's field equations, is said to be *observational space–time associated with* $\mathcal{D}(w_0, z^*)$ if it contains a hypersurface Σ such that $\Sigma \cup \{q\}$ is the past light cone $L^-(q)$ of an event q situated on a regular timelike curve C, and if all observational data as predicted by the model $(\mathcal{M}, q)$

---

[10]This distance is determined by observing the pencil of null geodesics emanating from an observation event q. If this pencil forms a solid angle $d\Omega$ at q and spans a space element dS at a certain point p along the pencil, then the *area distance* from q to p is $dS = r^2 d\Omega$.

are exactly those data which are elements of $\mathcal{D}(w_0, z^*)$. If there exist more than one observational space–time associated with a given $\mathcal{D}(w_0, z^*)$, one is met with an *observational indeterminacy* of the model: there are more than one cosmological models that are consistent with observational data contained in $\mathcal{D}(w_0, z^*)$.

Actual achievements of the Ellis program consist in demonstrating that the maximal data set $\mathcal{D}(w_0, z^*)$ is simultaneously the smallest set of observational data required to uniquely determine the geometry of the past light cone $L^-(q)$ backward to $z = z^*$ In other words, the knowledge of the ideal data $\mathcal{D}(w_0, z^*)$ is the necessary and sufficient condition for uniquely determining, with the help of Einstein's equations, the geometry on our past light cone (backward to z^*). Doubtlessly, it is an interesting result. One cannot see any reason why it should be *a priori* true. One could easily imagine the situation in which a model of the Universe would be either undetermined or overdetermined by the set of ideal observational data available in principle (the first possibility occurs where the field equations are not taken into account; see above).

Geometric structure of the past light cone $L^-(q)$ is only a part of a cosmological model. Now, one should extend the solution obtained on the light cone to the Cauchy domain of dependence inside the cone. It is a difficult mathematical problem which can be successfully solved if relationships between mathematical quantities are analytic. However, there are some reasons suggesting that this assumption is not necessary. Since the analycity assumption is very restrictive, it would be desirable to have a proof without it.

It is also possible to treat $\mathcal{D}(w_0, z^*)$ as the initial data and look for a solution to the future, rather than backwards. This is a still more difficult problem. From the hyperbolic character of Einstein's equations it follows that one cannot expect a "deterministic" solution starting from a set of initial data, unless one assumes that the solution is analytic or that it cannot be perturbed by anything which could not be contained in the initial conditions (see sec. 1). The fact that, for a cosmologist, to reconstruct the past is more important than to forecast the future is hardly a consolation.

The question of stability of observational space–times is an important problem which still awaits its full elaboration. Roughly speaking, an observational space–time $(\mathcal{M}_0, g_0)$ associated with the maximal data set $\mathcal{D}_0(w_0, z^*)$ is said to be *observationally stable* if a small perturbation $\mathcal{D}(w_0, z^*)$ of the set $\mathcal{D}_0(w_0, z^*)$ leads to observational space–times $(\mathcal{M}_i, g_i)$, associated with $\mathcal{D}(w_0, z^*)$ such that $(\mathcal{M}_i, g_i)$ only slightly differ from $(\mathcal{M}_0, g_0)$. To define precisely the meaning of "slightly differs from" would require introducing a topology on the space of solutions to Einstein's equations. The observational stability condition should be satisfied by realistic world models. The situation in which slightly different observational data would lead to drastically different world models would be disastrous for any research in cosmology.

Although the analysis of the observational situation in cosmology with the ideal observations replaced by the real ones is still to be carried out, it is not difficult to foresee its general result. If the precise knowledge of the ideal data $\mathcal{D}(w_0, z^*)$ is necessary and sufficient to reconstruct the geometry of our past light cone, and since there is no doubt that the postulate to be realistic will restrict the data (to the set $R(w_0, z^*)$, say), one can expect a significant observational indeterminacy of cosmological models. Analogously to the concept of observational space–times associated with $\mathcal{D}(w_0, z^*)$, one can introduce the concept of *realistic space–times associated with* $R(w_0, z^*)$. On the strength of the above analysis it is clear that the set of all realistic space–times associated with $R(w_0, z^*)$ consists of more than one space–time. This statement can be thought of as a formulation of the *observational indeterminacy principle* in cosmology. It turns out to be an unavoidable consequence of the extrapolation method used in cosmological research.

Observational indeterminacy in cosmology can be either acknowledged as a necessary element of our world picture, or neutralized with the help of different philosophical postulates such as the cosmological principle claiming that the earthly observer occupies a typical position in the Universe (see Ellis 1975; Heller 1978). If one chooses the second possibility, one has to define cosmology not only as a science about the Universe, but also as a science about the assumptions which must be made in order to render such a science possible. If we take into account the fact that even the observational Ellis' program is based on many (both explicit and tacit) assumptions (laws of physics are everywhere the same, space–time is modelled by the pair $(\mathcal{M}, g)$, etc.), it seems that the above definition of cosmology can hardly be avoided.

Chapter 6

Geometry of the Standard Cosmological Model

0. Introduction

In the preceding chapter we have discussed some fundamental problems of cosmology understood as a non–local physics. The discussion gave us a kind of program or suggested a style of research in cosmology. Unfortunately, the scope of the present book does not allow us to implement this challenge in a more ambitious way. Especially, to deal with cosmology as a theory of the ensemble of universes would require a separate monograph. Moreover, it would seem more reasonable to postpone writing such a monograph till the domain is worked out more fully. Nevertheless, the present book cannot be complete without at least touching the space of solutions of Einstein's equations. No wonder that in such a situation our choice goes to a subset (of the zero measure!) of this space, namely to the *Robertson–Walker solutions*. These solutions are distinguished: first, by the maximal degree of symmetry admitting the existence of the "open" global time; second, by the surprising property that there are exactly these simplest, in a sense, world models that correctly — albeit in an approximate way — describe the observed Universe. Because of the last property one sometimes speaks of the *standard cosmological model*.

The present chapter is intended as an example of doing cosmology according to the program outlined in the preceding chapter (inasmuch as a zero–measure set can serve as an example of anything!). Therefore, from the very beginning we turn to global mathematical methods. They allow us to compactify the material and pay off with precise definitions of such concepts as particle with a non–zero rest-mass, photon, observer, etc. which have to be put into the formalism in order to change mathematics into cosmology. The books by Sachs and Wu (1977) and by O'Neill (1983) were both good examples and a great help. However, in the final parts of the chapter, concerning observational aspects of cosmology, we postpone this formalistic approach. It turns out to be too narrow for embracing the observational practice. To treat a galaxy or a cluster of galaxies as a curve in space–time can be very precise from the mathematical point of view, but — if taken literally — does not allow one even meaningfully to speak about the angular diameter of a celestial object. Cosmology, in a very transparent way, reveals both the weakness and the strength of the empirical method: The weakness — since the reality, as it manifests itself in empirical results, turns out to be too rich to be translated into the available mathematical structures; and the strength — since simple mathematical structures surprisingly well approximate the reality.

The organisation of the present chapter is the following. In sec. 1, the concept of warped (Cartesian) product is introduced; it turns out to be a good tool significantly simplifying subsequent analyses. In sec. 2, the aforementioned definitions of particles and observers are presented, and some propositions connected with them are proved; they will serve us afterwards in shortening computations. The construction of the standard cosmological model (of the Robertson–Walker space–time) is carried out in sec. 3. The most important features of the evolution of the standard world model (singularities, spatial curvature, etc.) are expressed in the three theorems. The goal of sec. 4 is to formulate and to prove them. In cosmological formulations two sub-classes of the Robertson–Walker world models play an important role, namely dust– and radiation–filled universes. Solutions describing these models are presented in sec. 5. Observational aspects of the standard world model are (very) briefly discussed in sec. 6. Many comments and remarks can be made as far as the standard model is concerned; in sec. 7, we focus on the problem of idealisations that are incorporated into the construction of this model; we briefly mention solutions with the cosmological constant different from zero and world models with non–standard topologies (which are locally Robertson–Walker space–times); we touch upon a difficult problem concerning causally disconnected regions of space–time and the two attempts to solve it: the older one, known under the name of the mixmaster mechanism, and the newer one, the so–called inflationary scenario.

The present chapter is followed by a *Bibliographical Essay*; its aim is to introduce the reader, who might feel disappointed with the brevity of the chapter dealing with the standard universe, to a fascinating realm of books on cosmology and related topics.

1. Warped Products

Let us consider a Cartesian product $\mathcal{B} \times \mathcal{F}$ of two pseudo–Riemannian manifolds $(\mathcal{B}, g_B)$ and $(\mathcal{F}, g_F)$, where g_B and g_F are metric tensors on $\mathcal{B}$ and $\mathcal{F}$, correspondingly. $\pi^*(g_B) + \sigma^*(g_F)$, where $\pi: \mathcal{B} \times \mathcal{F} \to \mathcal{B}$, $\sigma: \mathcal{B} \times \mathcal{F} \to \mathcal{F}$ are projections, is a metric tensor on $\mathcal{B} \times \mathcal{F}$. Let $f > 0$ be a smooth function on $\mathcal{B}$.

1.1. Definition. Cartesian product $\mathcal{B} \times \mathcal{F}$, equipped with the metric tensor

$$g = \pi^*(g_B) + (f \circ \pi)^2 \sigma^*(g_F),$$

is called *warped product* and denoted by $\mathcal{M} = \mathcal{B} \times_f \mathcal{F}$.

If $f = 1$, the warped product is the usual Cartesian product. As it can be easily seen, warped product is a special case of a fibre bundle. Therefore, it is reasonable to call $\mathcal{B}$ a *base manifold* (or *base*, for short), and $\pi^{-1}(p) = p \times \mathcal{F}$, $p \in \mathcal{B}$, a fibre over p

(fig. 6.1). Fibres and cross–sections $\sigma^{-1}(q) = \mathcal{B} \times q$, $q \in \mathcal{F}$, are pseudo–Riemannian manifolds. f is said to be a *warping function*.

1.2. Example. All surfaces of revolution are warped products. Let $\mathcal{M}$ be a surface obtained by the revolution of a flat curve C around an axis in $\mathbf{R}^3$. The distance from the curve C to this axis is given by the function $f: C \to \mathbf{R}^1$. One has $\mathcal{M} = C \times {}_f\mathcal{S}^1$.

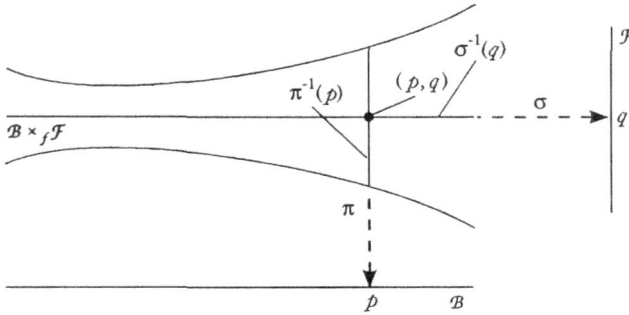

Fig. 6.1. Warped product $\mathcal{M} = \mathcal{B} \times {}_f\mathcal{F}$.

1.3. Definitions. Let $\mathcal{B} \times \mathcal{F}$ be a Cartesian product of manifolds $\mathcal{B}$ and $\mathcal{F}$, π and σ the projections as above, and f a function on $\mathcal{B}$. The function $\tilde{f} = f \circ \pi$ defined on $\mathcal{B} \times \mathcal{F}$ is said to be the *lift of the function* f to the product $\mathcal{B} \times \mathcal{F}$. Let $p \in \mathcal{B}$, $q \in \mathcal{F}$, and $v \in T_p(\mathcal{B})$. The unique vector $\tilde{v} \in T_{(p,q)}(\mathcal{B} \times \mathcal{F})$, such that $d\pi(\tilde{v}) = v$, is said to be the *lift of the vector* v to $(p,q) \in \mathcal{B} \times \mathcal{F}$.

In an analogous manner, one can *lift* functions and vectors from the manifold $\mathcal{F}$ to $\mathcal{B} \times \mathcal{F}$. The above definitions remain valid also with respect to warped products.

1.4. Definition. Let $(\mathcal{M}, g_M)$ and $(\mathcal{N}, g_N)$ be pseudo–Riemann manifolds. A diffeomorphism $\psi: \mathcal{M} \to \mathcal{N}$ such that $\psi^*(g_N) = cg_M$, where $c = \text{const} \neq 0$, is called a *homotetic mapping*, or a *homotety* for brevity (it is a special case of a *conformal mapping* which is defined by the condition: $c = h$ is a different from zero function on $\mathcal{M}$). $|c|^{\frac{1}{2}}$ is called a *scale factor*.

It is easy to prove the following.

1.5. Proposition. Let $\mathcal{M} = \mathcal{B} \times {}_f\mathcal{F}$ be a warped product,

(a) for every $q \in \mathcal{F}$, the mapping $\pi|_{(\mathcal{B} \times q)}$ is an isometry on $\mathcal{B}$;
(b) for every $p \in \mathcal{B}$, the mapping $\sigma|_{(p \times \mathcal{F})}$ is a homotety with the scale factor $1/f(p)$;
(c) for every $(p,q) \in \mathcal{M}$, the fibre $\pi^{-1}(p)$ and the cross–section $\sigma^{-1}(q)$ are orthogonal at (p,q). □

The properties expressed in this proposition allow one easily to reconstruct the geometry of the warped product $\mathcal{B} \times {}_f\mathcal{F}$ when the geometry is known of the manifolds $\mathcal{B}$ and $\mathcal{F}$ (see O'Neill 1983, pp. 204–211; Beem and Ehrlich 1981, pp. 55–79).

The concept of warped product was introduced by Bishop and O'Neill(1969).

2. Macroscopic Description of Particles and Observers

Let us begin with definitions the aim of which is to make the description more precise of an "averaged matter content" of the Universe traditionally used in cosmology. In what follows, all structures are defined on the space–time manifold $(\mathcal{M}, g)$.

2.1. Definition. A timelike future–oriented curve $\alpha\colon I \to \mathcal{M}$, such that $|\alpha'(\tau)| = 1$ for all $\tau \in I$, is said to be an *observer* in space–time $\mathcal{M}$; τ is his proper time.

2.2. Definition. A timelike future–directed curve $\alpha\colon I \to \mathcal{M}$, such that $g(\alpha', \alpha') \equiv \langle \alpha', \alpha' \rangle = -m^2$, is said to be a *particle with the rest-mass m*, or a *particle* for brevity.

2.3. Definition. If the curve α of definition 2.1 or 2.2 is a geodesic curve, the observer or the particle is called *freely falling*.

2.4. Definition. A future–directed zero geodesic $\gamma\colon I \to \mathcal{M}$ is said to be a *photon* or a *light beam*.

Images $\mathrm{Im}(\alpha)$, $\mathrm{Im}(\gamma) \subset \mathcal{M}$ of the curves $\alpha\colon I \to \mathcal{M}$ and $\gamma\colon I \to \mathcal{M}$ are one-dimensional manifolds; they are called *histories* (or *world lines*) of observers or particles.

2.5. Definitions. Timelike future–directed unit vector $v \in \mathcal{T}_q(\mathcal{M})$ is said to be an *instantaneous observer* at $q \in \mathcal{M}$. By performing the orthogonal decomposition of the tangent space $\mathcal{T}_q(\mathcal{M}) = \mathbf{R}u + u^\perp$, one obtains the *time axis* $\mathbf{R}u$ of the observer and its *rest–space* $u^\perp$.

One could imagine that an instantaneous observer is situated in $\mathcal{T}_q(\mathcal{M})$ and observes the Universe with the help of the *exponential mapping* $\exp_q\colon \mathcal{T}_q(\mathcal{M}) \to \mathcal{M}$. If $u \in \mathcal{T}_q(\mathcal{M})$, then there exists a unique geodesic curve α such that $a'(0) = u$; the exponential mapping is defined by $\exp_q(u) = \alpha(1)$. By the phrase "to observe the space–time $\mathcal{M}$ from the tangent space $\mathcal{T}_q(\mathcal{M})$ with the help of the exponential mapping" one should understand projecting structures from an open neighbourhood $U \subset \mathcal{M}$ of q by $\exp_q^{-1}$ and treating them as structures on $\mathcal{T}_q(\mathcal{M})$ (see Sachs and Wu 1977a, sec. 4.2).

2.6. Definition. A unit vector field, the integral curves of which are observers (of definition 2.1), is said to be a *reference frame* in the space–time $(\mathcal{M}, g)$. Observers in question will be called *observers in a given reference frame*.

Satisfactory agreement with actual observational data is achieved under a surprisingly strong simplification allowing one to treat the "material content of the Universe" as a perfect fluid. In physics, perfect fluid is defined as a continuous medium moving, at each point $q \in \mathcal{M}$, with a velocity v_q such that an observer (in an everyday meaning of this term), moving with the same velocity v_q, sees this medium as isotropically distributed in space. Till the end of the present section, all analyses will be carried out locally, i.e., in a neighbourhood of a point q at which the perfect fluid approximation remains valid. The above ideas are formalized in the following.

2.7. Definition. A *perfect fluid* in space–time $(\mathcal{M}, g)$ is the triple (U, ρ, p), where U is a reference frame, called the *co–moving reference frame*; $\rho \colon \mathcal{M} \to [0, \infty)$ is the so–called *energy density function*, and $p \colon \mathcal{M} \to [0, \infty)$ is the so–called *pressure* (or *stress*) *function* (it is assumed that ρ and p are of C^∞ class). *The energy-momentum tensor of the perfect fluid* is defined to be

$$T = (\rho + p)U^* \otimes U^* + pg, \tag{2.1}$$

where U^* is a one–form metrically equivalent to the vector field U.

2.8. Proposition. The components of the energy–momentum tensor of the perfect fluid (in the co–moving reference frame) are

$$T_{00} = \rho, \qquad T_{ij} = p\delta_{ij}, \quad (i, j = 1, 2, 3). \ \blacksquare$$

Proof. Let U be the co–moving frame. We shall use the basis (U, X, Y) where $X, Y \perp U$. Then $T(U, U) = \rho$; $T(U, X) = 0$; $T(X, X) = \langle X, Y \rangle$. $\square$

If $p = 0$, the perfect fluid is called *dust*; if $p = \frac{1}{3}\rho$, the perfect fluid is called *photon gas* or *radiation field*.

The local conservation laws requires $\operatorname{div} T = 0$.

2.9. Proposition. Let (U, ρ, p) be a perfect fluid. $\operatorname{Div} T = 0$ implies

(1) $$U\rho = -(\rho + p)\operatorname{div} U;$$

(2) $$-(\operatorname{grad} p)^\perp = (\rho + p)\nabla_U U. \ \blacksquare$$

Proof. Let $^*T$ be a tensor field of the type $(2, 0)$ metrically equivalent to the tensor field T. One has $(\operatorname{div} T = 0) \Rightarrow (\operatorname{div} {}^*T = 0)$, and one obtains

$$\operatorname{div} {}^*T = U(\rho + p)U + (\rho + p)\nabla_U U + (\rho + p)(\operatorname{div} U)U + \operatorname{grad} p = 0.$$

The vertical component of this equation $(\langle \operatorname{div} {}^*T, U \rangle = 0)$ gives equation (1), and its longitudinal component $(\langle \operatorname{div} {}^*T, \nabla_U U \rangle = 0)$ equation (2). $\square$

Equation (1) expresses the rate of change of the energy density as measured by an observer in the co–moving reference frame U. If in equation (2) the space component of the gradient of the pressure function $((-\operatorname{grad} p)^\perp)$ is treated as a force, the coefficient $(\rho + p)$ as a counterpart of mass, and $\nabla_U U$ as an acceleration, then equation (2) turns out to be Newton's second law.

The presented here formalised macroscopic description of particles and observers was originally proposed by Sachs and Wu (1977a, b), and its somewhat simplified form can be found in O'Neill (1983) (see also Heller 1982).

3. Construction of the Standard Cosmological Model

In a geometric approach to physics adopted also in the present book — in contradistinction to the purely mathematical texts — definitions are more important than theorems. Very often definitions express fundamental facts or experimental generalizations, whereas theorems are usually only their formal consequences. Through definitions the reality enters mathematics (see Sachs and Wu 1977a, sec. 1.5). The following definition constitutes both a mathematical stylization and a synthesis of many observational facts lying at the basis of the standard cosmological model (these facts will be briefly presented in sec. 6 of the present chapter).

3.1. Definition. Let S be a connected three–dimensional Riemann manifold with constant curvature $k = -1, 0$, or $+1$; I be an open interval of the space $\mathbf{R}_1^1$, where $\mathbf{R}_1^1$ denotes the space $\mathbf{R}^1$ with the signature 1; $R(t) > 0$, $t \in I$ be a smooth real function on I. The warped product $\mathcal{M} = I \times_R S$ is called a *Robertson–Walker space–time*.

A Robertson–Walker space–time is, therefore, a manifold $I \times S$ with the line element

$$ds^2 = -dt^2 + R^2(t)d\sigma^2 \tag{3.1}$$

where $d\sigma^2$ is a line element on the manifold S lifted to $I \times S$. Fibres $S(t)$, $t \in I$, are called *instantaneous spaces* or *space sections* (at an instant t) of the Robertson–Walker space–time. Every instantaneous space $S(t)$ has a constant curvature equal to $k/R^2(t)$. The warping function $R(t)$, in this context, will be called the *scaling function*, and its value $R(t_0)$, for an instant $t_0 \in I$, the *scale factor*.

The line element σ^2 can be reduced to the one of the following two forms,

$$d\sigma^2 = \frac{(dx^1)^2 + (dx^2)^2 + (dx^3)^2}{\left\{1 + \frac{k}{4}\left[(x^1)^1 + (x^2)^2 + (x^3)^2\right]\right\}^2} \tag{3.2a}$$

$$= \frac{dr^2 + r^2 d\theta^2 + r^2 \sin^2\theta d\varphi^2}{\left[1 + \frac{kr^2}{4}\right]^2} \tag{3.2b}$$

Let d/dt be a vector field on $I \subset \mathbf{R}^1$, and $U = \partial/\partial t$ its lifting to $I \times S$. The curve $I \times q$, $q \in S$, can be parametrized in the following way: $\alpha_q(t) = (t, q)$. Of course, it is an integral curve of the field U, and consequently an observer in the reference frame U (definition 2.6); such an observer will be called a *fundamental observer*. Thus, the function t ascribes to every fundamental observer his proper time; this function will be called the *cosmic time function*, or *time function*, for brevity.

It can be easily checked that $\langle U, U \rangle = g(U, U) = -1$, and $U \perp S(t)$, for every $t \in I$ (see proposition 1.5c), i.e., every instantaneous space $S(t)$ is a spacelike hypersurface in the space–time $\mathcal{M} = I \times_R S$.

Let us introduce two further concepts playing important roles in cosmology.

3.2. Definition. A space–time $\mathcal{M} = I \times_R S$ is said to be *spatially isotropic with respect to a point* $q \in S(t)$ if the point $(t, q) \in \mathcal{M}$ has a neighbourhood $\mathcal{N}$ such that, for any two unit vectors v and w tangent to $\mathcal{S}(t)$ at (t, q), there exists an isometry $\phi: \mathcal{N} \rightarrow \mathcal{N}$ such that $\phi = \mathrm{id} \times \phi|_{\mathcal{S}}$ and $d\phi(v) = w$.

3.3. Definition. A space–time $\mathcal{M} = I \times {}_R \mathcal{S}$ is said to be *spatially homogeneous* if, for $p, q \in \mathcal{S}(t)$, $t \in I$, there exists an isometry $\psi: \mathcal{U} \rightarrow \Theta$, where $\mathcal{U}$ and Θ are neighbourhoods of p and q respectively.

Less strictly but quite commonly, one simply says that the space $\mathcal{S}(t)$ is isotropic around a point $q \in \mathcal{S}(t)$ or homogeneous.

The concept of spatial isotropy formalises the intuitive statement that there are no privileged directions with respect to a point q. The concept of spatial homogeneity formalizes the intuitive statement that there are no privileged points in the considered space. These concepts can be more generally defined in terms of the Lie group theory (see Wolf 1967, Robertson and Noonan 1968). It is known from the Schur theorem that a space which is isotropic with respect to each of its points is also homogeneous and that a space is isotropic with respect to each of its points if and only if it is a space of constant curvature (see Wolf 1967, p. 57).

From these considerations we obtain the following.

3.4. Corollary. Instantaneous spaces of a Robertson–Walker space–time are isotropic around every of its points, and consequently they are also homogeneous. □

This corollary plays an important role in observational motivation of the the choice of a Robertson–Walker space–time for the standard cosmological model.

3.5. Proposition. For any Robertson–Walker space–time $\mathcal{M} = I \times {}_R \mathcal{S}$, in the reference frame $U = \partial/\partial t$, non–vanishing components of the Ricci tensor are

$$\Re_{00} = -3\frac{\ddot{R}}{R},$$

$$\Re_{11} = \Re_{22} = \Re_{33} = -2\frac{\dot{R}^2}{R^2} + 2\frac{k}{R^2} + \frac{\ddot{R}}{R}$$

(notice the change of the Ricci tensor symbol as compared with chapter 4) and the curvature scalar is

$$\Re = 6\left\{ \frac{\dot{R}^2}{R^2} + \frac{k}{R^2} + \frac{\ddot{R}}{R} \right\}. \quad \blacksquare$$

Proof by direct computation from the definition; one can also employ the theorem on the Riemann and Ricci tensors in warped products (see O'Neill 1983, pp. 209–211). □

Einstein's equations introduce dynamics into the cosmological model. In the following, for the sake of simplicity, we will use Einstein's equations in the form (5.7) of chapter 4, i.e., without the cosmological constant. This simplification is motivated by observational data which point out that cosmological constant effects at the present epoch are negligible within limits of observational accuracy (see Petrosian 1974). In the following, we shall use the *geometric system of units* in which the light velocity c

and the Newtonian constant of gravity G are assumed to be equal to one; in such a case the constant κ appearing in equation (5.7) of chapter 4 has the value 8π.

3.6. Theorem. Perfect fluid (U, ρ, p) in the Robertson–Walker space–time $\mathcal{M} = I \times_R S$ is described by the following equations

$$\frac{8\pi}{3}\rho = \frac{\dot{R}^2}{R^2} + \frac{k}{R^2},$$
(3.3)

$$-8\pi p = 2\frac{\ddot{R}}{R} + \frac{\dot{R}^2}{R^2} + \frac{k}{R^2}. \quad \blacksquare$$
(3.4)

Proof by direct computation from Einstein's equations $\mathfrak{R}_{\mu\nu} - \frac{1}{2}\mathfrak{R}g_{\mu\nu} = -8\pi T_{\mu\nu}$; proposition 3.5 can be used. By comparing equations (3.3) and (3.4) with formula (2.1) one can see that these equations describe a perfect fluid. □

Definition 3.1, together with theorem 3.6 constitutes the essence of the construction of the standard cosmological model.

3.7. Corollary.

$$-4\pi(\rho + 3p) = -3\frac{\ddot{R}}{R}. \quad \square$$
(3.5)

3.8. Corollary. Equation (1) of proposition 2.9 in a Robertson–Walker space–time assumes the form

$$U\rho = \dot{\rho} = -3(\rho + p)\frac{\dot{R}}{R}.$$
(3.6)

(Equation (2) of the same proposition is trivially satisfied since in the considered case $(\operatorname{grad} p)^{\perp} = \nabla_U U = 0$.) □

4. Evolution of the Standard Cosmological Model.

The structure of the cosmological model presented in the preceding section is rich enough to contain in itself information concerning the world evolution and its limitations. This information is deciphered in the following theorems.

4.1. Theorem. (O'Neill 1983, p. 348). Let $\mathcal{M} = I \times_R S$ be a Robertson–Walker space–time satisfying the condition: $\rho + 3p > 0$ (the *strong energy condition*). If for a certain $t_0 \in I$ the relationship $H_0 \equiv H(t_0) \equiv \dot{R}_0/R_0 > 0$ is satisfied, then the interval I has the initial point t_{in} such that $t_0 - H_0^{-1} < t_{\text{in}} < t_0$, and either (1) $\dot{R} > 0$ or (2) R attains maximum at $t_{\max}$, $t_{\max} > t_0$. If (2) takes place, then $I = (t_{\text{in}}, t_{\text{fin}})$. $\blacksquare$

Proof. From the condition $\rho + 3p > 0$ and relationship (3.5) it follows that $\ddot{R} < 0$, i.e., that the diagram of the function $R(t)$ is a convex curve. In particular, this diagram is everywhere (besides the tangency point) beneath the diagram of the function

$$Q(t) = R(t_0) + H_0 R(t_0)(t - t_0),$$

which is the tangent line to $R(t)$ at the point $R(t_0)$. The condition $H_0 > 0$ implies vanishing of the function $Q(t)$ at $t_0 - H_0^{-1}$ (see, fig. 6.2). Since, however, the diagram of the function $R(t)$ is a convex curve and $R(t)$ cannot assume negative values, $R(t)$ must break down at a certain instant t_{in} situated between $t_0 - H_0^{-1}$ and t_0 .

From the fact that $\ddot{R}(t) < 0$ on the entire I, it follows that either (1) $\dot{R} > 0$ on the entire I or (2) R attains maximum for a certain $t_{max} > t_0$, and afterwards $\dot{R} < 0$. In the last case, the reasoning analogous to that in the first part of the proof leads to the conclusion that I must have the final point $t_{fin} > t_0$. $\square$

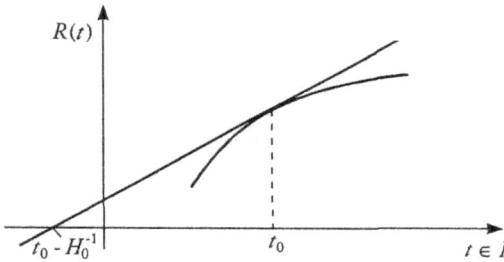

Fig. 6.2. Breaking down of the history of the Robertson–Walker world model at the instant t_{in}; $t_0 - H_0^{-1} < t_{in} < t_0$

Note that theorem 4.1 says nothing about the state of the Universe in the process of its approaching backward in time to the initial instant t_{in}; in particular, it does not say that $\rho \to \infty$ and $R \to 0$ when $t \to t_{in}$.

4.2. Theorem. Let $\mathcal{M} \times_R \mathcal{S}$ be a Robertson–Walker space–time with $H_0 > 0$ for a certain $t_0 \in I$. If $\rho > 0$, $-\frac{1}{3} < a \le p/\rho$, $a = $ const, and $R \to 0$ as $t \to t_{in}$, then $\dot{R} \to \infty$ and $\rho \to \infty$ when $t \to t_{in}$. $\blacksquare$

Proof. From the assumption $-\frac{1}{3} < a \le p/\rho$ it follows that $\rho + 3p > \varepsilon p > 0$ for a certain $\varepsilon > 0$. Therefore, theorem 4.1 applies; in particular $\ddot{R} < 0$.

Since in the interval (t_{in}, t_0) one has $\dot{R} > 0$, and on the strength of the above condition $\rho - \varepsilon \rho \ge -3p$, equation (3.4) changes into the inequality

$$\dot{\rho} \le -(2 + \varepsilon)\rho \frac{\dot{R}}{R},$$

which, as can be easily seen, is equivalent to

$$\frac{d}{dt}(\rho R^{2+\varepsilon}) \le 0,$$

i.e., in the interval (t_{in}, t_0) one has $\rho R^{2+\varepsilon} \ge \rho(t_0)R(t_0)^{2+\varepsilon}$. However, from the assumption one knows that $R \to 0$ if $t \to t_{in}$. Therefore, $\rho R^2 \to \infty$ and $\rho \to \infty$ as $t \to t_{in}$.

Equation (3.3) can be given the form

$$\frac{8\pi}{3}\rho R^2 = \dot{R}^2 + k, \qquad (4.1)$$

which immediately leads to the conclusion that also $\dot{R} \to \infty$ as $t \to t_{\text{in}}$. □

Therefore, the evolution of the standard cosmological model begins with the singularity, the so-called *initial singularity*. It is characterized by the fact that as time, going backwards, approaches the initial instant, $t \to t_{\text{in}}$, both the energy density ρ and the expansion rate $\dot{R}$ grow indefinitely. This fact justifies the name *Big Bang*.

4.3. Theorem. Let $\mathcal{M} = I \times_R \mathcal{S}$ be a Robertson–Walker space–time satisfying the conditions of theorems 4.1 and 4.2. (A) If $\dot{R} > 0$ on the entire interval $I = (t_{\text{in}}, \infty)$, then either $k = 0$ or $k = -1$. (B) If R attains maximum at $t_{\text{max}} \in I$, then $k = +1$ (and $I = (t_{\text{in}}, t_{\text{fin}})$), and if $R \to 0$, as $t \to t_{\text{fin}}$, then $\rho \to \infty$ and $\dot{R} \to -\infty$, as $t \to t_{\text{fin}}$.∎

Proof. (A) First, let us assume that $\dot{R} > 0$ on the entire interval $I = (t_{\text{in}}, \infty)$. As in the preceding theorem, $p > 0$ and $\rho + 3p > 0$, which implies $\rho + p > 0$. Therefore, equation (3.6) leads to the conclusion that $\dot{\rho} < 0$, i.e., ρ decreases indefinitely. Two cases ought to be considered:

(1) $R \to \infty$ as $t \to \infty$. From the proof of theorem 4.2 it is known that $\frac{d}{dt}(\rho R^{2+\epsilon}) \le 0$ (but now on the entire I) which implies that $\rho R^{2+\epsilon}$ is a bounded function for large $t \in I$. In other words, $\rho R^2 \to 0$ as $t \to \infty$. From equation (4.1) one gets either $k = 0$ or $k = -1$.

(2) However, the case $R \to b = \text{const}$ as $t \to \infty$ cannot be excluded. With this assumption, and remembering that $\ddot{R} < 0$, one obtains that $\dot{R} \to 0$. This, together with equation (4.1) leads to the conclusion that either $k = 0$ or $k = +1$. However, the case $k = +1$ is excluded. Indeed, let us assume, on the contrary, that $k = +1$. From equation (4.1) one gets $\rho R^2 \to 3/(8\pi)$, and consequently $\rho \ge \delta$ for a certain $\delta > 0$. Since $\dot{R} \to 0$ as $t \to \infty$, there exists a sequence $\{t_i\} \to \infty$, $t_i \in I$, such that $\{\ddot{R}(t_i)\} \to 0$. Therefore, equation (3.5) gives $\{(\rho + 3p)(t_i)\} \to 0$, but this contradicts the inequality $\rho + 3p > \epsilon p > 0$ (see, beginning of the proof of theorem 4.2).

Therefore, if $\dot{R} > 0$ on the entire I, then either $k = 0$ or $k = -1$. (It can be shown that the case $R \to b$ as $t \to \infty$ also excludes $k = 0$; see O'Neill 1983, p. 349.)

(B) Let us assume that R attains maximum at $t_{\text{max}} \in I$, i.e., $\dot{R}(t_{\text{max}}) = 0$. Equation (3.3) gives $\rho(t_{\text{max}}) = 3k/8\pi R^2(t_{\text{max}}) > 0$; hence $k = +1$. The rest of the proof is similar to that of theorem 4.2. □

4.4. Corollary. Let us define the so-called *critical density* $\rho_{\text{crit}} = 3H_0^2/8\pi$. If $\rho_0 = \rho_{\text{crit}}$, then $k = +1$ (index zero denotes the value of the corresponding parameter at a certain instant $t_0 \in I$, for example at the instant of performing observation). ∎

Proof. From equation (3.3) one gets $\rho_0 - \rho_{\text{crit}} = 3k/8\pi R_0^2$. □

5. Dust Filled and Radiation Filled Universes

Two special cases of perfect fluid Robertson–Walker space–times play an important role in cosmological applications, namely dust filled and radiation filled world models. We will consider them in turn.

5.1. Proposition. The scaling function $R(t)$ of a Robertson–Walker space–time $\mathcal{M} = I \times {}_R S$ filled with dust $(U, \rho, p = 0)$ satisfies the so–called *Friedman equation*

$$\dot{R}^2 = \frac{8\pi}{3} \frac{E}{R} - k, \tag{5.1}$$

where $E = \rho R^3 > 0$ is a conserved magnitude on the entire interval I. ∎

Proof. Equation (5.1) is equivalent to equation (3.3). Equation (3.6) with $p = 0$, after simple transformations, gives $d(\rho R^3)/dt = 0$. □

Solutions of the Friedman equation describe the evolution of the cosmological model. In the considered case, these solutions — for different values of the parameter k — are the following,

$$
\begin{array}{lll}
k = 0: & R = Ct^{2/3}, & C = 6\pi E; \\
k = +1: & t = \frac{4}{3}\pi E(\theta - \sin\theta); \\
& R = \frac{4}{3}\pi E(\theta - \cos\theta), & (0 < \theta < 2\pi); \\
k = -1: & t = \frac{4}{3}\pi E(\sinh\eta - \eta); \\
& R = \frac{4}{3}\pi E(\cosh\eta - 1) & (\eta > 0).
\end{array}
$$

These solutions are shown in fig. 6.3; they are often called *dust filled Friedman universes*, or *Friedman universes* for short. Their evolution starts with the Big Bang (the instant t_{in} of the Big Bang has been accepted as the beginning of time counting, i.e., $t_{in} = 0$). In the case $k = 0$ (the *flat model*), $R \to \infty$ and $\dot{R} \to 0$ as $t \to \infty$. This solution is also called the *Einstein–de Sitter model*. If $k = +1$, the solution is a branch of the cycloid. The maximum occurs at $t_{max} = (4/3)\pi^2 E$ and is $R_{max} = 8\pi E/3$. The final singularity takes place at $t_{fin} = 8\pi^2 E/3$. In the case $k = -1$, $R \to \infty$ and $\dot{R} \to 1$ as $t \to \infty$.

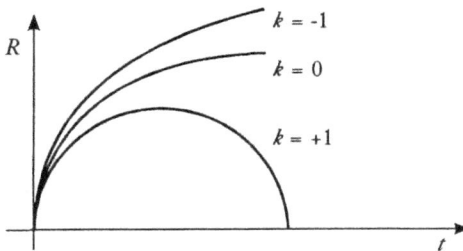

Fig. 6.3. Friedman's world models.

5.2. Proposition. The scaling function R(t) of a Robertson–Walker space–time $\mathcal{M} = I \times {}_R\mathcal{S}$ filled with radiation $(U, \rho, p = \rho/3)$ satisfies the equation

$$\dot{R}^2 = \frac{8\pi\varepsilon}{3R^2} - k, \qquad (5.2)$$

where $\varepsilon = \rho R^4$ is a conserved magnitude on the entire interval I. ∎

Proof. Equations (3.4) and (3.5), after straightforward manipulations taking into account that $p = \rho/3$, give equation (5.2). In this case, equation (3.6) turns out to be equivalent to $d(\rho R^4)dt = 0$. □

The solutions of equation (5.2) are the following,

$$k = 0: \qquad R = (2bt)^{\frac{1}{2}};$$
$$k = +1: \qquad R = (2bt - t^2)^{\frac{1}{2}};$$
$$k = -1: \qquad R = (2bt + t^2)^{\frac{1}{2}}.$$

where $b = (8\pi\varepsilon/3)^{\frac{1}{2}}$.

5.3. Corollary. If $t \to t_{\text{in}}$, all solutions behave as the flat solution, i.e., curvature effects become negligible. □

6. Observational Background of the Standard Cosmological Model

So far analyses of the present chapter had a purely mathematical character. Definitions of observers, particles, perfect fluid etc. (see, sec. 2) were very formal and rather loosely connected with physical or astronomical practice. Now, the formalism must be put into resonance with the results of measurements and observations. In other words, mathematical definitions, theorems and corollaries must be changed into what is usually called a cosmological model. These procedure immediately relaxes the entire formal fabric. Even a high idealisation of physical objects (e.g., of galaxies to points in space or to curves in space–time, see below) impresses on mathematics a stigma of empirical indeterminacy. Therefore, in the present section we postpone our previous mathematical style of exposition and change into a more "narrative language". Consequently, terms such as observer, particle, and so on, will be understood in their intuitive everyday meaning, and not in the sense of definitions 2.1–2.7.

First of all, one should be aware of the kind of vicious circle in which every empirical testing of cosmological models is entangled (in fact, it is a feature of the empirical method as such, but in cosmology it could be more dangerous than in other branches of the empirical sciences). On the one hand, without the help of a theory or of a model one does not know what and how to measure, and in which way the obtained "data" should be interpreted. On the other hand, with no observational data one could hardly create a theory or a model, not to mention testing of their validity. This vicious circle has to be somehow broken down. The method — honoured by practice and supported by surprising, albeit not quite understandable, successes — proceeds in the following way. At the starting point one accepts (usually tacitly) the simplest model; the simplest in this context means remaining in agreement with a scientific tradition or with intellectual habits. Conceptual tools such a model supplies allow one to gather and to interpret observational data. As long as consistency of interpreted data with the model can be maintained, there is no reason to look for another model. Only under the pressure of increasing inconsistencies and anomalies, scientists begin (not without a certain unwillingness) to revise the model, to change some of its assumptions, to revaluate some of its argumentation methods ... This process usually leads to a new model which guarantees, at least for the time being, agreement of the theory with observational data. If necessary, the history repeats itself.

Cosmology of the twentieth century began its evolution from a traditional and intuitive model of a static flat and Euclidean space and an absolute time. Scientists believed that such a model was implied by the Newtonian theory of gravity (or at least it did not contradict the Newtonian theory) and remains in agreement with astronomical observations.[1] Gradually, observational data discarded this model. The standard cosmological model emerged as a better approximation to the reality; within its framework a satisfactory agreement has been reached between the theory and observation. However, it cannot be thought of as "empirically proved" but only as a well–functioning model providing a consistent interpretation of empirical data. Exclusively in this sense, in the following we shall speak about observational verification of the standard world model.

First of all, one must observationally verify definition 3.1 which constitutes the core of the standard model. If this definition is to be referred to the physical reality, the space–time $\mathcal{M}$ of the Universe is supposed to be of the form of the warped product $\mathcal{M} = I \times_R \mathcal{S}$, where $\mathcal{S}(t)$ are instantaneous spaces of constant curvature, and the parameter $t \in I$ represents the cosmic time. The empirical motivation for the isotropy and homogeneity of instantaneous spaces is provided by (A) measurements of the microwave background radiation and (B) investigations of the spatial distribution of galaxies. The information that $R(t) \neq$ const is supplied by (C) red shift measurements in galactic spectra. The family of three–dimensional homogeneous and isotropic spaces[2] can be, in a natural way, parametrized by the parameter $t \in I$; this defines the projection $\pi: \mathcal{M} \to I$ which can be interpreted as the existence of the cosmic time in the model (see Appendix). In what follows we shall notice some difficulties which this apparently simple model encounters.

(A) The microwave background radiation has the following properties interesting from the cosmological point of view (their mere enumeration allows one to see to which extent establishing of these properties is the result of both theory and observation.

---

[1] As it is shown in the Appendix, such a model is in fact not consistent with Newton's theory.

[2] More precisely, these spaces are defined by the action of certain symmetry groups, see, for example, MacCallum 1973.

(1) Observations are carried out at wavelengths 0.1–10 cm. The radiation does not come from any identifiable discrete sources. Most probably, it was emitted "by the Universe" about 10^5–10^6 years after the Big Bang; before that date, called the *epoch*, or sometimes, more geometrically, the *surface, of the last scattering*, photon remained in equilibrium with other components of the primeval plasma. At the last scattering epoch the Universe became transparent for microwave photons; one speaks of photons decoupled from other forms of matter. If this interpretation is correct, the microwave background radiation carries the information about the Universe as regards how it was at the last scattering epoch. In other words, the background radiation allows us to penetrate backward up to 99.999 % of the evolution of the Universe.

(2) The background radiation spectrum has, up to good approximation, a Planck spectrum of the black body at temperature $T = 2.7$ K. This means that, at the epoch of emission of the background photons, the Universe was in the thermodynamical equilibrium.

(3) The background radiation possesses a high degree of isotropy. Expected deviations from isotropy can be of a threefold kind:

(a) anisotropy with a 24–hour period (*dipole anisotropy*); it is caused by the Earth motion with respect to the radiation field (once every 24 hours antenna crosses the direction in which the Earth moves with respect to the background radiation field);

(b) anisotropy with a 12–hour period (*quadrupole anisotropy*); it could serve as a test of the anisotropic expansion of the Universe at the last scattering epoch (twice every 12 hours antenna crosses a privileged direction of the world's expansion);

(c) anisotropy on small scales would constitute an empirical proof of small scale inhomogeneities of the Universe at the last scattering surface (the radiation was scattered on inhomogeneities).

The anisotropy with a 24–hour period has indeed been found. It turns out that — after taking into account corrections for the Earth's motion around the Sun and the Sun's motion around the centre of our Galaxy — one must ascribe to our Galaxy a speed of about $540 \, \mathrm{km \cdot s^{-1}}$ with respect to the radiation field (in the direction: $\alpha = 10^{\mathrm{h}},7$, $\delta = -22°$). It is worth noticing that the background radiation field is a physical realization of the idea of a co–moving reference frame.

The background radiation field has a high degree of isotropy of types (b) and (c). The upper limit of the temperature T anisotropy, at any scale, is $\Delta T/T < 10^{-3}$, and for angular scales less than $3°$, $\Delta T/T < 10^{-5}$. The last result comes as a surprise since, according to present theories of formation of galaxies and cluster of galaxies, pregalactic condensations of matter should manifest themselves in small scale anisotropies of the background radiation. The actually measured degree of the isotropy puts severe constraints on theories of galaxy formation.

To sum up, one must say that observational data concerning the microwave background radiation inform us about: (1) a high degree of its isotropy with respect to the Earthly observer (after taking into account the observer's motion with respect to the co–moving frame); (2) homogeneity of matter distribution on the last scattering surface (the lack of small scale inhomogeneities); (3) the isotropic expansion at the last scattering epoch (the lack of 12–hour anisotropies). The often overlooked fact is worth noticing; it is that the background photons carry information concerning not only the spatial isotropy of the matter distribution in the Universe, but also its spatial homogeneity (conclusion (2)). (More about cosmological significance of the microwave background radiation, see Peebles 1971, pp. 121–158; Weinberg 1972, pp. 506–228; Sciama 1971, pp. 176–203; Demiański 1985, pp. 274–279; Raine 1981b, pp. 44–61; Bajtlik 1984; Kolb and Turner 1990, pp. 14-15.) Recent observational results and their analysis can be found in the thesis by Bajtlik (1984).)[3]

(B) Investigations of the spatial distribution of galaxies is a difficult observational task. Till not long ago, distances to very few galaxies were known (with large measurement errors); without knowing the "third coordinate" one could only determine the projection of the galaxy distribution onto the celestial sphere. The reconstruction of the three–dimensional distribution from observational data (with no knowledge of the third coordinate) is not unique, and leads only to probable result. From among several methods used to reach the goal, the so-called *correlation function* method turned out to be the most effective one. Two–point correlation function measures the probability of finding (on the celestial sphere) two galaxies such that the distance between them is smaller than the average distance between two neighbouring galaxies. Three–, four–, ..., and n–point correlation functions are defined analogously (for precise definitions, see Peebles 1980, pp. 143–152). Time and effort consuming investigations, carried out with the help of this method, have led to the following results:

(1) For large angular distances the correlation function tends to zero. This means that the distribution of galaxies is homogeneous at large. Observations suggest that the world can be regarded as spatially homogeneous at distance scales larger than 10^8 ps.[4]

(2) On smaller scales a strong effect occurs of the clustering of galaxies. This effect remains in agreement with the well–established existence of smaller aggregations of galaxies (such as the Local Group), clusters, and superclusters of galaxies.

(3) The investigations do not show any characteristic scale of clustering. This means that there is no privileged unit (cluster of galaxies, say) which could be regarded as a fundamental building–block of the Universe. The average distribution of such building blocks would implement the observed large–scale distribution of matter. This suggests the hypothesis that there is clustering of galaxies on all scales: smaller aggregates tend to be clustered into larger ones, but this tendency is less and less pronounced as one goes to larger and larger distances.

---

[3]The above presented results have been remarkably improved by the very successful satellite COBE (COsmic Background Explorer) launched in November 1989. Preliminary measurements fit perfectly to a Planck blackbody radiation spectrum for a temperature of 2.735 ± 0.06 K. At the 1 % level there are no deviations from an ideal blackbody spectrum. Dipole and quadrupole anisotropies are not observed. See B. Schwarzschild, *Physics Today*, **43**, 1990, 17–20. (Note added in proof.)

[4]1 ps = 206 265 AU = 3.263 light years = 3.083×10^{13} km (AU stands for Astronomical Unit).

Many red shift measurements of galactic spectra have been recently obtained (see below (C)). By assuming that the red shifts provide information on distances to galaxies, the lacking third coordinate can be determined for a considerable number of galaxies, and three–dimensional charts of the sky can be constructed. This kind of research has confirmed the previous results but, quite unexpectedly, revealed a peculiar structure of clustering on smaller scales. It turns out that galaxies are clustering in elongated membrane structures the linear size of which is of order of magnitude larger than the transversal one. Between membranes crossing each other in an irregular manner, large domains extend devoid of galaxies. Within the membranes galaxies are distributed homogeneously with a tendency of clustering on all scales. (See Silk et al. 1983; more about distribution of galaxies in the Universe, Peebles 1980; Groth et al. 1977; Raine 1971b, pp. 1–33, 62–83; Kolb and Turner 1990, pp. 16–26.)

Important cosmological information can be obtained by comparing empirical results concerning the microwave background radiation and the distribution of galaxies. In this way, one can get, as it were, two photographs of the Universe: the background radiation conveys information about the structure of the last scattering surface, the distribution of galaxies–about the "present" cosmological epoch. It turns out that the evolution leads the Universe towards the destruction of homogeneity. This effect should be ascribed to a strongly unstable character of gravitational field which consists in a tendency to strengthen all casual matter inhomogeneities. Onto this general trend other physical mechanisms are superimposed the nature of which has not been fully understood. The origin and evolution of structures in the Universe is one of the most difficult problems of contemporary cosmology.

(C) Red shifts of the spectra of galaxies were the first significant observable of the relativistic cosmology, it played an exceptional role in its history. Now, we will show how red shifts are connected with theoretical concepts appearing in the standard model.

Let us consider Robertson–Walker line element (3.1)–(3.2b). Let us assume that at an instant t_1 of the cosmological time a photon is emitted from a galaxy situated at the point $(r_1, \theta_1, \varphi_1)$ in the direction r (i.e., $\theta = \varphi = 0$). The photon is registered on the Earth, situated at $(0,0,0)$, at the instant t_0. Taking into account that light moves along null geodesics, one obtains

$$\int_{t_1}^{t_0} \frac{dt}{R(t)} = \int_0^{r_1} \frac{dr}{1 + \frac{1}{4}r^2} \equiv f(r_1), \qquad t_0 > t_1. \tag{6.1}$$

Since the coordinates $(r_1, \theta_1, \varphi_1)$ of the galaxy are co–moving coordinates, $f(r_1)$ does not depend on time; it can be thought of as measuring a conventional distance corresponding to the time difference $t_0 - t_1$ (conventional since it is not directly measurable). If the subsequent photon is emitted from the galaxy at an instant $t_1 + \Delta t_1$ and registered on the Earth at $t_0 + \Delta t_0$ with $\Delta t_1, \Delta t_0 \ll t_0 - t_1$, one has

$$\int_{t_1+\Delta t_1}^{t_0+\Delta t_0} \frac{dt}{R(t)} = f(r_1),$$

or

$$\frac{\Delta t_o}{R(t_o)} = \frac{\Delta t_1}{R(t_1)}.$$

The last formula is often expressed in terms of the wavelength λ_1 (or frequency ν_1) of the emitted light and of the wavelength λ_0 (or frequency ν_0) of the registered light; in such a case we have

$$\frac{\lambda_1}{\lambda_0} = \frac{\nu_0}{\nu_1} = \frac{\Delta t_1}{\Delta t_o} = \frac{R(t_1)}{R(t_0)}.$$

Customarily, one defines the *red shift parameter* (called also *red shift*, for short) $z = (\lambda_0 - \lambda_1)/\lambda_1$ which gives

$$z = \frac{R(t_0)}{R(t_1)} - 1. \tag{6.2}$$

If the Universe expands, $R(t_0) > R(t_1)$ and z is positive, i.e., the registered light wave is shorter than the emitted one. If the Universe contracts, the situation is just the opposite, in which case one obtains blue shift. By expanding (6.1) and (6.2) with respect to the parameter $h = (t_0 - t_1)/R(t_0)$, which is assumed to be small (i.e. one could hope that the procedure is valid for not too distant galaxies), and then by eliminating this parameter from expansions of (6.1) and (6.2), one obtains

$$z = \dot{R}(t_0)f(r_1).$$

This gives a linear dependence of red shift on the "distance" $f(r_1)$. Distances, actually measured by astronomers (see Weinberg 1972, pp. 418–427) turn out to be proportional to $R(t_0)f(r_1)$, symbolically $d \cong R(t_0)f(r_1)$, i.e.,

$$z \cong \frac{\dot{R}(t_0)}{R(t_0)}d. \tag{6.3}$$

This is the famous *Hubble's law*, discovered observationally by this astronomer in 1929. It was this law that, before the discovery of the background radiation, constituted the main argument of the "expanding Universe".

Before the discovery of the microwave background radiation the so-called *classical cosmological tests* were expected to provide relevant observational information. The idea consists in identifying actually measured astronomical quantities (sometimes called *observables*) with theoretical magnitudes appearing in a model, and comparing dependencies between pairs of them, as predicted by various models, with those established experimentally. The most important dependencies are: red shift — angular size of objects, red shift — stellar magnitude, counting of objects up to a certain limiting magnitude[5] (see for instance, Weinberg 1972, pp. 441–458; Raine 1981b, pp. 153–166). However, so far the efficiency of these tests has proved to be rather limited.

---

[5]In cosmological tests magnitude often serves as a distance indicator.

They are based on a number of in principle unverifiable assumptions which have to fill in gaps in our knowledge, especially as far as the evolution of objects is concerned (such as galaxies, radiogalaxies, quasars). By maneuvering "free parameters" many different cosmological models can be fitted to actually observed dependencies. The most significant result obtained with the help of classical tests is the support given by them to the idea of the evolutionary Universe against the steady–state cosmology of Bondi and Gold (1948), and Hoyle (1948).

To sum up, one must say that the standard world model presented in the present chapter does not contradict any observational data available at present and, although none of known tests "proves" it directly, all tests together strongly suggest that the standard model — up to a good approximation — correctly describes the observed Universe.

We should notice another important circumstance as far as observational aspects of cosmology are concerned. Because of the final velocity of light signals, the so–called *horizon* may appear in the Universe limiting the field of vision of a given observer. At a moment t, an observer situated at the beginning of the co–moving coordinate system, $r = 0$, can receive signals emitted by galaxies (more strictly by particles at rest with respect to the co–moving coordinate system), the radial coordinate of which, r, is smaller than r_1, where r_1 is the coordinate of a galaxy such that the light signal emitted from it at t_1 reaches the observer (situated at $r = 0$) at t. t_1 can be computed from the formula (see formula (6.1))

$$\int_0^{r_1} \frac{dr}{1 + \frac{1}{4}r^2} = \int_{t_1}^{t_0} \frac{dt}{R(t)} \tag{6.4}$$

If the integral on the right–hand side of (6.4) converges for $t_1 \to 0$ (or in world models without the initial singularity, for $t_1 \to -\infty$), then there exist galaxies such that light signals emitted from them do not have enough time to reach the observer at $r = 0$; such a model is said to have a *particle horizon*.

It can also happen that the integral on the right–hand side of formula (6.4) converges for $t \to t_{max}$ (t_{max} can be equal to $+\infty$ or to the moment of the final singularity); if this is the case, there exist galaxies such that light signals emitted from them will never reach the observer situated at $r = 0$; such a model is said to have an *event horizon*. (More about horizons, see Rindler 1946; 1969, pp. 240–242; Weinberg 1972, pp. 489–491; Hawking and Ellis 1973, pp. 127–130.)

The existence of horizons imposes strong constraints upon observational possibilities of cosmology. If in our Universe there are horizons, and if we want to extrapolate our knowledge acquired within the horizons to regions situated outside them, we have to make strong assumptions that would justify such a procedure. Assumptions of this kind will always be of an empirically unverifiable character.

7. Comments and Remarks

A. After having made the first contact with the standard cosmological model, one is struck by a certain tension between the degree of idealization incorporated into this model and the agreement of its predictions with observational data. Definitions of particle, photon, etc. from the beginning of sec. 2 allow one to "theoretically identify" the material content of the standard model; however, to coordinate physical objects to particular mathematical structures becomes a very difficult task. For instance, in the standard model, density of particles with non–zero rest–mass satisfies the relationship $\rho_{\text{mat}} \sim R^{-3}(t)$, whereas for particles with zero rest–mass one has $\rho_{\text{rad}} \sim R^{-4}(t)$ (number of photons in the volume unit changes as R^{-3}, and the energy of each photon as R^{-1}). If, therefore, in the present Universe, there is even a slight amount of radiation, there had to be in its past an epoch (when R was sufficiently small) in which the radiation dominated the dynamics of the world. If the present Universe is correctly described by a dust–filled model with even a slight admixture of radiation, its sufficiently distant past must be described by a radiation–filled model.

In its "dust period" the model is filled with dust particles, i.e., with timelike geodesics (definition 2.2), but it is less clear with which physical objects a "theoretical dust particle" should be identified. From the theoretical point of view, dust particles are distributed throughout the space of the model in a homogeneous and isotropic manner, and consequently one must look for real objects on the level of which the spatial homogeneity and isotropy would be guaranteed. Galaxies for sure, and clusters of galaxies most probably, are not such objects. If a version of the hypothesis is true of clustering of galaxies on all scales, it might turn out that such objects simply do not exist. Having this in mind, successes of the standard world model in organizing observational material are really astonishing. (For a more detailed discussion, see Heller 1982.)

B. All solutions of Einstein's equations, considered in the present chapter, have been obtained with the assumption that the cosmological constant Λ is equal to zero (see chapter 4, sec. 5). By rejecting this assumption one obtains a much richer class of solutions. Counterparts of equations (3.3) and (3.4) with non–vanishing cosmological constant are the following,

$$\frac{8\pi}{3}\rho = \frac{\dot{R}^2}{R^2} + \frac{k}{R^2} - \frac{1}{3}\Lambda,$$

$$-8\pi p = 2\frac{\ddot{R}}{R} + \frac{\dot{R}^2}{R^2} + \frac{k}{R^2} - \Lambda.$$

Solutions of these equations, for different values of the parameters Λ and k, with the equation of state $p = 0$ (dust), are shown in fig. 6.4 and 6.5; the table gives their straightforward classification. (For more details, see Tolman 1934, pp. 394–419; Rindler 1977, pp. 235–238; Weinberg 1972, pp. 613–616.)

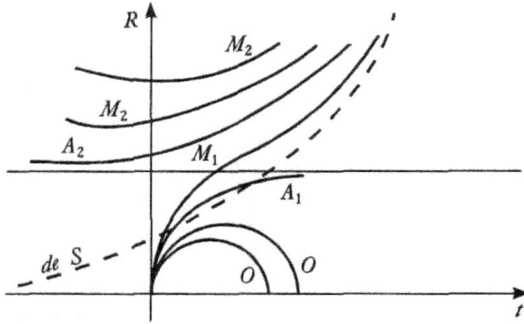

Fig. 6.4. World models with $k = +1$, $p = 0$, for different values
of the cosmological constant.

Einstein (1917) introduced the cosmological constant Λ to obtain a static world model; however, after finally convincing himself that the Universe really expands, he considered it to be superfluous (see Einstein 1958). Observational data testify that the influence of the cosmological constant upon the dynamic of the planetary system is negligible. The interest in cosmological models with the cosmological constant increased around 1967 when a surplus of number was discovered of quasars with red shifts about $z = 2$. It seemed that this suggested the existence of an epoch in which the rate of expansion slowed down almost to zero. Such an epoch appears only if the value of the cosmological constant Λ is slightly larger than Λ_E (the so-called *Lemaître's models*, see model $\mathcal{M}_1$ in fig. 6.4). When afterwards it had turned out that it was only a kind of a selection effect, models with $\Lambda = 0$ became again fashionable (see Petrosian 1974).

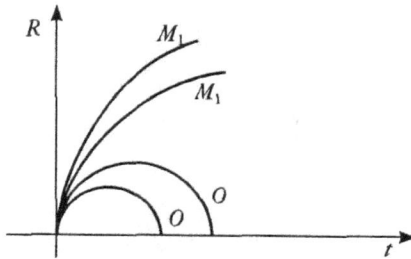

Fig. 6.5. World models with $k = 0$, $p = 0$, for different values
of the cosmological constant.

Table. World models with the cosmological constant.

$\Lambda\backslash k$		$= +1$	$= 0$	$= -1$
> 0	$> \Lambda_E$	M_1 — monotonically expanding models with the initial singularity. $M_1 \to$ de S as $t \to \infty$	Models M_1	Models M_2
	$= \Lambda_E$	$R < R_E$: A_1 — first asymptotic model. $A_1 \to E$ as $t \to \infty$		
		$R = R_E$: E — static Einstein's model.		
		$R > R_E$: A_2 — second asymptotic model. $A_2 \to E$ as $t \to \infty$		
	$< \Lambda_E$	$R < R_E$: O — oscillating models with the initial and final singularities		
		$R > R_E$: M_2 — monotonically expanding models without the initial singularity. $M_2 \to$ de S as $t \to \infty$		
$= 0$	$= 0$	Models O		
< 0	< 0		Models O	Models O

de S denotes de Sitter's world model; Λ_E and R_E are values of the cosmological constant and of the scale factor characteristic for the static Einstein's model, correspondingly.

The cosmological constant has returned to scientific papers in connection with grand unification theories. It was known for a long time that dynamical effects of the cosmological constant appear as "corrections" to the density and pressure: $\rho_{ef} = \rho + \Lambda/(8\pi)$, $p_{ef} = p - \Lambda/(8\pi)$ (see Weinberg 1972, p. 614).[6] The energy density of the vacuum presupposed by unifying theories is thought of as being connected with these corrections (see, for instance, Kazanas 1980, Abbot 1988).

C. Einstein's equations, being partial differential equations, determine the structure of space–time only locally. In particular, they do not contain any global information about space–time topology. For instance, for a given space–time $(\mathcal{M}, g)$, there exist a simply connected universal covering manifold[7] $\tilde{\mathcal{M}}$ equipped with the metric $\tilde{g}$ such that $(\mathcal{M}, g)$ can be obtained from $(\tilde{\mathcal{M}}, \tilde{g})$ by suitably identifying points in $\tilde{\mathcal{M}}$. If a solution of Einstein's equations is given, it is always possible to find its universal covering $(\tilde{\mathcal{M}}, \tilde{g})$, and to perform all possible identifications in it (if space–time has no symmetries its universal covering coincides with it). Space–times obtained in this

---

[6]To obtain a static solution from the field equation without the cosmological constants requires a negative pressure. The cosmological constant was introduced by Einstein in order to avoid this "non–physical" effect.

[7]Definitions of these concept can be found, for example, in a book by O'Neill (1983, pp. 441–445).

way are automatically solutions of Einstein's equations, but they can drastically differ from each other from the topological point of view. In sec. 3 of the present chapter, it has been said that the standard choice of topology for instantaneous spaces are hyperbolic space $\mathcal{H}^3$ for $k = -1$, Euclidean space $\mathbf{R}^3$ for $k = 0$, and spherical space $\mathcal{S}^3$ for $k = +1$. However, the problem could be stated in a more general way, namely, one could ask about all possible space–times which locally are expanding Robertson–Walker worlds with instantaneous spaces of constant curvature. The answer to this question gives globally different cosmological models that locally are non–distinguishable from each other (and from the standard model) as far as their observational aspects are concerned. Global differences between these models could manifest themselves in the possibility of observing the same object more than once on the celestial sphere (the so–called *ghost images*).

Mathematically, the question is reduced to the classical problem of *Clifford–Klein space forms* (see Wolf 1967, and in a more concise way O'Neill 1983, pp. 243–248). It was Ellis (1971) who considered this problem in a cosmological context. It turns out that there exist infinitely many space forms[8] locally reproducing Robertson–Walker universes. In the case $k = +1$, all these space forms are compact but in the cases $k = 0$ and $k = -1$, there exist both compact and non–compact space forms. This means that not all world models with $k = 0$ and $k = -1$ should be regarded as "open".

This problem is strictly connected with the extrapolation problem in cosmology which essentially reduces itself to the question: how spaces, determined locally, could be extended to non–local scales (see Heller 1978)? Some observational aspects of this problem have been discussed by Ellis (1983).

D. The existence of horizons in some world models (see end of sec. 6) also leads to difficult cosmological problems. Let us notice that two domains of space–time separated by a particle horizon never have interacted with each other. It can be shown that domains, distant from each other by no more than 75° on the celestial sphere, were separated by the particle horizon at the epoch at which the microwave background radiation interacted for the last time with other forms of matter filling the Universe (see, for example, Weinberg 1972, pp. 525–526). If one admits that physical interactions smoothing out the Universe cannot propagate with a speed greater than that of light, one cannot see any reason why the background radiation, coming from domains separated from each other by a particle horizon during the last scattering epoch, is now isotropic to such a high degree.

If one postpones looking for a physical mechanism explaining this puzzle, one has to assume very special initial conditions for the Universe which would *a priori* synchronize cosmological parameters (such as temperature of the background radiation) in regions separated by particle horizons from the very beginning.

A program aiming at explaining the question of the "smoothness of the Universe"

---

[8]A complete and connected Riemann manifold of constant curvature is said to be a (*Clifford–Klein*) *space form*.

was proposed by Misner (1969). The idea was to find such mechanisms with the help of which it would be possible "to deduce" the actually observed state of the Universe from *arbitrary* initial conditions (this program has been named the *mix-master cosmology*). To this end one must (1) find world models in which a light signal would be able to go many times around the world till the present epoch, and (2) identify physical mechanisms which could effectively smooth out the Universe. Different mechanisms were proposed: among others dissipative effects connected with the viscosity of neutrinos or with the quantum process of the pair particle–antiparticle creation in strong gravitational fields in the early stages of world's evolution. The heated discussion, which developed around the mix–master cosmology program, has shown that to satisfy condition (1) very special initial data are also required; therefore, it is not a good strategy to avoid special conditions. As far as condition (2) is concerned, it has turned out that the proposed mechanisms are either not efficient enough (neutrino viscosity) or too hypothetical (pair creation). In effect, the mix-master program is believed to be attractive but involved in so far insurmountable difficulties (see Barrow and Matzner 1977, Barrow 1978, Misner 1980, Raine 1981b, pp. 215–220).

New perspectives to solve the puzzle appeared in connection with the grand unification theories. Guth (1981) noticed that the phase transition from the early symmetric Universe to the Universe with the spontaneously broken symmetry can be associated with a short period of rapid (exponential) expansion enlarging the size of the Universe 10^{50} times (or more). "Inflation" of this kind is caused by the vacuum energy which, during the phase transition, dominates the cosmological dynamics. This scenario, called the *inflationary universe*, was afterward modified several times (see Linde 1982, Hawking and Moss 1982, Barrow and Turner 1982). Inflation seems to liquidate the horizon problem: because of the exponential expansion in its history, the Universe, observed at present, was never divided by particle horizons; it had originated from a tiny portion of the primordial plasma blown up to enormous size during the inflation period.

Great hopes were connected with inflationary scenarios. However, the enthusiasm which surrounded early versions of the model has recently slowed down. Some researches point out that physics, presupposed by inflation, is still dubious, and for solutions of some problems one is paid with new, and even more serious, problems (see Ellis and Stoeger 1988).

106

Bibliographical Essay

Reading of the present book does not give an idea of what the contemporary cosmology really is. The book's aim has been only to pose the cosmological problem and to discuss its fundamental assumptions. If the reader feels a little disappointed that he did not acquire the expected cosmological knowledge, the author is entitled to think that he has succeeded to reach his goals. However, wanting to help the reader in his further adventures with cosmology, the author feels obliged to add to this book a kind of guide throught the land of cosmological readings. It should be stressed that it is by no means a bibliography which could pretend to even partial completeness. The term "guide" seems to be a just expression; it is a guide, and — it should be added — a subjective one, to a great extent reflecting pathways travelled once by the author himself.

The guide contains only references to the literature in the field of relativistic cosmology; books dealing with other cosmologies (such as neo–Newtonian, Milne's or steady–state cosmology) are not included. With only one exception (Robertson's review article of 1933) the reader will not find in it any reference to scientific papers published in journals. Because of the great richness and variety of such papers any attempt to compose a guide through them, or a list of them, would be both groundless and impossible.

For reasons of clarity, cosmological literature will be divided into sections reflecting main stages of the evolution of cosmology in our century.

A. Classical Period (from the origin of relativistic cosmology to the Second World War).

First textbooks of general relativity appeared surprisingly early. To the earliest ones belong:

H. Weyl (1922), *Space — Time — Matter* (German edition in 1921);

A.S. Eddington (1923), *The Mathematical Theory of Relativity.*

As far as mathematical standards are concerned, both these books are equal to many later publications, and the depth of their philosophical insight surpasses many newest elaborations. Numerous comments in these books are devoted to the problem of the relativity of space and time; the problem was still new and claimed for elucidation. Only some sections deal with "the Universe as a whole"; they represent Einsteins's "cylindrical" world and de Sitter's "spherical" universe. The fact that the two world models exist fascinates the authors. For the first time space–time begins to reveal its global image.

About 1930, owing mainly to the works of Friedman, Lemaître and Robertson, all spatially homogeneous and isotropic cosmological models were known. The first complete review of them can be found in the paper:

H.P. Robertson (1933), *Relativistic Cosmology*,
published in the journal *Reviews of Modern Physics*. The article is important for
two reasons: first, a few generations of physicists learned from it basic facts of relativistic cosmology; second, appendix to the paper contains an almost complete list of
cosmological papers published in 1917–1932.

An extensive monographic textbook:

R.C. Tolman (1934), *Relativity, Thermodynamics and Cosmology*,
contains a detailed presentation of special and general relativity; in this aspect it
is certainly a kind of textbook. Chapters that present pioneering works of Tolman
himself are doubtlessly of a monographic character; they create the fundamentals of
relativistic thermodynamics. The cosmological part of the book, in its composition, is
similar to the paper by Robertson of 1933 but it puts more stress on the confrontation
of theoretical models with astronomical observations. Applications of relativistic
thermodynamics to cosmology are typical for this book. The clarity of Tolman's
work is exceptional. Even today scientists open it to check a formula or to find a
forgotten, but still useful, mathematical expression.

One should also mention another book which — although published after the
Second World War — by its spirit belongs to the pre–war period:

G. Lemaître (1972), *L'Hypothese de l'atome primitif* (first edition in 1946; English translation in 1950).

The book is a collection of popular essays, published earlier by the author on
various occasions. The value of the book lies in the fact that it shows the whole
cosmological vision of Lemaître. The hypothesis of the primeval atom was a first
attempt to create an overwhelming cosmological system (from the initial singularity,
through a cosmological theory of nucleosynthesis, to a theory of the origin of galaxies
and of clusters of galaxies). In this hypothesis, for the first time, geometric theory of
the world was supplemented with a reconstruction of physics of the early Universe.

The book:

O. Heckmann (1968), *Theorien der Kosmologie* (first edition in 1942)
closes this period. In the exposition style it follows Robertson's paper (1933). It also
contains a list of cosmological publications of 1933–1940.

B. Period of Stagnation (from the Second World War to the discovery of the background radiation in 1965).

First an explanation. The term "stagnation" implies a certain slowing down of
progress in cosmology, but it does not apply to the theory of relativity which had
many successes in this period especially as far as its mathematical aspects were concerned. Of course, all these successes were relevant also for cosmology. Speaking about
stagnation I have in mind a certain impasse in cosmological observations. In the fifti-
eth discussion focused on polemics between defenders of the relativistic paradigm

(which was then called the "theory of the expanding Universe") with the proponents of the steady–state cosmology, created in 1948 by Bondi and Gold, and Hoyle.

A note of pessimism can be heard as early as in 1945 in the Cosmological Appendix added by Einstein to the second edition of his *The Meaning of Relativity* (first edition in 1921). A small number of publications in the field of cosmology that are to be quoted in the present section also testifies about the stagnation. We should certainly mention

H. Bondi (1960), *Cosmology* (editions in 1952, 1960, 1961).

The book presents "on equal footing": neo–Newtonian cosmology, relativistic cosmology, Milne's cosmology, and steady–state cosmology. Cosmological conceptions of Eddington, Dirac, and Jordan are also briefly reported. A transparent discussion of the conceptual and methodological basis of cosmology makes this book still currently relevant.

The following book, on the one hand, is a kind of summing up of formal achievements of the pre–war cosmology but, on the other hand, contains a number of theoretical results obtained already in this period:

H.P. Robertson and T. W. Noonan (1968), *Relativity and Cosmology*.

In spite of the date of its publication, this book doubtlessly belongs to the period before the discovery of the background radiation. After the tragic death of Robertson (in a traffic accident), it was compiled, from papers and notes left by him, by his last student, Thomas W. Noonan. The book was very carefully edited;[1] it shows the Robertsonian style of doing cosmology, in which a detailed analysis of space–time symmetries is as important as solving the field equations.

C. Period of Dynamical Progress (after the discovery of the background radiation).

The discovery of the microwave background radiation in 1965 together with other achievements of radioastronomy (quasars, radiogalaxies, pulsars) and later on also of optical astronomy brought new momentum into relativistic physics and especially into cosmology. Relatively quickly, the standard model began to consolidate. From the beginning of the seventieth the number of books in cosmology continuously increases. Only some of them can be quoted. Because of their great variety, I shall classify them into certain groups.

1. *Textbooks of general relativity.* It is now a custom to devote a chapter or (more often as time passes) an entire part to cosmology. A handbook which has gained considerable publicity is

C.W. Misner, K. Thorn, J.A. Wheeler (1973), *Gravitation.*

---

[1]At the end of the book the reader can find summaries of all chapters. The summary ends with the summary of the summary.

It is a big volume (1279 pages). The style is highly didactic with the ambition to teach a beginner all the mysteries and intricacies indispensable to start research in this field. (In my opinion, didactic goals of this book are a bit exaggerated.) Many pages are devoted to cosmology, but a serious drawback (in my opinion again) is that the authors, motivated by philosophical reasons, almost exclusively focused on "closed" world models (with $k = +1$).

The more recent book:

B. Schutz, *A First Course in General Relativity*
bears some similarities to Misner, Thorn and Wheeler's volume, but is much more concise, easier, and with no tricky gadgets. It introduces the reader to understanding the content of the theory presented and to manipulating with its computational machinery. This book is my personal recommendation for an ambitious freshman.

From among many other handbooks I would recommend the following to a reader who is prepared to make a more intimate acquaintance with cosmology:

S. Weinberg (1972), *Gravitation and Cosmology: Principles and Applications of the General Theory of Relativity.*

The fifth part of this book deals with cosmology. In my view, only two features separate the book from the ideal: first, the author unnecessarily avoids the modern coordination free notation; second, the book was written several years ago, and in a few points it would require modernization. Nevertheless, the principles of physics do not change so rapidly and Weinberg's book is still an excellent source to learn relativistic physics and cosmology.

Let us also quote:

W. Rindler (1968, 1977), *Essential Relativity: Special, General, and Cosmological.*

R. Adler, M. Bazin, M. Schiffer (1965), *Introduction to General Relativity.*

The first of these books, providing more imaginative pictures, focuses on the problem of the relativity of space and time (Mach's principle), and from the second the reader can learn many useful mathematical techniques.

L.D. Landau, E.M. Liphschitz (1973), *The Field Theory* (in Russian).

This edition of the famous book has a significantly enlarged cosmological part as compared with previous editions. A serious drawback, from the cosmological point of view, is the absence of observational aspects of cosmology. A chapter devoted to cosmology can be found in the book

M. Demiański (1985), *Relativistic Astrophysics.*

The following book is of an exceptional character,

R.S. Sachs and H. Wu (1977b), *General Relativity for Mathematicians.*

It adopts a quasi mathematical style of presentation. However, mathematical rigors break down when a physical reality has to be put into formulae. In the part dealing with cosmology, the Einstein–de Sitter world model is mainly exploited (it is a spatially isotropic and homogeneous world model with $k = p = \Lambda = 0$).

Very useful for every relativity student is the collection of exercises:

A.P. Lightman, W.H. Press, R.H. Price and S.A. Teukolsky (1975), *Problem Book in Relativity and Gravitation.*

Chapter 19 is on cosmology (solutions to problems are included).

2. *Handbooks of cosmology.* a) Elementary level:

(1) *Understanding Space and Time*, Block 5: *Cosmology* (1979),

(2) M. Berry (1976), *Principles of Cosmology and Gravitation,*

(3) M. Rowan-Robinson (1977), *Cosmology,*

(4) D.W. Sciama (1975), *Modern Cosmology,*

(5) J. Heidmann (1973), *Introduction à la cosmologie,*

(6) P.T. Landsberg, D.A. Evans (1977), *Mathematical Cosmology — An Introduction.*

The above books are enumerated, roughly speaking, in the order from the easier to the more difficult. I would advise one to make the first encounter with cosmology with the help of (1) or (2). (1) is a publication of the British "Open University", (2) is an excellent textbook for beginners. (3) deals more with the observational side of cosmology, and (4) nicely presents mutual dependencies of cosmology and astrophysics. (5) is a sample of the French approach to cosmology. By using (6) one can acquire a proficiency in "mathematical cosmology" without going beyond the fundamental level of higher mathematics. Although the book deals, in principle, with the neo-Newtonian cosmology only, it gives a good picture of cosmological research.

b) Advanced level:

(1) D.J. Raine (1981b), *The Isotropic Universe,*

(2) P.J.E. Peebles (1971), *Physical Cosmology,*

(3) J.B. Zeldovitch and I.D. Novikov (1975), *Structure and Evolution of the Universe* (in Russian).

(4) A.D. Dolgov, I.B. Zeldovitch and M.B. Sagin (1988), *Cosmology of the Early Universe* (in Russian).

(5) M.P. Ryan and L.C. Shepley (1975), *Homogeneous Relativistic Cosmologies.*

These books are also enumerated in the order of increasing difficulty. (1) is a good introduction to, and overview of, cosmology. (2) and (3) are attempts at reconstructing physical processes in the evolving Universe. (3) is a mosaic of various cosmological problems rather than a textbook of cosmology, albeit a mosaic providing the general picture. (4) originated from courses given by Zeldovitch at the Moscow University, and was composed by his co-workers after the death of this eminent cosmologist; it gives a very penetrating account of recent problems and results in cosmology, such as: inflation, grand unification, quantum gravity, creation of the Universe as a quantum tunneling process, ... (5) is mathematically elegant and explores theoretical aspects of Bianchi cosmology (homogeneous but not necessarily isotropic world models). This book contains a rich catalogue of cosmological publications till 1974, which can be considered as a continuation of similar lists composed by Robertson (1933) and Heckmann (1968) (however, now without any ambition of completeness).

3. Monographs.
(1) P.J.E. Peebles (1980), *The Large–Scale Structure of the Universe,*
(2) S.W. Hawking, G.F.R. Ellis (1973), *The Large–Scale Structure of Space–Time,*
(3) J.K. Beem, P.E. Ehrlich (1981), *Global Lorentzian Geometry,*
(4) D. Kramer, H. Stephani, E. Herlt, M. MacCallum (1980), *Exact Solutions of Einstein's Field Equations.*
(1) deals with the problem of the origin and evolution of structures in the Universe (such as galaxies, clusters of galaxies, . . .). Works which gave birth to book (2) have initiated many new research programs in relativity: singularity theorems, causal structure of space–time, global structure of solutions to Einstein's field equations, and so on. This monograph closes a certain stage of discussions concerning the classical (non–quantum) singularity problem. (3) elaborates principally the same problems as (2) but in a more mathematical style. (4) is an indispensable source for studying solutions of Einstein's equations.

4. Books on the XXth century history of cosmology.
(1) J.D. North (1965), *The Measure of the Universe,*
(2) J. Merleau–Ponty (1965), *Cosmologie du XXe siecle,*
(3) O. Godart and M. Heller (1985), *Cosmology of Lemaître.*
(4) E.A. Tropp, W.Ya. Frenkel, A.D. Chernin (1988), *Alexander Alexandrovitch Friedman* (in Russian).

5. Books on philosophical aspects of relativity and cosmology.
(1) J.D. Barrow and F.J. Tipler (1985), *The Anthropic Cosmological Principle,*
(2) D.J. Raine and M. Heller (1981), *The Science of Space–Time,*
(3) R. Toretti (1983), *Relativity and Geometry,*
(4) M. Friedman (1983), *Foundations of Space–Time Theories,*
(5) J.D. Barrow (1988), *The World Within the World,*
(6) M. Heller (1986), *Questions to the Universe.*

6. Popular books on cosmology. Since a great number of such books appears each year, I choose, more or less at random, only a few:
(1) J. Silk (1980), *The Big Bang. Creation and Evolution of the Universe,*
(2) E.R. Harrison (1981), *Cosmology — The Science of the Universe,*
(3) S. Weinberg (1977), *The First Three Minutes,*
(4) R. Geroch (1978), *General Relativity from A to B,*
(5) A.G. Pacholczyk (1984), *The Catastrophic Universe,*
(6) J. Silk and J.D. Barrow (1983), *The Left Hand of Creation,*
(7) J.S. Trefil (1983), *The Moment of Creation.*
My personal recommendation would be (1), (6) and (7).

Appendix

Evolution of Dynamical Space–Time Theories

0. Introduction

General relativity is something more in contemporary physics than one of the many theories describing a certain domain of the world. A distinguished role of Einstein's theory of gravity can be reduced to two circumstances: first, this theory provides a space–time stage to many other physical theories and, second, it does so not only *locally*, in the neighbourhood of an investigated event, but also globally when, for instance, it deals with inextendible space–times, i.e., with such space–times which cannot be embedded into "larger" ones. Doubtlessly, the last function of general relativity appeals to aesthetic feelings of physicists; however, its significance in our system of knowledge is both more prosaic and more important: without it the correct statement of many concrete problems of physics and astronomy would be impossible. Contemporary observational astronomy controls such extensive regions of the Universe that with no help of global (or at least large scale) theory of space–time there could be no question of distilling any information carried by signals travelling through space and time. We have also seen in previous chapters how cosmology becomes unavoidable as far as the program of unification of physics is concerned (see chapter 3, sec. 0).

Science always tended towards a synthesis of its various theories. In previous paradigms, there were dynamical theories of mechanics that played the role of preparing spatio–temporal stage for eventual cosmological synthesis. This was the case with both Aristotelian and Newtonian images of the world. In contemporary science this function has been taken over, and much sharpened, by general relativity.

After having read all previous chapters, only a slight mathematical preparation would be enough in order to see, in a more detailed way, how the evolution of views on the space–time structure has led to the present distinguished position of general relativity in science. This could serve us as an additional argument on behalf of this theory — the argument from a "logic of development" of science. Since, on the one hand, analyses of this kind do not fit the main line of reasoning adopted in this book and, on the other hand, it would be a great negligence to omit an occasion to present the place of general relativity in the evolution of our views on the Universe altogether, we will do this in the present Appendix.

To do so, we must prepare certain concepts from the mathematical theory of categories. This theory allows one to organise large parts of contemporary mathematics in the spirit similar to that of the Erlangen Program proposed by F. Klein as a framework for unifying different geometric theories. Some mathematicians share the opinion that this announces a revolution in mathematics analogous to that initiated by Cantor's work (see Semadeni and Wiweger 1978, p. 13). No wonder, therefore, that the application of the categorial style of thinking to the analysis of physical theories not only turns out to be an attractive possibility, but also begins to bring fruits in the form of significant sharpening of some concepts and of revealing logical dependence between present physical theories and their predecessors (see Trautman 1973b).

Finally, one should stress that analyses carried out in the present appendix are of no strict historical character; they are rather a stylization of the history of science: they try to look at former theories in the light, and with the help, of modern mathematical tools. Such an approach enables one better to see a logic of the evolution of science, and consequently, more responsibly to predict perspectives of further development.

Such analyses have been carried out by many authors (see bibliographical note at the end of the appendix). Of course, we shall make ample use of them (especially of our own study, see Raine and Heller 1981); however, we are in a better position than all previous authors inasmuch as we can base our analysis upon an advanced mathematical machinery prepared throughout the entire book. Owing to this, we are able to significantly shorten the "explanatory text" and provide a deeper insight into the nature of analysed problems.

After presenting elementary concepts of the mathematical category theory (in sec. 1), we shall present the structure and dynamics of space–time in the Aristotelian theory (sec. 2), in classical mechanics without gravitational field (sec. 3), and in classical mechanics with gravitational field (sec. 4). The evolution of these space–time structures naturally leads to special and general relativity theories (sec. 5). Some branching from the main evolutionary trunk, namely certain ideas coming from Berkeley, Leibniz and Mach, together with prospects for future progress, will be presented as *Comments and Remarks* closing the Appendix (sec. 6).

1. Elementary Repetition of the Category Theory

1.1. Definition. We say that a *category* A is given if there are

(1) a class A^0 (not necessarily a set[1]), elements of which are called *objects*;
(2) a function which to every ordered pair of objects $\langle \mathcal{X}, \mathcal{Y} \rangle$, $\mathcal{X}, \mathcal{Y} \in A^0$ ascribes a set Mor $(\mathcal{X}, \mathcal{Y})$, elements of which are called *morphisms* from $\mathcal{X}$ to $\mathcal{Y}$; additionally
(3) a *composition* of morphism

$$\text{Mor}\,(\mathcal{X}, \mathcal{Y}) \times \text{Mor}\,(\mathcal{Y}, \mathcal{V}) \longrightarrow \text{Mor}\,(\mathcal{X}, \mathcal{V})$$

is defined by

$$(f, g) \longmapsto h = g \circ f,$$

where $f \in \text{Mor}\,(\mathcal{X}, \mathcal{Y})$, $g \in \text{Mor}\,(\mathcal{Y}, \mathcal{V})$, $h \in \text{Mor}\,(\mathcal{X}, \mathcal{V})$; the composition of morphisms is assumed to satisfy the following conditions,

---

[1] In the sense of the axiomatization of the set theory proposed by Gödel (see, Semadeni and Wiweger 1978, pp. 21–22).

(a) *associativity*: $j \circ (g \circ f) = (j \circ g) \circ f$, where $j \in \mathrm{Mor}\,(\mathcal{V},\mathcal{U})$, and f and g are as above;

(b) the existence of *morphism identity*: for every object $\mathcal{X} \in A^0$ there exists one and only one morphism $\mathrm{Mor}\,(\mathcal{X},\mathcal{X})$, denoted by Id_X, such that for every object $\mathcal{Y} \in A^0$ one has

$$f \circ \mathrm{Id}_x = f, \quad \text{for every } f \in \mathrm{Mor}\,(\mathcal{X},\mathcal{Y}),$$
$$\mathrm{Id}_x \circ g = g, \quad \text{for every } g \in \mathrm{Mor}\,(\mathcal{Y},\mathcal{X}).$$

1.2. Example. A category the objects of which are any sets and morphisms — any mappings is called Ens.

1.3. Example. A category the objects of which are any topological spaces, and morphisms — continuous mappings is called Top.

1.4. Example. A category the objects of which are arbitrary, but finitely dimensional $\mathcal{C}^k$ differential manifolds and morphisms — arbitrary $\mathcal{C}^k$ diffeomorphisms is called $\mathcal{C}^k\mathrm{Diff}$.[2]

1.5.–1.7. Examples. Analogously, one speaks about categories Vect, Gr, and Ab, the objects of which are vector spaces, groups, Abelian groups, and morphisms — any linear mappings and any homomorphisms, correspondingly.

1.8. Example. Let us recall that an *affine space* is a triple $\mathcal{A} = (\mathcal{E},\mathcal{V},+)$, where $\mathcal{E}$ is any set, $\mathcal{V}$ a vector space, and $+$ a mapping $+\colon \mathcal{E} \times \mathcal{V} \to \mathcal{E}$ such that $\mathcal{V}$ acts as a transitive and free group of transformations of $\mathcal{E}$ (for more details consult Schwartz, vol. I 1979, pp. 185–199). Let $\mathcal{A}_1 = (\mathcal{E}_1,\mathcal{V}_1,+)$ and $\mathcal{A}_2 = (\mathcal{E}_2,\mathcal{V}_2,+)$ be affine spaces. Let us define a mapping $f\colon \mathcal{E}_1 \to \mathcal{E}_2$ such that there exists a corresponding mapping $\tau f\colon \mathcal{V}_1 \to \mathcal{V}_2$ satisfying the condition

$$f(v + p) = \tau f(v) + f(p),$$

for any $p \in \mathcal{E}_1$ and $v \in \mathcal{V}_1$. The mapping f is said to be an *affine morphism*. The class of affine spaces together with affine morphisms as morphisms forms a category called Aff.

1.9. Example. A category, the objects of which are any metric spaces and morphisms — any contractions, i.e., mappings $\alpha\colon \mathcal{X} \to \mathcal{Y}$ such that

$$\rho_Y(\alpha(x_1),\alpha(x_2)) \le \rho_X(x_1,x_2),$$

where ρ_X and ρ_Y are metrics on spaces $\mathcal{X}$ and $\mathcal{Y}$ correspondingly, $x_1, x_2 \in \mathcal{X}$, is called Metr.

1.10. Example. A category the objects of which are fibre bundles and morphisms — bundle morphisms (see chapter 3, sec. 2), is called Bun.

1.11. Definition. A category B is a *subcategory* of a category A if the objects and morphisms of B are also objects and morphisms of A, and if the compositions

---

[2]In the case of $\mathcal{C}^\infty\mathrm{Diff}$, we shall simply write Diff.

of morphisms in B are restrictions of the corresponding compositions of morphisms in A.

For instance, Ab is a subcategory of Gr.

1.12. Definition. A morphism $f \in \mathrm{Mor}\,(\mathcal{X}, \mathcal{Y})$ is said to be an *isomorphism* if there exists another morphism $g \in \mathrm{Mor}\,(\mathcal{Y}, \mathcal{X})$ such that $f \circ g = \mathrm{Id}_Y$ and $g \circ f = \mathrm{Id}_X$.

For instance, bijections are isomorphisms in Top, isometries — in Metr, group isomorphisms — in Gr and Ab. Having a category A, one can define a category $\oint A$, the objects of which are those of A, and morphisms — the isomorphisms of A.

In the category theory, morphisms play, in a sense, a more important role than objects. One can construct a category axiomatically in such a way that its primitive concepts, besides logical constants and the equality sign, are (1) "α is a morphism", and (2) "composition of morphisms $\alpha \circ \beta$ is defined and equal γ". The axioms guarantee the associativity of compositions and the existence of identity morphisms. Since the concept of objects does not appear in the axioms, one speaks of an *objectless category theory*. It can be shown that a category defined at the beginning of the present section is a model of the axiomatic objectless category theory (see Semadeni and Wiweger 1978, pp. 44–46).

An important concept in the category theory is that of functor.

1.13a. Definition. Let A and B be two categories, A^0 and B^0 — the classes of their objects, and $\mathrm{Mor}\,(A)$ and $\mathrm{Mor}\,(B)$ — the classes of their morphisms. A *covariant functor* from the category A to the category B is a mapping

$$C: (A, A^0, \mathrm{Mor}\,(A)) \longrightarrow (B, B^0, \mathrm{Mor}\,(B)),$$

satisfying the following conditions:

1) if $f \in \mathrm{Mor}\,(\mathcal{X}, \mathcal{Y}) \in \mathrm{Mor}\,(A)$, $\mathcal{X}, \mathcal{Y} \in A^0$, then $C(f) \in \mathrm{Mor}\,(C(\mathcal{X}), C(\mathcal{Y})) \in \mathrm{Mor}\,(B)$, $C(\mathcal{X}), C(\mathcal{Y}) \in B^0$;
2) if $f, g \in \mathrm{Mor}\,(A)$ and f can be composed with g, then $C(g) \circ C(f) = C(g \circ f) \in \mathrm{Mor}\,(B)$;
3) $C(\mathrm{Id}_X) = \mathrm{Id}_{C(X)}$, $\mathcal{X} \in A^0$, $C(\mathcal{X}) \in B^0$, $\mathrm{Id}_X \in \mathrm{Mor}\,(A)$, $\mathrm{Id}_{C(X)} \in \mathrm{Mor}\,(B)$.

The following diagram elucidates the structure of the above definition:

$$A: \mathcal{X} \xrightarrow{g} \mathcal{Y} \xrightarrow{f} \mathcal{Z};$$
$$B: C(\mathcal{X}) \xrightarrow{C(f)} C(\mathcal{Y}) \xrightarrow{C(g)} C(Z)$$

The symbol $C(f)$ is often replaced by the symbol f_*. To simplify the notation we shall also write $C: A \to B$.

1.13b. Definition. *A contravariant functor* from the category A to the category B is obtained by introducing to definition 1.13a modifications visualised in the following diagram:

$$A: \mathcal{X} \xrightarrow{f} \mathcal{Y} \xrightarrow{g} \mathcal{Z};$$
$$B: C(\mathcal{X}) \xleftarrow{C(f)} C(\mathcal{Y}) \xrightarrow{C(g)} C(\mathcal{Z})$$

The symbol $\mathcal{C}(f)$ is often replaced by the symbol f^*.

1.14. Example. A mapping

$$\mathcal{C}\colon (C^k\mathrm{Diff}, (X^n, x), f) \longrightarrow (C^{k-1}\mathrm{Diff}, T_X(X), f'(x)),$$

where X^n is any n–dimensional C^k manifold, $T_X(X^n)$ — the tangent space to X^n at a point $x \in X^n$, and f' — the derivative mapping of the mapping f transforming a C^k manifold into a C^k manifold, is a covariant functor from $C^k\mathrm{Diff}$ to $C^{k-1}\mathrm{Diff}$.

The category theory has enormous ability to order mathematical structures and make precise concepts which has so far been used intuitively. The notion of naturality belongs to such concepts. Eilenberg and MacLane considered the formalization of this notion as a main motive of creating the category theory (this can be seen in the title of their pioneering work; see Eilenberg and MacLane 1945). Even if further developments of the theory transcended this goal, it remains true for the present Appendix. Strict definition of the naturality concept is essential to grasp logical dependencies between different space–time structures.

1.15. Definition. Let C_1 and C_2 be two functors of the same variance (i.e., either both covariant or both contravariant) from a category A to a category B:

$$C_1\colon A \longrightarrow B,$$
$$C_2\colon A \longrightarrow B.$$

A *natural mapping* (called also a *morphism functor*) of C_1 into C_2 is a mapping

$$\mathcal{N}\colon A^0 \longrightarrow \mathrm{Mor}\,(B),$$

ascribing to every object $A \in A^0$ a morphism of the category B:

$$A \longmapsto \mathcal{N}(A) \in \mathrm{Mor}\,(B),$$

in such a way that

$$\mathcal{N}(A)\colon C_1(A) \longrightarrow C_2(A).$$

It is also assumed that, for every $f\colon A_1 \to A_2$, $f \in \mathrm{Mor}\,(A)$, $A_1, A_2 \in A^0$, the following diagram is commutative (it is assumed that all the required compositions are meaningful)

$$
\begin{array}{ccc}
C_1(A_1) & \xrightarrow{\;\;C_1(f)\;\;} & C_1(A_2) \\
\downarrow{\scriptstyle \mathcal{N}(A_1)} & & \downarrow{\scriptstyle \mathcal{N}(A_2)} \\
C_2(A_1) & \xrightarrow{\;\;C_2(f)\;\;} & C_2(A_2)
\end{array}
$$

(all in this diagram occurs in the category B). If, additionally, for every $A \in A^0$ the morphism $\mathcal{N}(A)$ is isomorphism in B, then the functors C_1 and C_2 are said to be *naturally equivalent*.

1.16. Example. Let us consider a contravariant functor $*: \mathrm{Vect} \to \mathrm{Vect}$, defined by $f^*(g) = g \circ f$ for any mapping $f: \mathcal{A} \to \mathcal{B}$, $g: \mathcal{B} \to \mathbf{R}$ (f^* is called a *pull back* of the function g by f; see chapt. 1, sec. 4). Analogously, we define f^{**} (a pull back of a pull back), and a covariant functor $**: \mathrm{Vect} \to \mathrm{Vect}$. Now, let $A = B = \mathrm{Vect}$, $\mathcal{C}_1 = \mathrm{Id}$, $\mathcal{C}_2 = {}^{**}$. The mapping $\mathcal{N}(\mathcal{V}): \mathcal{V} \to \mathcal{V}^{**}$, $\mathcal{V} \in \mathrm{Vect}^0$ is a natural transformation as it can be easily seen from the commutative diagram, valid for any $f: \mathcal{V} \to \mathcal{U}, \mathcal{U} \in \mathrm{Vect}^0$

$$
\begin{array}{ccc}
\mathcal{C}_1(\mathcal{V}) = \mathcal{V} & \xrightarrow{\;\mathcal{C}_1(f)=f\;} & \mathcal{C}_1(\mathcal{U}) = \mathcal{U} \\[2mm]
{\scriptstyle \mathcal{N}(\mathcal{V})}\Big\downarrow & & \Big\downarrow{\scriptstyle \mathcal{N}(\mathcal{U})} \\[2mm]
\mathcal{C}_2(\mathcal{V}) = \mathcal{V}^{**} & \xrightarrow{\;\mathcal{C}_2(f)=f^{**}\;} & \mathcal{C}_2(\mathcal{U}) = \mathcal{U}^{**}
\end{array}
$$

If we limit our considerations to the category of finitely–dimensional vector spaces, the functors $\mathcal{C}_1 = \mathrm{Id}$ and $\mathcal{C}_2 = {}^{**}$ are naturally equivalent (see Trautman 1973b, p. 185).

Definitions of natural transformation and natural equivalence can be "naturally" extended to the case of functors of many variables (see Semadeni and Wiweger 1978 (pp. 134–136).

2. Aristotelian Space–Time

Dynamical theories investigate motions of bodies under the influence of forces. The first goal of a dynamical theory is to provide a criterion allowing one to determine whether any force has appeared. The role of such a criterion is usually played by the definition of "free motions", i.e., motions which are going on without any intervention of forces. A deviation from a free motion testifies to the appearance of a force. The formulation of such a criterion is the content of the *first law of dynamics*. If one assumes that motions under no influence of forces are determined by the structure of space–time — and this assumption seems to be very natural — then the first law of dynamics contains in itself information about this structure. Geometry, interpreted in such a way, becomes a part of the given dynamical theory, and, as it belongs to physics, is subject to empirical verification.

In Aristotle's dynamics[3] the first law can be formulated as follows:

There exists the natural rest,

or, equivalently,

---

[3]It is less known that Aristotle formulated a complete dynamics, and attempted at doing it in a quantitative manner (see Aristotle's *Physics*, especially the final parts of book VII). From our present point of view the Aristotelian "physics" is a mixture of philosophical views, purely verbal analyses, empirical statements (often erroneous ones), and a natural knowledge based on them. One could distillate from this a part which would correspond to what is today called physics. It goes without saying that the basic principles of the Aristotelian physics have been experimentally falsified.

a body, not acted upon by any force, remains in the absolute rest.[4]

This law implies the structure of space–time in the following way. The absolute rest allows one to meaningfully speak about the same place at different times (a body in the state of absolute rest **always** occupies the same place), and consequently it determines a projection π_Σ from the four–dimensional space–time $\mathcal{A}$ into a three–dimensional manifold Σ. The existence of the Aristotelian absolute time makes it possible to meaningfully speak of the absolute simultaneity of events (even if they are distant from each other), and consequently it defines a projection π_T of the manifold $\mathcal{A}$ into the "axis of absolute time" T. The existence of the projections π_Σ and π_T is equivalent to the existence of the absolute space and the absolute time, respectively. It follows that the Aristotelian space–time $\mathcal{A}$ is a product bundle $(\mathcal{A} = \Sigma \times T, \Sigma, \pi_\Sigma)$.[5] One could easily show that this structure is natural for the Aristotelian space–time or, more strictly, that between suitably defined category of Aristotelian space–times and the category of product bundles there exists a natural transformation. Proof of this fact is similar to the proof which will be carried out in detail for space–time of classical mechanics.

Ascribing to the sets Σ, T and $\mathcal{A}$ the structure of smooth manifolds is a stylization of the history, but a stylization not too far exaggerated, if one takes into account the fact that the contemporary differential manifold notion directly comes from the Aristotelian concept of *continuum* for which ideas of space and time in turn constituted its prototypes.

The structure of the Aristotelian space–time is invariant with respect to the direct product of dilatation, translation and rotation groups (acting on Σ) by an affine group (acting on T).

The appearance of absolute velocity with respect to the absolute space testifies to the intervention of a force. In Aristotle's views, this force is proportional to the velocity produced by it. This statement can be thought of as to form the second law of the Aristotelian dynamics. However, we shall skip this problem as marginal for our discussion.[6]

3. Space–Time of Classical Mechanics Without Gravitational Field

Reconstruction of the space–time structure presupposed by classical mechanics is especially instructive. It not only shows that the structures, discovered later on by

---

[4]One should remember that Aristotle assumed the existence of the so–called natural motions (e.g. motions of heavy bodies towards the centre of the Earth). Of course, we could include such motions into our reconstruction of the Aristotelian dynamics, but this would highly complicate the picture without throwing any new light on questions we are interested in.

[5]One can also say that the Aristotelian space–time $\mathcal{A}$ is the Cartesian product $\mathcal{A} = \Sigma \times T$.

[6]More precise formulation of the second law of the Aristotelian dynamics would require a far–reaching historical stylization. It would mainly consist in translating some terms, used by Aristotle in their every–day meanings, into their counterparts elaborated much later in order to correctly express laws of dynamics. For instance, the Aristotelian term "efficient factor" should be replaced by "force", "heaviness" by "mass", etc. (See also Toulmin and Goodfield, 1961, pp. 93–100.)

relativity theory, existed and were well functioning from the very beginning of modern physics, but also demonstrates how quickly a physical theory gains independence from the philosophical views of its inventor and begins to lead an autonomous life on its own. Newton ascribed to space and time, which appeared in mechanics created by himself, Aristotelian properties, that is to say he treated them as absolute ones. It turns out that this remains correct as far as time is concerned, but is false with regard to space. One should distinguish the structure of space–time as postulated by Newton himself and the structure of space–time as presupposed by classical mechanics. The former is often called *Newton's space–time* (see Ehlers 1973), the latter *Galileo space–time* (see Kopczyński and Trautman 1981, Trautman 1970, 1973b). In the following we shall speak simply of *space–time of classical mechanics*. This seems to be justified since there is a necessity to distinguish the space–time structures of classical mechanics *without gravitational field and with gravitational field*. Now, we are going to analyse the first of these two cases, postponing the second one to the next section.

The first law of the Newtonian dynamics establishes a democracy among all inertial reference frame: none of such frames is privileged in any way. The standard of the absolute rest disappears from the theory. The Aristotelian projection π_Σ ceases to be a natural structure. Every observer, connected with a reference frame, has his private standard of rest ("with respect to his own inertial frame"); that is to say, his own projection corresponds to the Aristotelian projection π_Σ. However, absolute simultaneity and absolute acceleration remain valid in the classical world picture. Absolute simultaneity defines "spaces of equal time", and consequently leaves the Aristotelian projection π_Σ unchanged. Absolute acceleration connects the velocities at different points of space–time with one another: vanishing of acceleration allows one to meaningfully speak of "the same velocity at space–time points separated from one another by non–zero distances" (but not of the zero absolute velocity in different inertial frames). This leads to the so–called *teleparallelism*, i.e., to the concept of parallel vectors at points separate from one another in space–time.[7] Therefore, one must assume that space–time of classical mechanics (without gravitational field) is a four–dimensional differential manifold equipped with an affine connection which makes meaningful parallel transport of vectors (teleparallelism). It is natural to consider this connection in terms of a principal fibre bundle over space–time. Since any local inertial frame of reference can be identified with a basis at a given space–time point, the principal fibre bundle in question should be concretized to the bundle of frames over space–time. Let us consider this structure in more detail.

An affine space $\mathcal{A} = (\mathcal{E}, \mathcal{V}, +)$ (see example 1.8) equipped with a bilinear map-

---

[7]Let us turn our attention to the fact that the absolute character of acceleration in classical mechanics does not imply the existence of an absolute space. From the historical point of view, the appearance of "absolute accelerations" constituted the main Newton's argument on behalf of the existence of absolute space. This argument was the focus of heated discussions around the conceptual structure of classical mechanics (see Raine 1981a, p. 1157).

ping $h: \mathcal{V}^* \times \mathcal{V}^* \rightarrow \mathbf{R}^1$, such that h is a symmetric positively defined mapping of rank $n - 1$, where n is the dimension of the vector space $\mathcal{V}$, will be called *space–time of classical mechanics without gravitational field* and denoted by $\mathcal{N}$. We shall write $\mathcal{N} = (\mathcal{E}, \mathcal{V}, +; h)$.

Let $\mathcal{N}_1 = (\mathcal{E}_1, \mathcal{V}_1, +; h_1)$ and $\mathcal{N}_2 = (\mathcal{E}_2, \mathcal{V}_2, +; h_2)$ be two such space–times. An affine morphism f will be called *Galileo's morphism* if $h_2 = h_1 \circ (\tau f)^*$, where the asterisk denotes the dual functor. The class of space–times of classical mechanics without gravitation with Galileo morphisms as morphisms is a category which will be denoted by Gal; it can be thought of as a subcategory of the category $\not\!\phi$Diff. A Galileo's morphism, which is an automorphism, is called a *Galileo's transformation*.

Let $\mathcal{S} \subset \mathcal{V}^*$ be defined as a subspace for which $h = 0$, and $\mathcal{S}\perp$ as a subspace of vectors orthogonal to $\mathcal{S}$. Let $\pi_T: \mathcal{E} \rightarrow \mathcal{E}/\mathcal{S} \equiv T$ be a canonical projection. Now, N can be presented as a fibre bundle $(\mathcal{E}, T, \pi_T, \mathcal{S})$, where $\mathcal{S}$ is a typical fibre. A functor $\sigma: \text{Gal} \rightarrow \text{Bun}$ can be defined from the category of space–times of classical mechanics without gravity to the category of bundles

$$\sigma(\mathcal{N}) = (\mathcal{E}, T, \pi_T, \mathcal{S}),$$
$$\sigma(f) = (f, a),$$

where $f: \mathcal{E}_1 \rightarrow \mathcal{E}_2$ is a Galileo's morphism and $a: T_1 \rightarrow T_2$ a uniquely determined mapping satisfying the condition: $\Pi_{T_1} \circ f = a \circ \pi_{T_2}$.

Now, let us define another functor $\omega: \text{Gal} \rightarrow \text{Bun}$ to every space–time $\mathcal{N} \in \text{Gal}^0$ ascribing a product fibre $(T \times \mathcal{S}, T, \pi_T)$. It turns out that there is no natural transformation between functors σ and ω. This statement is mathematical counterpart of the intuitive statement that space–time of classical mechanics (without gravity) is not a Cartesian product of absolute time and absolute space.

The existence of teleparallelism in space–time determines horizontal cross-sections of the frame bundle over space–time. Indeed, all frames, resulting from a parallel transport of an arbitrarily chosen frame at a certain point of space–time, smoothly select a single point of space–time in each fibre of the frame bundle. Intuition suggests that in this way the natural product structure, lost in space–time, can be recovered in the fibre bundle of frames over this space–time. We will see that this is indeed the case.

Let us define the following two functors: the functor

$$\mathcal{B}: \not\!\phi\text{Aff} \longrightarrow P\text{Bun}$$

from the category of affine spaces to the category of principal fibre bundles, associating with every object $\mathcal{A} = (\mathcal{E}, \mathcal{V}, +) \in \not\!\phi\text{Aff}^0$ a frame bundle $\mathcal{B}(\mathcal{E}) \in P\text{Bun}$, and the functor

$$\mathcal{C}: \not\!\phi\text{Aff} \longrightarrow P\text{Bun},$$

associating with every $\mathcal{A} = (\mathcal{E}, \mathcal{V}, +)$ a product bundle $\mathcal{C}(\mathcal{A}) = \mathcal{E} \times \mathcal{B}(\mathcal{V})$. These functors are naturally equivalent as it can be easily seen from the following commutative diagram for every $f : \mathcal{E}_1 \to \mathcal{E}_2$:

$$
\begin{array}{ccc}
\mathcal{B}(\mathcal{E}_1) & \xrightarrow{\;\;\mathcal{B}(f)\;\;} & \mathcal{B}(\mathcal{E}_2) \\[2mm]
\mathcal{N}(\mathcal{E}_1) \Big\downarrow & & \Big\downarrow \mathcal{N}(\mathcal{E}_2) \\[2mm]
C(\mathcal{E}_1) = \mathcal{E}_1 \times \mathcal{B}(\mathcal{V}_1) & \xrightarrow{\;\;C(f)\;\;} & C(\mathcal{E}_2) = \mathcal{E}_2 \times \mathcal{B}(\mathcal{V}_2)
\end{array}
$$

where $\mathcal{C}(f) = f \times \mathcal{B}(\tau f)$, and $\mathcal{N}$ is a natural transformation of the functor $\mathcal{B}$ into the functor $\mathcal{C}$. Therefore, in classical mechanics without gravity the product structure is naturally transferred from space–time to the frame bundle over space–time (see Trautman 1973b).

4. Space–Time of Classical Mechanics with Gravitational Field

At the basis of the Newtonian theory of gravity lies the statement which deserves the name *Galilean principle of equivalence*; it asserts that in a gravitational field all bodies fall with the same acceleration. As is well-known, this principle allows one to treat freely falling reference frames as *locally* non–accelerating, that is to say, reference frames in which laws of the Newtonian dynamics are locally valid (see Raine and Heller 1981, p. 83–84). However, the presence of a gravitational field causes focusing (or antifocusing) of space–time trajectories of freely falling bodies. For instance, the trajectories of bodies freely falling on Earth should meet at the point corresponding to the centre of the Earth. This effect is connected with the appearance of tidal forces. These forces liquidate the existence of teleparallelism: now, parallel transport depends on the path along which it is performed. We express this by saying that the affine connection over space–time of classical mechanics with gravity is non–integrable. This means that in the frame bundle over space–time there are no global cross–sections (in the case without gravity such cross–sections were implied by teleparallelism which now does not exist), and consequently the connection is not–flat (see chapt. 3, sec. 5). In other words: the presence of gravity manifests itself in the existence of a non–vanishing curvature form on the frame bundle over space–time.

The non–existence of global cross–sections in the frame bundle over space–time suggests that this bundle, in the presence of a gravitational field, cannot be expressed in the form of a Cartesian product (globally). Indeed, in this case, there are no naturally equivalent functors associating with the space–time a trivial frame bundle. (See Raine 1981a, Raine and Heller 1981, chapter 6.)

5. The Special and General Theories of Relativity

Further logic of the evolution of space–time strucures is almost automatic. Space–time of special relativity (see chapter 2, sec. 3) has no product structure. Moreover, it has no natural fibration into global spaces of equal time (i.e., it has no projection π_T which would be independent of the frame of reference) that existed in space–time of classical mechanics.

Space–time of special relativity — in contradistinction to space–time of classical mechanics — carries a Lorentz structure and its affine connection is consistent with this structure. It is a flat connection which makes it possible to express the bundle of (orthonormal) frames in the form of a Cartesian product.

In general relativity (see chapter 4) the frame bundle over space–time has a non-vanishing curvature form which implies that this bundle is not, in any natural way, a trivial one. The essential difference between the space–time structure of classical mechanics with gravitational field and that of general relativity consists in that the latter is equipped with a Lorentz metric. In the architecture of the frame bundle this difference manifests itself in the existence of the soldering form which does not exist in classical mechanics.

The existence of a linear connection in the frame bundle $\mathcal{B}(\mathcal{M})$ over space–time $\mathcal{M}$ is responsible for the fact that the frame bundle $\mathcal{B}(\mathcal{B}(\mathcal{M}))$ over the bundle $\mathcal{B}(\mathcal{M})$ is — in a natural way — a Cartesian product. This fact might suggest that the next step in the evolution of space–time theories could consist in depriving the bundle $\mathcal{B}(\mathcal{B}(\mathcal{M}))$ of its product structure and in the physical interpretation of this procedure (see Trautman 1970a, 1973b).

It is worth noticing that, when discussing the naturality problem in the case of general relativity, one should take into account the *category of principal fibre bundles with linear connections*. A *connection morphism* is a morphism of principal bundles such that the image under this morphism of any horizontal curve is a horizontal curve (see Sulanke and Wintgen 1977, p. 133).

The diagram on the adjacent page shows the logic of evolution of space–time structures (it is a modified version of the diagram appearing in the two works of Trautman quoted above).

6. Comments and Remarks

A. The logic of evolution of space–time theories, shown in the present Appendix, is a consequence of the strong stylization of the history of physics. It was the reconstruction of space–time structures of various theories in terms of modern differential geometry that has allowed us to see the logical connections between subsequent stages of the evolutionary process. The view that such a reconstruction reflects, at least some, real features of development of science seems well founded. On the other hand, however, there can be no doubt that the true history of science was much more complicated and much less logical: it was like a tree with the main stem and many

$\mathcal{B}(\mathcal{B}(\mathcal{M}))$

$\mathcal{B}(\mathcal{M})$

$\mathcal{M}$

$\mathcal{B}(\mathcal{M})$

$\mathcal{M}$

$\mathcal{B}(\mathcal{N})$

$\mathcal{N}$

$\mathcal{B}(\mathcal{N})$

$\mathcal{N}$

T

$\mathcal{A} = T \times \Sigma$

T

Σ

Space-time $\mathcal{A}$ of Aristotle has a product structure

Space-time $\mathcal{N}$ of classical mechanics without gravity has no product structure, but it admits fibration onto spaces of equal time

Frame bundle $\mathcal{B}(\mathcal{N})$ has a product structure

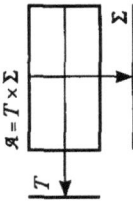

$\mathcal{N}$ is an affine manifold

Space-time $\mathcal{N}$ of classical mechanics with gravity

Frame bundle $\mathcal{B}(\mathcal{N})$ has no product structure

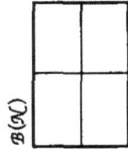

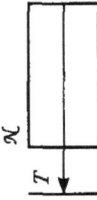

Space-time $\mathcal{M}$ of special relativity does not admit any natural fibration

Frame bundle $\mathcal{B}(\mathcal{M})$ has a product structure

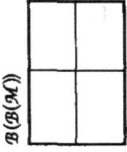

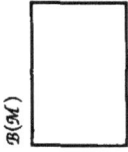

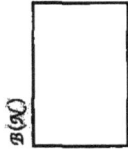

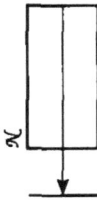

Space-time $\mathcal{M}$ of general relativity

Frame bundle $\mathcal{B}(\mathcal{M})$ has no product structure

Frame bundle $\mathcal{B}(\mathcal{B}(\mathcal{M}))$ has a product structure

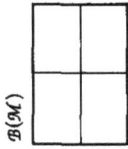

$\mathcal{M}$ is a Lorentz manifold

side–branches rather than a logical chain the parts of which smoothly change from one into another.

For example, the ideas of Berkeley and Leibniz, later on developed by Mach, constituted an important side–branch which could be thought of as a concurrence with respect to the Newtonian main stem (see Raine and Heller 1981, chapter 4; Barbour 1983). The essence of these ideas is contained in the postulate that the entire structure of space–time, and consequently the whole of dynamics, should be reduced to relationships between bodies without assuming any "absolute" manifold (see chapter 1, sec. 6, B). An attempt to construct such a dynamics has been undertaken by Barbour and Bertotti (1977). These authors believe that laws of the new dynamics should be invariant with respect to the following transformation group called the *Leibniz group*,

$$r \longmapsto \mathcal{A}(t)r + g(t),$$
$$t \longmapsto f(t),$$

where $\mathcal{A}(t)$ is an orthogonal matrix, and f and g are any smooth functions of time (additionally, one assumes that f is a monotonic function). The existence of the "universal time" t (the existence of the projection π_T) is justified by the Leibniz postulate of the "coexistence of events". The Leibniz group not only admits the rescaling of time, consisting in choosing new units and a new beginning of time counting (as did the Galileo group in classical mechanics), but also more general time transformations. The Leibniz group does not distinguish uniform motions; it also admits any suitably smooth rotations of instantaneous spaces. The Galileo group is a subgroup of the Leibniz group. (See also Liebscher and Yourgrau 1979).

B. In the philosophy of science there is a heated polemics in which the following views fight one another (a) the evolution of science goes on more or less logically, and the "time arrow" of this evolution is determined by the continuous accumulation of results (Popper, Lakatos); (b) strictly speaking, there is no evolution, but periods of "normal development" are separated from each other by all–changing revolutions; "normal periods" from before and after a revolution are "incommensurable" (Kuhn); (c) one should stop speaking about the progress in science since science is not ruled by logic but by laws of sociology and psychology of scientists (Wilson) or by the sole rule "anything goes" (Feuerabend). The polemics is often referred to in the literature as dispute on the rationality of science or of its evolution. The ideas briefly signaled in (c), which try to reduce all rationality criteria to social conditions changing from epoch to epoch, have recently grown in number and popularity (for a review of this dispute see for instance, Życiński 1988). The analyses, carried out in the present Appendix, throw some light on this problem. In spite of the strong historical stylization, which was indispensable in order to translate older space–time theories into the language of the present scientific paradigm, one cannot ignore the very fact that earlier theories **can be translated** into the modern language, and that in this process surprising logical dependencies are revealed within both the structures of particular theories and the structure of the entire evolutionary chain. Of course, it is true that in this way our present criteria have been adapted to the former theories in such a manner that the whole procedure could succeed. Nevertheless, the result of this analysis should be considered as philosophically significant.

Bibliographical Note. Elements of the mathematical category theory could be learned, for instance, from Dodson (1980) or Semadeni and Wiweger (1972, 1978). Applications of this theory to differential geometry can be found in Sulanke and Wintgen (1972), and to the analysis of space–time structures in the works by Trautman (1970a,b, 1973b). Mathematical aspects of the classification of space–times (in terms of bundles which can be associated with the frame bundle over space–time) are presented in Trautman's paper (1976).

Reconstruction of space–time theories in the language of modern differential geometry has been carried out in the following works: Penrose 1968; Ehlers 1973; Trautman 1968, 1970a, 1971, 1973b; Kopczyński and Trautman 1981, Thirring 1978 (chapter 6); Raine and Heller 1981; Toretti 1983; Friedman 1983. Toretti's book contains some historical comments; Friedman focuses on philosophical aspects. One should take into account the different terminological conventions adopted by various authors, which sometimes could give the misleading impression that results of the analyses lead to contradictory results.

126

References

Abbot, L., 1988, The Mystery of the Cosmological Constant, *Scientific American* **258**, 82–88.

Abraham R., Marsden J., 1979, *Foundations of Mechanics*, Benjamin, New York (revised edition).

Adler, R., Bazin, M., Schiffer, M., 1965, *Introduction to General Relativity*, McGraw–Hill, New York, St. Louis ...

Aharoni, J., 1959, *The Special Theory of Relativity*, Clarendon Press, Oxford.

Arnold, W. I., 1974, *Mathematical Methods of Classical Mechanics*, Nauka, Moscow.

Auslander, L., Mac Kenzie, R. E., 1963, *Introduction to Differentiable Manifolds*, MacGraw–Hill, New York, San Francisco ...

Banach, S., 1932, *Theorie des opérations linéaires*, Warszawa.

Barbour, J. B., 1974, Relative–Distance Machian Theories, *Nature (Phys. Sci.)* **249**, 328–329 (corrected in *Nature* **250**, 1974, 606).

Barbour, J. B., 1983, Philosophical Principles and the Problem of Motion, *Philosophy in Science* **1**, 71–87.

Barbour, J., Bertotti, B., 1977, Gravity and Inertia in a Machian Framework, *Nuovo Cim.* **38B**, 1–27.

Barrow, J. D., 1978, Quiescent Cosmology, *Nature* **272**, 211–215.

Barrow, J. D., 1983, Anthropic Definitions, *Q. Jl. Royal Astron. Soc.* **24**, 146–153.

Barrow J. D., 1988, *The World Within the World*, Clarendon Press, Oxford.

Barrow, J. D., Matzner, R. A., 1977, The Homogeneity and Isotropy of the Universe, *Mont. Not. Royal Astron. Soc.* **181**, 719–727.

Barrow, J. D., Tipler, F. J., 1986, The Anthropic Cosmological Principle, Clarendon Press, Oxford.

Barrow, J. D., Turner, M. S., 1982, The Inflatory Universe — Birth, Death and Transfiguration, *Nature* **298**, 801–805.

Bass, R. W., Witten, L., 1957, Remark on Cosmological Models, *Rev. Mod. Phys.* **29**, 452–453.

Beem, J. K., Ehrlich, P. E., 1981, *Global Lorentzian Geometry*, Marcel Dekker, New York–Basel.

Bergamnn, P. G., 1976, *Introduction to the Theory of Relativity*, Dover Publ., New York.

Berstein, H. J., Philips, A. V., 1981, Fibre Bundles and Quantum Theory, *Scientific American* **245**, 94–109.

Berry, M., 1976, *Principles of Cosmology and Gravitation*, Cambridge University Press.

Bishop, R. L., Crittenden, R. J., 1964, *Geometry of Manifolds*, Academic Press, New York–London.

Bishop, R. L., Goldberg, S. I., 1980, *Tensor Analysis on Manifolds*, Dover Publ., New York.

Bishop, R. L., O'Neill. B., 1969, Manifolds of Negative Curvature, *Trans. Amer. Math. Soc.* **145**, 1–49.

Bondi, H., 1960, *Cosmology*, Cambridge University Press.

Bondi, H., Gold, T., 1948, The Steady-State Theory of the Expanding Universe, *Mont. Not. Royal Astron. Soc.* **108**, 252–270.

Bourbaki, N., 1969, *Elements d'histoire des mathematiques*, Hermann, Paris.

Brickell, F., Clark, R. S., 1970, *Differentiable Manifolds — An Introduction*, Van Nostrand, London, New York ...

Brill, D., 1982, Linearization Stability, in: *Space-time and Geometry*, ed.: R. A. Matzner, L. C. Shepley, University of Texas Press, Austin, 59–81.

Carmeli, M., 1977, *Group Theory and General Relativity*, McGraw–Hill, New York, London ...

Carr, J. B., 1982, The Anthropic Principle, *Acta Cosmol.* **11**, 143–151.

Carr, J. B., Rees, M. J., 1979, The Anthropic Principle and The Structure of the Physical World, *Nature* **278**, 605–612.

Cartan, E., 1922, Sur une generalisation de la notion de courbure de Riemann et les Espaces a Torsion, *Compt. Rend. Acad. Sci. (Paris)* **174**, 593–595.

Cartan, E., Sur les varietes a connexion affine et la theorie de la relativité genearalisée 1923, I: *Ann. Ec. Norm. Sup.* **40**, 325–412;

1924, I (suite): ibid. **41**, 1–25;

1925, II: ibid. **42**, 17–88.

Carter, B., 1974, Large Number Coincidences and the Anthropic Principle in Cosmology, in: *Confrontation of Cosmological Theories with Observational Data, IAU Symposium*, ed.: M. S. Longair, Reidel, Dordrecht–Boston, 291–298.

Choquet-Bruhat, Y., DeWitt-Morette,C., Dillard-Bleick, M., 1982, *Analysis, Manifolds and Physics* (revised edition), North–Holland, Amsterdam.

Clarke, C. J. S, 1970, On the Global Isometric Embedding of Pseudo–Riemannian Manifolds, *Proc. Royal Soc. London*, **A314**, 417–429.

Clarke, C. J. S., 1973, Local Extensions in Singular Space–Times, *Commun. math. Phys.* **32**, 205–214.

Clarke, C. J. S., 1975, The Classification of Singularities, *GRG* **6**, 35–40.

Clarke, C. J. S., 1976, Space–Time Singularities, *Commun. math. Phys.* **49**, 17–23.

Clarke, C. J. S., Schmidt, B. G., 1977, Singularities: The State of the Art, *GRG* **8**, 129–137.

Collins, C. B., 1974, Tilting at Cosmological Singularities, *Commun. math. Phys.* **39**, 131–151.

Collins, C. B., 1977, The Role of Mathematics in Gravitational Physics — From the Sublime to the Subliminal? *GRG* **8**, 717–721.

Collins, C. B., Hawking, S. W., 1973, Why is the Universe Isotropic? *Astrophys. J.* **180**, 317–334.

Daniel, M., Viallet, C. M., 1980, The Geometrical Setting of Gauge Theories of the Yang–Mills Type, *Rev. Mod. Phys.* **52**, 175–197.

Demaret, J., Barbier, C., 1981, Le principe anthropique en cosmologie, *Rev. Quest. Sci.* **151**, 181–222; 461–509.

Demiański, M., 1985, *Relativistic Astrophysics*, Polish Scientific Publishers — Pergamon Press, Warszawa, Oxford ...

Dicke, R. H., 1961, Dirac's Cosmology and Mach's Principle, *Nature* **192**, 440–441.

Dicke, R. H., Peebles, P. J. E., Roll, P. G., Wilkinson, D. T., 1965, Cosmic–Black–Body Radiation, *Astrophys. J.* **142**, 114–119.

Dixon, W. G., 1978, *Special Relativity — The Foundation of Macroscopic Physics*, Cambridge University Press.

Dodson, C. T. J., 1980, *Categories, Bundles and Spacetime Topology*, Shiva Publishing, Orpington.

Dolgov, A. D., Zeldovich, I. B., Sagin, M. B., 1988, *Cosmology of the Early Universe*, Moscow University Press (in Russian).

Drechsler, W., Mayer, M. E., 1977, *Fiber Bundle Techniques in Gauge Theories*, Springer, Berlin — Heidelberg — New York.

Eddington, A. S., 1923, *The Mathematical Theory of Relativity*, Cambridge University Press.

Eguchi, T., Gikey, P. B., Hanson, A. J., 1980, Gravitation, Gauge Theories and Differential Geometry, *Phys. Rep.* **66**, 215–393.

Ehlers, J., 1973, The Nature and Structure of Space–Time, in: *The Physicist's Conception of Nature*, ed.: J. Mehra, Reidel, Dordrecht, 71–91.

Ehlers, J., Pirani, F. A. E., Schild, A., 1972, The Geometry of Free Fall and Light Propagation, in: *General Relativity — Papers in Honour of J. L. Synge*, ed.: L. O'Raifeartaigh, Clarendon Press, Oxford, 65–83.

Eilenberg, S., MacLane, S., 1945, General Theory of Natural Equivalencies, *Trans. Amer. Math. Soc.* **58**, 231–294.

Einstein, A., 1905, Zur Elektrodynamik bewegter Körper, *Ann. Phys.* **17**, 891–921; English translation: *The Principle of Relativity*, Dover Publ., 1923, 35–65 and in: Miller 1981, 391–415.

Einstein, A., 1917, Kosmologische Betrachtungen zur allgemeinen Relativitätstheorie, *Sitzungs-ber. preuss. Akad. Wiss.* **1**, 142–152; English translation: *The Principle of Relativity*, Dover Publ., 1923, 177–188.

Ellis, G. F. R., 1971, Topology and Cosmology *GRG*, **2**, 7–21.

Ellis, G. F. R., 1975, Cosmology and Verifiability, *Q. Jl. R. Astr. Soc.* **16**, 245–264.

Ellis, G. F. R., 1980, Limits to Verification in Cosmology, *Ann. New York Acad. Sci.* **336**, 130–160.

Ellis, G. F. R., 1984, Relativistic Cosmology: Its Nature, Aims and Problems, in: *General Relativity and Gravitation, Invited Papers and Discussion Reports*, Padua, July 3–8, 1983, ed.: B. Bertotti, F. de Felice, A. Pascolini, Reidel, Dordrecht — Boston, 215–288.

Ellis, G. F. R., Maartens, R., Nel, S. D., 1978, The Expansion of the Universe, *Mont. Not. Royal Astron. Soc.* **184**, 439–465.

Ellis, G. F. R., Maartens, R., Nel, S. D., Stoeger W. R., Whitman A. P., Ideal Observational Cosmology, *Physic Reports* **124**, 315–417.

Ellis, G. F. R., Perry, J. J., 1979, Towards a "Correctionless" Observational Cosmology, *Mont. Not. Royal Astron. Soc.* **187**, 357–370.

Ellis, G. F. R., Schmidt, B. G., 1977, Singular Space–Times, *GRG* **11**, 915–953.

Ellis, G. F. R., Sciama, D. W., 1972, Global and Non–Global Problems in Cosmology, in: *General Relativity — Papers in Honour of J. L. Synge*, ed.: L. O'Raifeartaigh, Clarendon Press, Oxford, 35–59.

Ellis, G. F. R., Stoeger, W., 1988, Horizons in Inflationary Universes, *Class. Quantum Grav.* **5**, 207–220.

Engelking, R., 1975, *General Topology*, Polish Scientific Publishers, Warsaw (in Polish).

Fisher, A., Marsden, J. E., 1979, The Initial Value Problem and the Dynamical Formulation of General Relativity, in: *General Relativity — An Einstein Centenary Survey*, eds.: S. W. Hawking, W. Israel, Cambridge University Press, 138–211; 850–859.

Fisher, A., Marsden, J. E., Moncrief, V., 1980, Symmetry Breaking in General Relativity, in: *Essays in General Relativity — A Festschrift for Abraham Taub*, ed.: F. J. Tipler, Academic Press, New York, London ..., 79–96.

Flanders, H., 1963, *Differential Forms with Applications to the Physical Sciences*, Academic Press, New York — London.

Friedman, M., 1983, *Foundations of Space–Time Theories — Relativistic Physics and Philosophy of Science*, Princeton University Press.

Friedrich, H., Steward, J. M., 1982, *Characteristic Initial Data and Wave Front Singularities in General Relativity*, Preprint, Max–Planck–Institut, München.

Geroch, R., 1970, Singularities, in: *Relativity* ed.: S. Fickler, M. Carmeli, L. Witten, Plenum, New York, 259–291.

Geroch, R., 1971, Space–Time Structure from a Global Viewpoint, in: *General Relativity and Cosmology, Enrico Fermi XLVII Course*, Academic Press, Chicago — London.

Geroch, R., 1978, *General Relativity from A to B*, The University of Chicago Press, Chicago — London.

Geroch, R., Horowitz, G. T., 1979, Global Structure of Space–Time, in: *General Relativity — An Einstein Centenary Survey*, ed.: S. W. Hawking, W. Israel, Cambridge University Press, 212–293; 859–860.

Godart, O., Heller, M., 1985, *Cosmology of Lemaître*, Pachart, Tucson.

Goenner, H., 1970, Mach's Principle and Einstein's Theory of Gravitation, in: *Boston Studies in the Philosophy of Science*, vol. VI: Ernst Mach — Physicist and Philosopher, Reidel, Dordrecht, 200–215.

Golda, Z., Heller, M., Szydlowski, M., 1983, Structurally Stable Approximations to Friedmann-Lemaître World Models, *Astrophys. Space Sc.* **90**, 313–326.

Groth, E. J., Peebles, J. E., Seldner, M., Soneira, R. M., 1977, The Clustering of Galaxies, *Scientific American* **237**, 5, 76–98.

Gruszczak, J., Heller, M., Multarzyński, P., 1988, A Generalization of Manifolds as Space–Time Models, *J. Math. Phys.* **29**, 2576–2580.

Gruszak, J., Heller, M., Multarzyński, P., 1989, Physics With and Without the Equivalence Principle, *Foundations of Physics* **19**, 607–618.

Guth, A. H., 1981, Inflationary Universe: A Possible Solution to the Horizon and Flatness Problems, *Phys. Rev.* **23D**, 347–356.

Harrison, E. R., 1981, *Cosmology — The Science of the Universe*, Cambridge University Press.

Hartle, J. B., Hawking, S. W., 1983, Wave Function of the Universe, *Phys. Rev.* **D28**, 2960–2975.

Hawking, S. W., 1971, Stable and Generic Properties in General Relativity, *GRG* **1**, 393–400.

Hawking, S. W., 1979, The Path–Integral Approach to Quantum Gravity, in: *General Relativity — An Einstein Centenary Survey*, eds.: S. W. Hawking, W. Israel, Cambridge University Press, 746–789; 894–895.

Hawking, S. W., 1982, The Boundary Conditions of the Universe, in: *Astrophysical Cosmology*, eds.: H. A. Bruck, G. V. Coyne, M. S. Longair, Pontificiae Academiae Scientiarum Scripta Varia, vol. 48, 563–574.

Hawking, S. W., 1984a, The Quantum State of the Universe, *Nuclear Physics* **B239**, 257–276.

Hawking, S. W., 1984b, Quantum Cosmology, in: *Relativity, Groups and Topology II*, eds.: B. S. DeWitt, R. Stora, North–Holland, Amsterdam, Oxford . . . , 332–379.

Hawking, S. W., Ellis, G. F. R., 1973, *The Large Scale Structure of Space–Time*, Cambridge University Press.

Hawking, S. W., Moss, I. G., 1982, Supercooled Phase Transitions in the Very Early Universe, *Phys. Lett.* **110B**, 35–38.

Heckmann, O., Schücking, I., 1959, Newtonsche und Einsteinsche Kosmologie, in: *Handbuch der Physik*, Band LII, Astrophysik IV: Sternsysteme, Springer, Berlin, Gottingen . . . , 489–519.

Hehl, F. W., Spin and Torsion in General Relativity
1973, I: Foundations, *GRG* **4**, 333–349, 1974
1974, II: Geometry and Field Equations, *GRG* **5**, 491–516.

Heidmann, J., 1973, *Introduction à la Cosmologie*, Press Universitaires de France.

Heller, M., 1975a, The Influence of Mach's Thought on Contemporary Relativistic Physics, *Organon* **11**, 271–283.

Heller, M., 1977, Space–Time Structures, *Acta Cosmol.* **6**, 109–128.

Heller, M., 1978, Local — Large Scale — Global: On Certain Methodological Questions of Cosmology, *Acta Cosmol.* **7**, 83–99.

Heller, M., 1981a, The Manifold Model for Space–Time, *Acta Cosmol.* **10**, 33–51.

Heller, M., 1981b, Relativistic Model for Space–Time, *Acta Cosmol.* **10**, 53–69.

Heller, M., 1981c, The Origin of Time, in: *The Study of Time IV*, ed.: J. T. Fraser, N. Lawrence, D. Park, Springer, New York, Heidelberg . . . , 90–93.

Heller, M., 1982, Matter and Radiation Models in the Standard Universe, *Acta Cosmol.* **11**, 11–25.

Heller, M., 1988, Between Newton and Einstein, in: *Newton and the New Direction in Science, Proc. of the Cracow Conference, 25–28 May 1987*, eds.: G. V. Coyne, M. Heller, J. Życinski, Specola Vaticana, 155–173.

Heller, M., Multarzyński, P., Sasin, W., 1989, Algebraic Approach to Space–Time Geometry, *Acta Cosmol.* **16**, 53–85.

Hicks, N. J., 1965, *Notes on Differential Geometry*, Van Nostrand, Princeton.

Higgs, P. W., 1964, Broken Symmetries and The Masses of Gauge Bosons, *Phys. Rev. Lett.* **13**, 508–509.

Higgs, P. W., 1966, Spontaneous Symmetry Breakdown without Massless Bosons, *Phys. Rev.* **145**, 1156-1163.

Hoyle, F., 1948, A New Model for the Expanding Universe, *Mont. Not. Royal Astron. Soc.* **108**, 372-382.

Hubble, E., 1929, A Relation between Distance and Radial Velocity among Extra-Galactic Nebulae, *Proc. Nat. Acad. Sc.* **15**, 168-173.

Hubble, E., 1936, *The Realm of the Nebulae*, Yale University Press.

Hubble, E., 1937, *The Observational Approach to Cosmology*, Clarendon Press, Oxford.

Husemoller, D., 1975, *Fibre Bundles*, 2 edition, Springer, New York.

Kazanas, D., 1980, Dynamics of the Universe and Spontaneous Symmetry Breaking, *Astrophys. J.* **241**, L59-L63.

Kibble, T. W. B., 1961, Lorentz Invariance in Gravitational Field, *J. Math. Phys.* **2**, 212-221.

Kobayashi, S., Nomizu, K., 1963, *Foundations of Differential Geometry*, Interscience, New York — London.

Kramer, D., Stephani, H., Herlt, E., MacCallum, M., 1980, *Exact Solutions of Einstein's Field Equations*, ed.: E. Schmutzer, Cambridge University Press.

Kristian, J., Sachs, R. K., 1966, Observations in Cosmology, *Astrophys. J.* **143**, 379-399.

Landau, L., Lifszic, E., 1973, *The Field Theory*, Nauka, Moscow, (in Russian).

Landsberg, P. T., 1977, *Mathematical Cosmology — An Introduction*, Clarendon Press, Oxford.

Lemaître, G., 1927, Un univers homogène de masse constante et de rayon croissant, renadnt compte de la vitesse radiale des nébuleuses extra-galactiques, *Ann. Soc. Sc. Bruzelles* **47A**, 49-59; English translation: *Mont. Not. Royal Astron. Soc.* **91**, 1931, 483-490.

Lemaître, G., 1950, *The Primeval Atom — An Essay on Cosmology*, Van Nostrand, Toronto — New York — London.

Lemaître, G., 1972, *L'hypothese de l'Atome Primitif*, Éd. Culture et Civilisation, Bruxelles.

Liebscher, D-E., Yourgrau, W., 1979, Classical Spontaneous Breakdown of Symmetry and Induction of Inertia, *Annalen der Phys. (Leipzig)*, **36**, 20-24.

Lightman, A. P., Press, W. H., Price, R. H., Teukolsky, S. A., (1975), *Problem Book in Relativity and Gravitation*, Princeton Univ. Press.

Linde, A. D., A New Inflationary Universe Scenario: a Possible Solution of the Horizon, Flatness, Homogeneity, Isotropy, and Primordial Monopole Problems, *Phys. Lett.* **108B**, 398-395.

Longair, M., (ed.), 1974, *Confrontation of Cosmological Theories with Observational Data (IAU Symposium No 63)*, Reidel, Dordrecht — Boston.

Lubkin, E., 1963, Geometric Definition of Gauge Invariance, *Ann. Phys. (New York)* **23**, 233-283.

MacCallum, M. A. H., 1973, Cosmological Models from a Geometric Point of View, in: *Cargése Lectures*, vol. 6., ed.: E. Schatzman, Gordon and Breach, New York, 61-174.

Marsden, J. E.,1981, *Lectures on Geometric Methods in Mathematical Physics*, Society for Industrial and Applied Mathematics, Philadelphia.

Maurin, K., 1976, *Analysis*, Polish Scientific Publishers — Reidel, Warszawa — Dordrecht: 1976, part 1: *Elements*, 1980, part 2: *Integration, Distributions, Harmonic Functions, Tensor and Harmonic Analysis*.

Marleau-Ponty, J., 1965, *Cosmologie du XXe siècle*, Éd. Gallimard.

Miller, A. I., 1981, *Albert Einstein's Special Theory of Relativity — Emergence (1905) and Early Interpretation (1905-1911)*, Addison-Wesley, London, Amsterdam . . .

Milne, E. A., 1935, *Relativity, Gravitation and World Structure*, Clarendon Press, Oxford.

Minkowski, H., 1907, Das Relativitätsprinzip, a paper delivered on 5 of November 1907 during the meeting of the Math. Gesel. Göttingen, published in: *Ann. Phys.*, **47**, 1915, 927-938.

Minkowski, H., 1908a, Die Grundgleichungen für die elektromagnetischen Vorgänge in bewegten Körpern, *Göttinger. Nachr.* 53-111.

Minkowski, H., 1908b, Raum und Zeit, a paper delivered on 21 of September 1908 in Cologne, published in: *Phys. Z.*, **20**, 1909, 104–111; English Translation: *The Principle of Relativity*, Dover Publ., 73–96.

Misner, C. W., 1969, Mix–Master Universe, *Phys. Rev. Lett.* **22**, 1071–1074.

Misner, C. W., 1980, Symmetry Paradoxes and other Cosmological Comments, in: *Some Strangeness in the Proportion: A Centennial Symposium to Celebrate the Achievement of Albert Einstein*, Addison–Wesley, 405–415.

Misner, C. W., Thorne, K., Wheeler, J. A., 1973, *Gravitation*, Freeman, San Francisco.

Moncrief, V., Space–Time Symmetries and Linearization Stability of the Einstein Equations, *J. Math. Phys.*:
1975; I: **16**, 493–498,
1976, II: **17**, 1893–1902.

Nash, C., Sen, S., 1983, *Topology and Geometry for Physicists*, Academic Press, London, Orlando ...

O'Neill, B., 1966, *Elementary Differential Geometry*, Academic Press, New York — London.

O'Neill, B., 1983, *Semi–Riemannian Geometry with Applications to Relativity*, Academic Press, New York — London.

North, J. D., 1965, *The Measure of the Universe*, Clarendon Press, Oxford.

Pacholczyk, A. G., 1984, *The Catastrophic Universe*, Pachart, Tucson.

Peebles, P. J. E., 1971, *Physical Cosmology*, Princeton University Press.

Peebles, P. J. E., 1980, *The Large–Scale Structure of the Universe*, Princeton University Press.

Penrose, R., 1968, Structure of Space–Time, in: *Battelle Rencontres* 1967, eds.: C. M. DeWitt, J. A. Wheeler, Benjamin 121–235.

Penrose, R., 1980, Null Hypersurface Initial Data for Classical Fields of Arbitrary Spin and for General Relativity, *GRG* **12**, 225–264.

Penrose, R., Rindler, W., 1984, *Spinors and Space-Time*, vol. 1: *Two-Spinor Calculus and Relativistic Fields*, Cambridge University Press.

Penzias, A. A., Wilson, R. W., 1965, A Measurement of Excess Antenna Temperature at 4 080 Mc/s., *Astrophys. J.* **142**, 419–421.

Petrosian, V., 1974, Confrontation of Lemaitre Models and the Cosmological Constant with Observations, in: *Confrontation of Cosmological Theories with Observational Data (IAU Symposium No 63)*, Ed.: M. Longair, Reidel, Dordrecht — Boston, 31–46.

Pietrov, A. Z., 1966, *New Methods in General Relativity*, Nauka, Moscow (in Russian).

Raine, D. J., 1975, Mach's Principle in General Relativity, *Mont. Not. Royal Astron. Soc.* **171**, 507–528.

Raine, D. J., 1981a, Mach's Principle and Space–Time Structure, *Rep. Prog. Phys.* **44**, 1152–1195.

Raine, D. J., 1981b, *The Isotropic Universe*, Adam Hilger, Bristol.

Raine, D. J., Heller, M., 1981, *The Science of Space-Time*, Pachart, Tucson.

Reinhardt, M., 1973, Mach's Principle — A Critical Review, *Zeitschr. für Naturforschung*, 28a, 529–537.

Richtmyer, R. D., *Principles of Advanced Mathematical Physics*, Springer, New York — Berlin:
1978: vol. 1,
1981: vol. 2.

Riemann, G. F. B., 1868, Über die Hypothesen, welche der Geometrie zu Grunde liegen (Habilitationsschrift 1854). *Abhandl. kgl. Ges. Wiss. zu Göttingen* **13**.

Riesz, F., Sz–Nagy, B., 1972, *Leçons d'analyse fonctionelle*, Akadémiai Kiadó, Budapest.

Rindler, W., 1956, Visual Horizons in World Models, *Mont, Not. Royal Astron. Soc.* **116**, 662–677.

Rindler, W., 1969, *Essential Relativity: Special, General, and Cosmological*, Van Nostrand, New York, Cincinnati ...

Rindler, W., 1977, *Essential Relativity — Special, General, and Cosmological, Springer*, New York, Heidelberg, Berlin.

Robertson, H. P., 1933, Relativistic Cosmology, *Rev. Mod. Phys.* 5, 62–90.

Robertson, H. P., Noonan, T. W., 1968, *Relativity and Cosmology*, Saunders, Philadelphia, London ...

Roman, P., 1975, *Some Modern Mathematics for Physicists and Other Outsiders*, vol. 1 and 2, Pergamon, New York, Toronto ...

Rowan–Robinson, M., 1977, *Cosmology*, Clarendon Press, Oxford.

Ross, G. G., 1981, Unified Fields Theories, *Rep. Prog. Phys.* **44**, 655–718.

Ryan, M. P., Shepley, L. C., 1975, *Homogeneous Relativistic Cosmologies*, Princeton University Press.

Sachs, R. K., Wu, H., 1977a, General Relativity and Cosmology, *Bull. Amer. Math. Soc.* **83**, no. 6, 1101–1164.

Sachs R. K., Wu, H., 1977b, *General Relativity for Mathematicians*, Springer, New York.

Salam, A., 1968, in: *Elementary Particle Theory*, ed.:
N. Svartholm, Almquist and Wiksell Förlag, Stockholm.

Sandage, A. R., 1961, The Ability of the 200-inch Telescope to Discriminate between Selected World Models, *Astrophys. J.* **133**, 335–392.

Schmidt, B., 1972, Differential Geometry from a Modern Standpoint, in: *Relativity, Astrophysics and Cosmology*, ed.: W. Israel, Reidel, Dordrecht — Boston, 289–322.

Schutz, B. F., 1984, *Geometrical Methods of Mathematical Physics*, Cambridge University Press 1967, *Analyse mathématique*, Hermann, Paris, vol. 1 and 2.

Schutz, B. F., 1985, *A First Course in General Relativity*, Cambridge University Press.

Schwinger, J., The Theory of Quantized Fields, *Phys. Rev.*:
I: 1951, **82**, 914–927,
II: 1953, **91**, 713–728,
III: 1953, **91**, 728–740.

Sciama, D. W., 1962, On the Analogy between Charge and Spin in General Relativity, in: *Recent Developments in General Relativity*, Pergamon Press — Polish Scientific Publishers, 415–439.

Sciama, D. W., 1971, *Modern Cosmology*, Cambridge University Press.

Shapiro, I. I., 1979, *Experimental Tests of the General Theory of Relativity and Gravitation*, vol. 2, ed.: A. Held, Plenum Press, New York — London, 469–489.

Silk, J., Barrow, J. D., 1983, *The Left Hand of Creation*, Basic Books.

Silk, J., Szalay, A. S., Zeldovich, Y. B., 1983, The Large–Scale Structure of the Universe, *Scientific American*, **249**, 56–64.

Silk, J., 1980, *The Big Bang — The Creation and Evolution of the Universe*, Freeman and Comp.

Smale, S., 1980, *The Mathematics of Time*, Springer, New York — Heidelberg — Berlin.

Spanier, E. H., 1966, *Algebraic Topology*, McGraw-Hill, New York — San Francisco.

Steenrod, N., 1951, *The Topology of Fibre Bundles*, Princeton University Press.

Sulanke, R., Wintgen, P., 1972, *Differential Geometrie und Faserbündel*, VEB Deutscher Verlag der Wissenschaften, Berlin.

Synge, J. L., 1965, *Relativity: The Special Theory*, North-Holland, Amsterdam.

Szydlowski, M., Heller, M., Golda, Z., 1984, Structural Stability Properties of Friedman Cosmology, *GRG* **16**, 877–890.

Taylor, E. P., Wheeler, J. A., 1966, *Spacetime Physics*, Freeman and Comp., San Francisco — London.

Taylor, E. P., 1978, *Gauge Theories of Weak Interactions*, Cambridge University Press.

Thirring, W., 1978, *A Course in Mathematical Physics*, vol. 1: *Classical Dynamical Systems*, Springer, New York — Wien.

Thom, R., 1977, *Stabilité structurelle et morphogénèse*, Interéditions.

Tolman, R. C., 1934, *Relativity, Thermodynamics and Cosmology*, Clarendon Press, Oxford.

Tonnelat, M. A., 1971, *Histoire du principe de relativité*, Flammarion, Paris.

Toretti, R., 1983, *Relativity and Geometry*, Pergamon Press, Oxford, New York ...

Trautman, A., 1964, Foundations and Current Problems of General Relativity, in: *Lectures on General Relativity*, Brandeis Summer Institute in Theoretical Physics, Prentice–Hall, Englewood Cliffs, 1–248.

Trautman, A., 1970a, Fibre Bundles Associated with Space–Time, *Rep. Math. Phys.* **1**, 29–62.

Trautman, A., 1970b, Riemannian Bundles, *Bull. Acad. Pol. Sc., Sér. sc. math., astr. et phys.* **18**, 667–672.

Trautman, A., On the Einstein–Cartan Equations, *Bull. Acad. Pol. Sc., Sér, sc. math., astr. et phys.*:

I: 1972a, **20**, 185–190,

II: 1972b, **20**, 503–506,

III: 1972c, **20**, 859–896,

IV: 1973a, **21**, 345–346.

Trautman, A., 1973b, Theory of Gravitation, in: *The Physicist's Conception of Nature*, ed.: J. Mehra, Reidel, 179–198.

Trautman, A., 1973c, On the Structure of Einstein–Cartan Equations, *Symposia Mathematica* **12**, 139–162.

Trautman, A., 1975, Recent Advances in Einstein–Cartan Theory of Gravity, *Ann. N. Y. Acad. Sc.* **262**, 241–245.

Trautman, A., 1976, A Classification of Space–Time Structures, *Rep. Math. Phys.* **10**, 297–310.

Trautman, A., 1979, The Geometry of Gauge Fields, *Czech. J. Phys.* **B29**, 107–116.

Trautman, A., 1980, Fiber Bundles, Gauge Fields, and Gravitation, in: *General Relativity and Gravitation*, vol. 1, ed.: A. Held, Plenum Press, New York — London, 287–308.

Trautman, A., 1981, Geometrical Aspects of Gauge Configurations, *Acta Phys. Austriaca*, Suppl. XXIII, 401–432.

Trautman, A., 1984, *Differential Geometry for Physicists — Stony Brook Lectures*, Bibliopolis, Napoli.

Trefil, J. S., 1983, *The Moment of Creation*, Scribner's Sons, New York.

Tropp, E. A., Frenkel, W. Ya., Chernin, A. D., 1988, *Alexander Alexandrovitch Friedman — Life and Work*, Nauka, Moscow (in Russian).

[Understanding ...], 1979, *Understanding Space and Time*, The Open University Press.

Utiyama, R., 1956, Invariant Theoretical Interpretation of Interaction, *Phys. Rev.* **101**, 1597–1607.

Wallace, A. H., 1968, *Differential Geometry — First Steps*, W. A. Benjamin, New York — Amsterdam.

Weinberg, S., 1967, A Theory of Leptons, *Phys. Rev. Lett.* **19**, 1264–1266.

Weinberg, S., 1972, *Gravitation and Cosmology: Principles and Applications of the General Theory of Relativity*, E Wiley and Sons, New York, London ...

Weinberg, S., 1977, *The First Three Minutes*, A. Deutsch, London.

Weizsäcker von C. F., 1972, *Die Einheit der Natur*, Carl Hauser Verlag.

Weyl, H., 1918, Gravitation und Elektrizität, *Sitzungsber. Preuss. Acad. Wiss.* 465–480.

Weyl, H., 1919, Eine neue Erweterung der Relativitätstheorie, *Ann. Phys. Leipzig* **59**, 101–133.

Weyl, H., 1922, *Space–Time–Matter*, Dover Publ.

Weyl, H., 1955, *The Concept of a Riemann Surface*, Addison–Wesley, Reading Mass. German original published in 1913.

Whitney, H., 1936, Differentiable Manifolds, *Ann. Math.* **37**, 645–680.

Will, C. M., 1979, The Confrontation between Gravitation Theory and Experiment, in: *General relativity — An Einstein Centenary Survey*, ed.. S. W., Hawking, W., Israel, Cambridge University Press, 24, 89, 833–846.

Will, C. M., 1979, The Confrontation between Gravitation Theory and Experiment, in: *General relativity — An Einstein Centenary Survey*, ed.: S. W., Hawking, W., Israel, Cambridge University Press, 24–89, 833–846.

Wolf, J. A., 1967, *Spaces of Constant Curvature*, McGraw–Hill, New York, St. Louis . . .

Yang, C. N., Mills, R. L., 1954, Conservation of Isotopic Spin and Isotopic Gauge Invariance, *Phys. Rev.* **96**, 191–195.

Yasskin, Ph. B., 1979, *Metric-Connection Theories of Gravity*, University of Maryland, Dept. of Phys. and Astron., Thesis.

Zeldovitch, Ya, B., 1975, *The Structure and Evolution of the Universe*, Nauka, Moscow (in Russian).

Życinski, J., 1988, *The Structure of the Metascientific Revolution*, Pachart, Tucson.

Symbol Index

Author and Subject Index

Polyadic
Transcendental
Number Theory

Polyadic Transcendental Number Theory

Vladimir G Chirskii
Lomonosov Moscow State University, Russia

World Scientific

NEW JERSEY • LONDON • SINGAPORE • BEIJING • SHANGHAI • HONG KONG • TAIPEI • CHENNAI • TOKYO

Published by

World Scientific Publishing Europe Ltd.

57 Shelton Street, Covent Garden, London WC2H 9HE

Head office: 5 Toh Tuck Link, Singapore 596224

USA office: 27 Warren Street, Suite 401-402, Hackensack, NJ 07601

Library of Congress Cataloging-in-Publication Data

Names: Chirskii, V. G., author.

Title: Polyadic transcendental number theory / Vladimir G. Chirskii,
    Lomonosov Moscow State University, Russia.

Description: New Jersey : World Scientific, [2025] | Includes bibliographical references and index.

Identifiers: LCCN 2024007634 | ISBN 9781800615885 (hardcover) |
    ISBN 9781800615892 (ebook for institutions) | ISBN 9781800615908 (ebook for individuals)

Subjects: LCSH: Transcendental numbers. | Number theory.

Classification: LCC QA247.5 .C45 2025 | DDC 512.7/3--dc23/eng20240720

LC record available at https://lccn.loc.gov/2024007634

British Library Cataloguing-in-Publication Data

A catalogue record for this book is available from the British Library.

For any available supplementary material, please visit
https://www.worldscientific.com/worldscibooks/10.1142/Q0468#t=suppl

Desk Editors: Nambirajan Karuppiah/Rosie Williamson/Shi Ying Koe

Typeset by Stallion Press
Email: enquiries@stallionpress.com

Polyadic transcendental number theory has NOT a long history. Its main focus involves the transfer of the profound ideas of the Siegel–Shidlovskii method to a new interesting class of power series.

This book is dedicated to the memory of my teacher, Professor A. B. Shidlovskii (1915–2007).

Preface

This book describes the recent results of polyadic transcendental number theory. A modification of the Siegel–Shidlovskii method and the concept of a global relation introduced by E. Bombieri are proposed to obtain the results. The Hermite–Padé approximations of generalised hypergeometric functions constructed by Yu. V. Nesterenko also play an important role.

The author is grateful to the corresponding member of the Russian Academy of Sciences, Professor Yu. V. Nesterenko, and to Professor V. Kh. Salikhov for useful discussions on the results of the works that form the basis of the content of this book.

Some extracts in this book are reproduced from earlier articles written by the author. With kind permission from Pleiades Publishing, Ltd., extracts have been reproduced from the following articles: Product Formula, Global Relations and Polyadic Integers, *Russian Journal of Mathematical Physics*, Vol. 26, No. 3, 2019, pp. 286–305; Arithmetic Properties of Generalized Hypergeometric F-series, *Russian Journal of Mathematical Physics*, Vol. 27, No. 2, 2020, pp. 175–184; Arithmetic Properties of the Values of Generalized Hypergeometric Series with Transcendental Polyadic Parameters, *Doklady Mathematics*, Vol. 106, No. 2, 2022, pp. 386–397; Transcendence of p-adic Values of Generalized Hypergeometric Series with Polyadic Transcendental Parameters, *Doklady Mathematics*, Vol. 107, No. 2, 2023, pp. 109–111.

With kind permission from the Steklov Mathematical Institute of Russian Academy of Sciences, extracts have been reproduced from the following articles: On the Arithmetic Properties of Generalized Hypergeometric Series with Irrational Parameters, *Izvestiya Mathematics*, Vol. 78, No. 6, 2014, pp. 1244–1260; Arithmetic Properties of Polyadic Series with Periodic Coefficients, *Izvestiya Mathematics*, Vol. 81, No. 2, 2017, pp. 444–461.

About the Author

Vladimir Grigoryevich Chirskii graduated from Lomonosov Moscow State University in 1972 and enrolled in the postgraduate course of the Faculty of Mechanics and Mathematics at the same university. His supervisor was the famous scientist Professor A. B. Shidlovskii, one of the creators of the Siegel–Shidlovskii method in the theory of transcendental numbers. From 1975 to the present, V. G. Chirskii has been primarily working at the Faculty of Mechanics and Mathematics, Lomonosov Moscow State University. He is a doctor of physical and mathematical sciences (equivalent to doctor habilitatus) and is currently a professor at the Department of Mathematical Analysis, Lomonosov Moscow State University. The area of V. G. Chirskii's scientific research involves the theory of transcendental numbers in direct products of fields of p-adic numbers. This research focus began to develop with his work in the early 1990s. He has published over a hundred scientific papers on number theory and the applications of mathematical methods to problems in economics and more than twenty books, including those written for schoolchildren and teachers. Among his students are several candidates pursuing physical and mathematical sciences (equivalent to a Ph.D.) and one doctor of physical and mathematical sciences (equivalent to a doctor habilitatus). In addition to his role at Moscow State University, V. G. Chirskii headed the Department of Number Theory at the Moscow State Pedagogical University for several years. He also offers lectures at the Russian Academy of National Economy and Public Administration.

Contents

Introduction

Let $\mathbb{F}$ be a ring or a field containing a subfield $\mathbb{K}$. An element $\alpha \in \mathbb{F}$ is called *algebraic* over the field $\mathbb{K}$ if there exists a non-zero polynomial with coefficients from $\mathbb{K}$ such that the equality $P(\alpha) = 0$ is satisfied in $\mathbb{F}$ (0 is the zero element of $\mathbb{F}$). The *degree* of an algebraic element α is the smallest of the degrees of polynomials $P(x)$ such that $P(\alpha) = 0$. If $\alpha \in \mathbb{F}$ is not algebraic over the field $\mathbb{K}$, then it is called *transcendental* over $\mathbb{K}$.

In this introduction, we briefly outline the main results from that part of the theory of transcendental numbers, the development of which constitutes the material of this book. Of course, this information does not give an idea of all the directions of the theory of transcendental numbers. Books such as that by Nesterenko and Feldman (1998) that contain a fairly detailed description of the most important methods of this theory. The purpose of this book is to describe the generalisation and development of one of these methods — the Siegel–Shidlovskii method — for the case of areas with non-Archimedean valuations.

1. Liouville Numbers

The first example of a transcendental number was found by Liouville (1844). Namely, he proved the following:

For any algebraic number α with degree $n > 1$, there exists a constant $c(\alpha) > 0$ such that the inequality

$$\left| \alpha - \frac{p}{q} \right| > \frac{c(\alpha)}{q^n}$$

holds for all rational numbers $\frac{p}{q}$, $q > 0$.

This theorem provides us with an example of a transcendental number, namely $\alpha = \sum_{n=0}^{\infty} 10^{-n!}$. Indeed, for any k, we put

$$p_k = \sum_{n=0}^{k} 10^{k!-n!}, \quad q_k = 10^{k!}.$$

Now, p_k and q_k are relatively prime integers, and we evidently have

$$\left| \alpha - \frac{p_k}{q_k} \right| = \sum_{n=k+1}^{\infty} 10^{-n!} < 10^{-(k+1)!} \sum_{n=0}^{\infty} 10^{-n}$$

$$= \frac{10}{9} 10^{-(k+1)!} < (10^{k!})^{-k} = q_k^{-k}.$$

Thus, this number does not satisfy the inequality of Liouville's theorem and is transcendental.

2. Transcendence of e and π

In 1873, Hermite created an analytic method that allowed him to prove the transcendence of one of the main classical constants, the number e. Developing Hermite's method, von Lindemann proved the transcendence of π in 1882. This remarkable result led to a negative solution to the well-known problem of the quadrature of a circle, which had remained unsolved for thousands of years. Lindemann also sketched a much more general result, which was later demonstrated by K. Weierstrass: a theorem that completely solved the question of the transcendence and algebraic independence of the values of the exponential function at algebraic points (Weierstrass, 1885).

This theorem can be formulated as follows:

If $\alpha_1, \ldots, \alpha_m$ are algebraic numbers that are linearly independent over $\mathbb{Q}$, then the numbers

$$e^{\alpha_1}, \ldots, e^{\alpha_m}$$

are algebraically independent.

The Hermite–Lindemann method is based on the following important properties of the exponential function:

1. The function $f(z) = e^z$ satisfies the addition theorem

$$f(x + y) = f(x)f(y).$$

2. The function $f(z) = e^z$ is a solution of the differential equation $y' = y$.

3. Siegel's Method

After the development of the Hermite–Lindemann method, a natural problem arose: its extension to classes of functions satisfying more general differential equations.

The first significant success in the development of the classical Hermite–Lindemann method belongs to K. Siegel. In 1929, he proposed a new method for proving the transcendence and algebraic independence of values at algebraic points of analytic functions of a certain class (which contained e^z).

The main result, confirmed by Siegel in his 1929 article, refers to the function

$$K_\lambda(z) = \sum_{n=0}^{\infty} \frac{(-1)^n}{n!(\lambda + 1)\cdots(\lambda + n)} \left(\frac{z}{z}\right)^{2n}, \quad \lambda \neq -1, -2, \ldots,$$

satisfying a linear differential equation of the second order:

$$y'' + \frac{2\lambda + 1}{z}y' + y = 0.$$

The function $K_\lambda(z)$ differs from the well-known Bessel function $J_\lambda(z)$ only by a multiplier

$$\frac{1}{\Gamma(\lambda+1)} \left(\frac{z}{z}\right)^{2\lambda}.$$

Here, $\Gamma(z)$ denotes Euler's gamma function and $K_0(z) = J_0(z)$.

Siegel proved the following:

If λ is a rational number other than half of an odd number and if $\xi \neq 0$ is an algebraic number, then the numbers $K_\lambda(\xi), K'_\lambda(\xi)$ are algebraically independent.

If each of the rational numbers $\lambda_1, \ldots, \lambda_m$ is not equal to half of an odd number, if for $i \neq j$ the numbers $\lambda_i \pm \lambda_j$ are not integers, and if $\xi_1, \ldots, \xi_n$ are algebraic numbers with different squares, then the $2mn$ numbers $K_{\lambda_i}(\xi_k), K'_{\lambda_i}(\xi_k)$ are algebraically independent.

The method proposed by Siegel can be applied to one class of integer functions, which he called *E-functions*. An analytical function

$$\sum_{n=0}^{\infty} \frac{c_n}{n!} z^n$$

is called an *E-function* if the following conditions are met:

1. $c_n \in \mathbb{K}$, $n = 0, 1, \ldots$, $\mathbb{K}$ is an algebraic number field of a finite degree over the field $\mathbb{Q}$.
2. For any $\varepsilon > 0$, we have

$$\overline{|c_n|} = O(n^{\varepsilon n}), \quad n \to \infty.$$

Here, $\overline{|a|}$ denotes the height of the algebraic number a, i.e., the largest of the absolute values of the number itself and its conjugates in $\mathbb{K}$.

3. There exists a sequence of positive integers q_n such that $q_n c_k \in \mathbb{Z}_\mathbb{K}$, $k = 0, 1, \ldots, n$ and $n = 0, 1, 2, \ldots$ (here, $\mathbb{Z}_\mathbb{K}$ denotes the ring of integers of the field $\mathbb{K}$). Further,

$$\overline{|q_n|} = O(n^{\varepsilon n}), \quad n \to \infty.$$

Examples of E-functions are polynomials with algebraic coefficients and functions $\exp z$, $\sin z$, $\cos z$.

It is not difficult to verify that E-functions form a commutative ring of functions closed with respect to the operations of differentiation, integration from 0 to z, and the replacement of the argument z by λz, where λ is an algebraic number. If the coefficients of E-functions belong to a field $\mathbb{K}$, then they are called $\mathbb{K}E$-functions. This case is highlighted when $\mathbb{K} = \mathbb{I}$ is an imaginary quadratic field over $\mathbb{Q}$.

In 1949, Siegel presented his method in the form of a general theorem on the algebraic independence of values at algebraic points of a set of E-functions satisfying a system of linear homogeneous differential equations with coefficients which are rational functions. This theorem reduces the proof of the statement about the algebraic independence of values at algebraic points of a set of E-functions to the verification of some sufficient analytical condition, which he called the *normality condition* for a set of products of degrees of the functions under consideration.

Siegel himself managed to verify the fulfilment of the normality condition for sets of E-functions, each of which satisfies a linear differential equation of the first or second order. The verification of this condition in a much more general case was carried out in an article by Beukers *et al.* (1988).

4. Shidlovskii's Fundamental Theorems

In 1954, Shidlovskii published a theorem in which the normality condition of a system of functions was replaced by a less restrictive irreducibility condition of this system. This condition was also sufficient, not necessary and sufficient.

In 1955, Shidlovskii published a criterion for the algebraic independence of a set of values at the algebraic points of E-functions that make up the solution of a system of linear differential equations. For convenience, we recall the concepts of algebraically dependent and algebraically independent elements. Let $\mathbb{F}$ be a ring or a field containing a subfield $\mathbb{K}$. The elements $\alpha_1, \ldots, \alpha_m$ from $\mathbb{F}$ are *algebraically dependent* over a field $\mathbb{K}$ if there exists a non-zero polynomial

$P(x_1, \ldots, x_m)$ with coefficients from $\mathbb{K}$ such that $P(\alpha_1, \ldots, \alpha_m) = 0$ holds in $\mathbb{F}$. Otherwise, they are called *algebraically independent*. If we consider only homogeneous polynomials in these definitions, we obtain the definitions of *homogeneous algebraic dependence* and *homogeneous algebraic independence*.

Let the analytical functions $f_1(z), \ldots, f_m(z)$ constitute a solution of a system of linear homogeneous differential equations of the first order:

$$y_k' = \sum_{i=1}^{m} Q_{k,i}(z) y_i, \quad k = 1, \ldots, m. \tag{1}$$

The functions $Q_{k,i}(z)$, $i, k = 1, \ldots, m \in \mathbb{C}(z)$ ($\mathbb{C}(z)$ is the field of rational functions). Let $T(z)$ be a polynomial that is the least common denominator of all rational functions $Q_{k,i}(z)$. Let us formulate the first main theorem by Shidlovskii.

Let the set of E-functions $f_1(z), \ldots, f_m(z)$ constitute a solution of a system of linear homogeneous differential equations (1) and be uniformly algebraically independent over $\mathbb{C}(z)$. Let ξ be an algebraic number such that $\xi\, T(\xi) \neq 0$. Then, the numbers $f_1(\xi), \ldots, f_m(\xi)$ are uniformly algebraically independent.

In the case where the considered E-functions constitute the solution of a linear inhomogeneous system of differential equations,

$$y_k' = Q_{k,0}(z) + \sum_{i=1}^{m} Q_{k,i}(z) y_i, \quad k = 1, \ldots, m, \tag{2}$$

the result bearing the name of the second fundamental theorem is valid.

Let the set of E-functions $f_1(z), \ldots, f_m(z)$ constitute a solution of a system of linear differential equations (2) and be algebraically independent over $\mathbb{C}(z)$. Let ξ be an algebraic number such that $\xi\, T(\xi) \neq 0$. Then, the numbers $f_1(\xi), \ldots, f_m(\xi)$ are algebraically independent.

These theorems have natural consequences related to linear differential equations of order m.

Let the E-function $f(z)$ be a solution of a linear homogeneous differential equation of order m,

$$P_m(z)y^{(m)} + \cdots + P_1(z)y' + P_0(z)y = 0, \quad m \geq 2,$$
$$P_k(z) \in \mathbb{C}(z), \quad k = 0, \ldots, m,$$

that does not satisfy any homogeneous algebraic differential equation with coefficients from $\mathbb{C}(z)$ of an order less than m. Let ξ be an algebraic number, $\xi P_k(\xi) \neq 0$. Then, the numbers

$$f(\xi), f'(\xi), \ldots, f^{(m-1)}(\xi)$$

are homogeneously algebraically independent.

Let the E-function $f(z)$ be a solution of a linear inhomogeneous differential equation of order m,

$$P_m(z)y^{(m)} + \cdots + P_1(z)y' + P_0(z)y + Q(z) = 0, \quad m \geq 2,$$
$$P_k(z) \in \mathbb{C}(z), \quad k = 0, \ldots, m,$$

that does not satisfy any algebraic differential equation with coefficients from $\mathbb{C}(z)$ of an order less than m. Let ξ be an algebraic number, $\xi P_k(\xi) \neq 0$. Then, the numbers

$$f(\xi), f'(\xi), \ldots, f^{(m-1)}(\xi)$$

are algebraically independent.

In 1970, Galochkin proposed generalisations of these theorems for the case where the values of E-functions were considered at transcendental points, admitting a good approximation by algebraic numbers.

Shidlovskii also considered the case where the considered E-functions are algebraically dependent over $\mathbb{C}(z)$.

If the maximal number of algebraically independent (homogeneously algebraically independent) elements over a field $\mathbb{K}$ of a set $U \subset \mathbb{F}$ of elements of a field or ring $\mathbb{F}$ is equal to l, then this number is called the transcendence degree (*homogeneous transcendence degree*) over $\mathbb{K}$ of this set and is denoted by $\mathrm{trdeg}_{\mathbb{K}}U$ ($\mathrm{homtrdeg}_{\mathbb{K}}U$).

We now formulate the third main theorem by Shidlovskii.

Let the set U of E-functions $f_1(z), \ldots, f_m(z)$ *constitute a solution of a system of linear differential equations* (2) (*or* (1)) *and the transcendence degree* (*homogeneous transcendence degree*) *of this set over* $\mathbb{C}(z)$ *be equal to l. Let* ξ *be an algebraic number such that* $\xi\, T(\xi) \neq 0$. *Then, the transcendence degree* (*homogeneous transcendence degree*) *of the set of numbers* $f_1(\xi), \ldots, f_m(\xi)$ *is also equal to l.*

These theorems are contained in Shidlovskii (1989).

5. Measures of Transcendence and Algebraic Independence

In the theory of transcendental numbers, it is customary to distinguish between two types of results: qualitative and quantitative. The first includes theorems on irrationality and transcendence as well as linear and algebraic independence. The results of the second type allow us to obtain estimates from below for the values of linear forms or polynomials in sets of numbers.

Let's move on to the exact formulations of these concepts. In other words, we define the concepts of measures of irrationality, transcendence, and linear and algebraic independence.

Let $\alpha_1, \ldots, \alpha_m$ be real or complex numbers. A *measure of the linear independence* of these numbers is a function

$$L(\alpha_1, \ldots, \alpha_m, H) = \min |h_1\alpha_1 + \cdots + h_m\alpha_m|,$$

where H is a positive integer, $h_1, \ldots, h_m$ are integers satisfying inequalities

$$|h_t| \leq H, \quad i = 1, \ldots, m, \quad |h_1| + \cdots + |h_m| > 0,$$

and the minimum is taken over all sets of numbers $h_1, \ldots, h_m$ satisfying these inequalities.

It is easy to see that the linear independence of a set of numbers means that the measure of the linear independence of this set of numbers is positive for any H.

Let $\alpha_1, \ldots, \alpha_m$ be real or complex numbers. The *algebraic independence measure* of these numbers is a function in s and H of the form

$$\Phi(\alpha_1, \ldots, \alpha_m, s, H) = \min |P(\alpha_1, \ldots, \alpha_m)|,$$

where s and H are positive integers and the minimum is taken over all non-zero polynomials $P(x_1, \ldots, x_m)$ with integer coefficients whose degree in the aggregate of variables does not exceed s.

Here is a slightly simplified formulation of Theorem 4 from Shidlovskii (1989).

Let the set of $\mathbb{I}E$-functions $f_1(z), \ldots, f_m(z)$ constitute a solution of a system of linear differential equations (2) and be algebraically independent over $\mathbb{C}(z)$. Let ξ be an algebraic number such that $\xi\, T(\xi) \neq 0$. Then, the numbers $f_1(\xi), \ldots, f_m(\xi)$ are algebraically independent.

Thus, there exists a constant C_0 depending on the functions $f_1(z), \ldots, f_m(z)$ and the numbers ξ and s such that for any ε, $0 < \varepsilon < \frac{1}{2}$, the inequality

$$\Phi(\alpha_1, \ldots, \alpha_m, s, H) > C_0 H^{1 - \frac{(s+m-1)!}{s!(m-1)!} - \varepsilon}$$

holds.

The theory of E-functions has been extensively developed in the works of Shidlovskii, his students, and many other researchers. The main results of this theory are described in detail in a monograph by Shidlovskii (1989). The subsequent results will be discussed in the relevant sections.

In 1929, Siegel pointed out that his proposed method could be used to study the arithmetic properties of another class of analytical functions, which he called G-functions (Siegel, 1929).

6. *G*-Functions

A function

$$f(z) = \sum_{n=0}^{\infty} c_n z^n$$

is called a *G-function* if the following conditions hold:

1. $c_n \in \mathbb{K}$, $n = 0, 1, \ldots, \mathbb{K}$ is an algebraic number field of a finite degree over the field $\mathbb{Q}$.

2. There exist constants γ, C such that

$$\overline{|c_n|} \leq \gamma C^n, \quad n = 0, 1, 2, \ldots.$$

Here, $\overline{|a|}$ denotes the height of the algebraic number a, i.e., the largest of the absolute values of the number itself and its conjugates in $\mathbb{K}$.

3. There exists a sequence of positive integers q_n such that $q_n c_k \in \mathbb{Z}_{\mathbb{K}}$, $k = 0, 1, \ldots, n$, $n = 0, 1, 2, \ldots$ (here, $\mathbb{Z}_{\mathbb{K}}$ denotes the ring of integers of the field $\mathbb{K}$) with some constants δ, D such that

$$\overline{|q_n|} = \delta D^n, \quad n = 0, 1, 2, \ldots.$$

As in the case of E-functions, it is not difficult to verify that the G-functions form a commutative ring of functions closed with respect to the operations of differentiation, integration from 0 to z, and replacement of the argument z by λz, where λ is an algebraic number.

Unlike E-functions, which are entire analytic functions, G-functions other than polynomials have a finite radius of convergence of the Taylor series.

Significantly fewer papers are devoted to G-functions than to E-functions. The results of Galochkin (1974) are noteworthy. In 1984, Chudnovsky proved, using the Hermite–Padé approximations, the following theorem:

Let the G-functions $f_1(z), \ldots, f_m(z)$ constitute a solution of a system of linear homogeneous differential equations (1) *and be linearly independent together with 1 over $\mathbb{C}(z)$. Then, for any $\varepsilon > 0$ and any non-zero number $r = \frac{a}{b}$, $a \in \mathbb{Z}$, $b \in \mathbb{N}$ such that for some constant C_1,*

$$b^{\varepsilon} \geq C_1 |a|^{(m+1)(m+\varepsilon)},$$

the numbers $1, f_1(r), \ldots, f_m(r)$ are linearly independent over $\mathbb{Q}$. Moreover, for any integers $h_1, \ldots, h_m$ satisfying the condition

$$H = \max(|h_1|, \ldots, |h_m|) > C_2,$$

the inequality

$$|h_0 + h_1 f_1(r) + \cdots + h_m f_1(r)| > H^{-m-\varepsilon}$$

holds. Here, the symbols $C_1 = C_1(f_1(z), \ldots, f_m(z), \varepsilon)$, $C_2 = C_2(f_1(z), \ldots, f_m(z), \varepsilon, r)$ *denote constants depending on the values written in parentheses. These constants can be effectively calculated.*

It is possible to consider not only the real or complex values of G-functions but also their p-adic values, as investigated by Flicker (1977).

7. p-Adic Numbers

Here, we give only the necessary information about p-adic numbers for further presentation. You can get acquainted with these numbers in more detail in very good books (e.g., Mahler, 1981; Koblitz, 1984).

Let's define and list the main properties of p-adic numbers.

A field a or ring $\mathbb{F}$ is called normalised if a *valuation* $\|a\| \in \mathbb{R}$ is defined for each of its elements, having the following properties:

1. $\|0\| = 0$, and if a is not the zero element of $\mathbb{F}$, then $\|a\| > 0$.
2. $\|ab\| = \|a\| \, \|b\|$.
3. $\|a + b\| \leq \|a\| + \|b\|$.

The field of rational numbers $\mathbb{Q}$ has a valuation

$$\|a\| = |a|,$$

which is also called *Archimedean*. In addition, for any prime number p, the so-called p-adic valuation of the field of rational numbers is defined as follows. For an arbitrary non-zero integer a, we define the value $\mathrm{ord}_p a$ as the degree to which a prime number p enters the factorisation of the number a. For any rational number $c = \frac{a}{b}$, we put

$$\mathrm{ord}_p \frac{a}{b} = \mathrm{ord}_p a - \mathrm{ord}_p b.$$

This definition is obviously correct. Now, we define the p-adic valuation of a rational number x by the equality

$$|x|_p = \begin{cases} p^{-\mathrm{ord}_p x}, & x \neq 0, \\ 0, & x = 0. \end{cases} \qquad (3)$$

Ostrovsky's famous theorem states the following:

Every non-trivial valuation in the field of rational numbers is equivalent to either the p-adic valuation for some prime number p or the usual absolute value.

An important *product formula* immediately follows from equality (3): for $a \in \mathbb{Q}$, $a \neq 0$,

$$|a| \prod_p |a|_p = 1, \qquad (4)$$

where the product is taken over all prime numbers p.

The theory of transcendental numbers in the p-adic domain has undergone less development than the theory of transcendental numbers in the complex domain. We note the works of Mahler (1935) and Adams (1966), which contain some overviews.

8.   Valuations of Algebraic Extensions of $\mathbb{Q}$

Let $\mathbb{K}$ be an algebraic number field of a finite degree κ over the field $\mathbb{Q}$. Let v be a valuation of $\mathbb{K}$. Denote by $\mathbb{K}_v$ the completion of $\mathbb{K}$ with respect to v. Any valuation v of $\mathbb{K}$ extends some valuation of $\mathbb{Q}$. If v extends the p-adic valuation (denoted by $v \mid p$), then $\mathbb{K}_v$ is an algebraic extension of $\mathbb{Q}_p$ of a finite degree κ_v, and for any prime p, we have

$$\sum_{v \mid p} \kappa_v = \kappa, \qquad (5)$$

where the summation on the left-hand side of (5) runs over all valuations $v \mid p$. Let's insert the case $v \mid p$:

$$|p|_v = p^{-\frac{\kappa_v}{\kappa}}. \qquad (6)$$

Suppose now that v extends the usual absolute value. Then, either $\mathbb{K}_v \subset \mathbb{R}$ and, in this case, $\kappa_v = 1$, or $\mathbb{K}_v \not\subset \mathbb{R}$ and, in this case, $\kappa_v = 2$, and so (5) still holds.

The *product formula* is also true: for $x \in \mathbb{K}$, $x \neq 0$,

$$\prod_v |x|_v = 1, \tag{7}$$

where the product is taken over all valuations v of $\mathbb{K}$.

9. Global Relations

In 1981, Bombieri introduced the notion of a global relation, which is defined as follows.

Let $P(x_1, \ldots, x_m)$ be a non-zero polynomial with coefficients from $\mathbb{K}$.

Let the power series $f_1(z), \ldots, f_m(z)$ have coefficients from the field $\mathbb{K}$, and let $\xi \in \mathbb{K}$. The relation

$$P(f_1(\xi), \ldots, f_m(\xi)) = 0 \tag{8}$$

is called *global* if it holds in all fields $\mathbb{K}_v$ where $f_1(\xi), \ldots, f_m(\xi)$ converge.

We formulate a somewhat simplified version of the theorem from the work of Bombieri (1981).

Let the G-functions $f_1(z), \ldots, f_m(z)$ constitute a solution of a system of linear differential equations (2) and be linearly independent over $\mathbb{C}(z)$. Then, there is a constant C such that all algebraic points ξ for which some linear global relation holds satisfy the inequality

$$\sum_v \ln(\max(1, |\xi|_v)) \leq C,$$

where the summation is performed over all valuations of an algebraic field $\mathbb{K}$ of a finite degree over $\mathbb{Q}$, obtained by joining to the field $\mathbb{Q}$ all coefficients of the series under consideration and the number ξ.

Global relations for G-functions were also investigated in the works of André (1989, 1996).

10.   *F*-Series: Motivation

In previous sections, we discussed E- and G-functions, which were defined as

$$\sum_{n=0}^{\infty} \frac{c_n}{n!} z^n, \quad \sum_{n=0}^{\infty} c_n z^n,$$

respectively. The properties of the coefficients c_n were quite similar. The following idea seems quite natural. Let's consider series of the form

$$\sum_{n=0}^{\infty} c_n \cdot n! \cdot z^n.$$

The conditions for the coefficients are formulated as follows. If these series are different from polynomials, then they have a zero radius of convergence in the field of complex numbers. However, under natural conditions for coefficients, they have convergence radii greater than 1 in all fields of p-adic numbers. This makes it possible to consider the concept of a global relation for series of this type.

The following chapters will be devoted to this concept. First, we consider the concept of a polyadic number.

We need some new definitions.

Suppose that $\alpha \in \mathbb{K}$ and that the series

$$\sum_{n=0}^{\infty} c_n \cdot n! \cdot z^n$$

converges in the field $\mathbb{K}_v$ for all valuations v of the field $\mathbb{K}$ that extend the p-adic valuation for all primes p, except possibly a finite number of them. As noted above, this allows us to consider this series as an element of an infinite direct product of the fields $\mathbb{K}_v$. This direct product has a natural ring structure, operations on which correspond to those on each coordinate. For the element $\mathfrak{a}$ of this direct product, we denote as $\mathfrak{a}^{(v)}$ its coordinate in the field $\mathbb{K}_v$.

If there exists a $P(x)$-polynomial with rational coefficients other than the identical zero such that $P(\mathfrak{a}) = 0$ (in other words,

$P(\mathfrak{a}^{(v)}) = 0$ in each field $\mathbb{K}_v$ of this direct product), then we say that $\mathfrak{a}$ is an *algebraic element*.

If the element $\mathfrak{a}$ is not algebraic, then it is called *transcendental*. The transcendence of the element means that for any $P(x)$-polynomial with rational coefficients other than the identical zero, there exists a prime number p and a valuation v of the field $\mathbb{K}$ that extends the p-adic valuation such that $P(\mathfrak{a}^{(v)}) \neq 0$ in the field $\mathbb{K}_v$.

Let's call an element $\mathfrak{a}$ *infinitely transcendental* if for any $P(x)$-polynomial with rational coefficients other than the identical zero, there exists an infinite set of primes p, for each of which there is a valuation v of the field $\mathbb{K}$ extending the p-adic valuation such that $P(\mathfrak{a}^{(v)}) \neq 0$ in the field $\mathbb{K}_v$.

An element $\mathfrak{a}$ is called *globally transcendental* if, for any $P(x)$-polynomial with rational coefficients other than the identical zero, the inequality $P(\mathfrak{a}^{(v)}) \neq 0$ holds in all fields $\mathbb{K}_v$ of the direct product under consideration.

By analogy, it is possible to define the concepts of *irrationality, infinite and global irrationality, linear and algebraic independence, infinite linear and algebraic independence, and global linear and algebraic independence* for an infinite product of rings. We will carefully formulate these definitions when there is a need for them.

Chapter 1

Polyadic Numbers

1.1. Definition and Motivation

We introduce on the ring of integers $\mathbb{Z}$ the smallest topology invariant with respect to shifts and such that the ideals $m\mathbb{Z}$ and $m \in \mathbb{Z}$ of the ring $\mathbb{Z}$ form the base of the system of neighbourhoods of zero. In this case, the addition and multiplication operations are continuous, and the ring of integers with the introduced topology has the structure of a topological ring. On this topological ring, one can introduce a metric by assuming

$$\rho(x, y) = \sum_{m=1}^{\infty} \frac{1}{2^m} \left\lfloor \frac{x - y}{m} \right\rfloor,$$

where the symbol $\lfloor t \rfloor$ denotes the distance from the number t to its nearest integer. An infinite sequence of integers $\{a_1, a_2, \ldots, a_n, \ldots\}$ is called *fundamental* if for any positive integer k, there exists a positive integer N such that for any positive integers $n, m, n > N, m > N$, $a_n \equiv a_m \pmod{k!}$ holds. The resulting metric space is not complete (for example, the sequence $1!, 1!+2!, 1!+2!+3!, \ldots$) but has no limits in the constructed topological ring.

The fundamental sequences form a ring. Understanding the limit in the sense of the topology introduced above, we call a sequence of integers $\{a_1, a_2, \ldots, a_n, \ldots\}$ a *zero sequence* if

$$\lim_{n \to \infty} a_n = 0.$$

Fundamental sequences are called *equivalent* if their difference is a zero sequence. The ring of polyadic integers is defined as a ring of equivalent fundamental sequences. The ring of *polyadic integers* admits a slightly different description. The ring system $\mathbb{Z}/m!\mathbb{Z}$ forms a projective system with respect to the maps

$$\mathbb{Z}/(m+1)!\mathbb{Z} \to \mathbb{Z}/m!\mathbb{Z}.$$

The ring of polyadic integers is the inverse limit of

$$\varprojlim_{m} \mathbb{Z}/m!\mathbb{Z}.$$

Here is a brief (and incomplete) overview of problems related to the theory of polyadic numbers. The elements λ of the ring of polyadic integers have a canonical representation in the form

$$\lambda = \sum_{m=1}^{\infty} a_m m!, \quad a_m \in N, \quad 0 \le a_m \le m.$$

It should be noted that any positive integer M admits a unique representation in the form

$$M = \sum_{m=1}^{N} a_m m!, \quad a_m \in N, \quad 0 \le a_m \le m.$$

This representation is called *factorial*.

These representations are relevant to solving some practical problems. Using the Haar–Lebesgue measure μ on the ring of polyadic numbers, integration can be determined (see, for example, Postnikov, 1971).

Novosyolov (1961) proved that the sequence of positive integers is uniformly distributed in the ring $\mathbb{Z}$.

This means that for any continuous function $f(x)$ on this ring, the following equality holds:

$$\lim_{N\to\infty} \frac{1}{N} \sum_{n=1}^{N} f(n) = \int f(x)d\mu.$$

Postnikov (1971) reviewed a number of other important results on the properties of the ring of polyadic numbers and functions defined on this ring.

Note the role of the ring of polyadic numbers in a number of algebraic problems. For more information, see the book by Fomin (2013).

For the issues considered in this book, it is important that the ring of polyadic integers is a direct product over all primes p of rings of integer p-adic numbers Z_p. Let's prove this statement.

The *canonical decomposition of a polyadic number* λ has the form

$$\lambda = \sum_{m=1}^{\infty} a_m m!, \quad a_m \in N, \quad 0 \le a_m \le m.$$

This series converges in any field of p-adic numbers $\mathbb{Q}_p$. Denoting the sum of this series in the field $\mathbb{Q}_p$ with the symbol $\lambda^{(p)}$, we get that any polyadic number λ can be considered an element of the direct product of rings of integer p-adic numbers $\mathbb{Z}_p$ over all primes p. The converse statement is also true, meaning that the ring of polyadic integers coincides with this direct product.

Theorem 1.1. *Let $p_1 < p_2 < \cdots$ be an ordered set of all primes. For any given p_i-adic integers Λ_i, $i = 1, 2, \ldots$, there exists a polyadic number λ such that $\lambda^{(p_i)} = \Lambda_i$, $i = 1, 2, \ldots$. In other words, the ring of polyadic numbers coincides with the direct product of the rings $\mathbb{Z}_p$ of p-adic integers over all primes p.*

Proof of Theorem 1.1. To prove this theorem, assume that

$$\Lambda_i = \sum_{t=0}^{\infty} a_{ti} p_i^t, \quad \Lambda_i^{(k)} = \sum_{t=0}^{k} a_{ti} p_i^t, \quad i = 1, 2, \ldots. \tag{1.1}$$

Consider an arbitrary positive integer M, and denote by

$$p_1 < p_2 < \cdots < p_{k(M)}$$

all prime numbers not exceeding the number M. Applying the Chinese remainder theorem, we get that there are natural numbers $x_1^{(M)}, \ldots, x_{k(M)}^{(M)}$ such that

$$x_l^{(M)} \equiv 1 (\operatorname{mod} p_l^M), \quad x_l^{(M)} \equiv 0 (\operatorname{mod} p_r^M),$$

$$r, l = 1, 2, \ldots, k(M), \quad r \ne l. \tag{1.2}$$

For any M, the number a_M is defined by the equality

$$a_M = \Lambda_1^{(M)} x_1^{(M)} + \cdots + \Lambda_{k(M)}^{(M)} x_{k(M)}^{(M)}. \qquad (1.3)$$

We prove that the sequence of numbers a_M is fundamental and that its limit is the desired polyadic number. To do this, fix the number M and consider the numbers $m > M, n > M$. They correspond to the numbers

$$a_m = \Lambda_1^{(m)} x_1^{(m)} + \cdots + \Lambda_{k(M)}^{(m)} x_{k(M)}^{(m)} + \cdots + \Lambda_{k(m)}^{(m)} x_{k(m)}^{(m)},$$

$$a_n = \Lambda_1^{(n)} x_1^{(n)} + \cdots + \Lambda_{k(M)}^{(n)} x_{k(M)}^{(n)} + \cdots + \Lambda_{k(n)}^{(n)} x_{k(n)}^{(n)}.$$

The difference of these numbers has the form

$$
\begin{aligned}
a_m - a_n = {} & (\Lambda_1^{(m)} x_1^{(m)} - \Lambda_1^{(n)} x_1^{(n)}) + \cdots + (\Lambda_{k(M)}^{(m)} x_{k(m)}^{(m)} - \Lambda_{k(M)}^{(n)} x_{k(M)}^{(n)}) \\
& + \Lambda_{k(M)+1}^{(m)} x_{k(M)+1}^{(m)} + \cdots + \Lambda_{k(m)}^{(m)} x_{k(m)}^{(m)} \\
& - \Lambda_{k(M)+1}^{(n)} x_{k(M)+1}^{(n)} - \cdots - \Lambda_{k(n)}^{(n)} x_{k(n)}^{(n)}.
\end{aligned}
\qquad (1.4)
$$

Let $l \leq k(M)$. If $r > k(M)$, then, according to (1.2),

$$x_r^{(M)} \equiv 0 (\mathrm{mod}\, p_l^M), \quad r = k(M)+1, \ldots, \max(k(m), k(n)), \quad r \neq l. \qquad (1.5)$$

If $s \leq k(M)$, but $s \neq l$, then, according to (1.2),

$$\Lambda_s^{(m)} x_s^{(m)} \equiv 0 (\mathrm{mod}\, p_l^M), \quad \Lambda_s^{(n)} x_s^{(n)} \equiv 0 (\mathrm{mod}\, p_l^M);$$

therefore,

$$\Lambda_s^{(m)} x_s^{(m)} - \Lambda_s^{(n)} x_s^{(n)} \equiv 0 (\mathrm{mod}\, p_l^M).$$

Consider, further, the difference

$$\Lambda_l^{(m)} x_l^{(m)} - \Lambda_l^{(n)} x_l^{(n)}.$$

According to (1.2),

$$x_l^{(m)} \equiv 1 (\mathrm{mod}\, p_l^m) \equiv 1 (\mathrm{mod}\, p_l^M), \quad x_l^{(n)} \equiv 1 (\mathrm{mod}\, p_l^n) \equiv 1 (\mathrm{mod}\, p_l^M).$$

By definition (1.1) of the numbers $\Lambda_i^{(k)} = \sum_{t=0}^{k} a_{ti} p_i^t$, one has

$$\Lambda_l^{(m)} x_l^{(m)} = \sum_{t=0}^{m} a_{ti} p_i^t x_l^{(m)} \equiv \sum_{t=0}^{M} a_{ti} p_i^t (\operatorname{mod} p_l^M),$$

$$\Lambda_l^{(n)} x_l^{(n)} = \sum_{t=0}^{n} a_{ti} p_i^t x_l^{(n)} \equiv \sum_{t=0}^{M} a_{ti} p_i^t (\operatorname{mod} p_l^M).$$

Therefore,

$$\Lambda_l^{(m)} x_l^{(m)} - \Lambda_l^{(n)} x_l^{(n)} \equiv 0 (\operatorname{mod} p_l^M). \tag{1.6}$$

Thus, according to (1.2)–(1.6), for any $m > M$, $n > M$, the number $a_m - a_n$ is divisible by each of the numbers p_l^M, $l = 1, \ldots, k(M)$.

Therefore, $a_m - a_n$ is divisible by the number $M!$, since the degree to which a prime number p, less than M, is included in the factorisation of the number $M!$ is equal to

$$\frac{M - S_M}{p - 1},$$

where S_M denotes the sum of digits in the p-adic expansion of the number M, and this value is less than M. Thus, the sequence of numbers a_n is fundamental. Let it be a polyadic number λ. It remains to prove that equality $\lambda^{p_i} = \Lambda^i$ holds for any prime number p_i. This immediately follows from the fact that the partial sum a_M of a series λ in the field $\mathbb{Q}_{p_i}$ of p_i-adic numbers is equal to $\Lambda_i^{(M)}$. The limit of these partial sums is equal to Λ_i. The theorem is proved. $\square$

The arithmetic properties of polyadic numbers have been studied relatively little.

The name Euler is associated with the series

$$\sum_{n=0}^{\infty} n!.$$

The properties of this series will be discussed in Chapter 4.

The *Kurepa hypothesis* (Kurepa, 1971) refers to partial sums of this series and states that for any prime number $p > 2$, the number

$\sum_{n=0}^{p-1} n!$ is not divisible by p. This means that $|\sum_{n=0}^{\infty} n!|_p = 1$ for any p. Kurepa's hypothesis has not yet been proven.

Note that the following statement is true:

For any polynomial $p(x) \in \mathbb{Z}[x]$, the following equality holds:

$$\sum_{n=0}^{\infty} p(n) \cdot n! = A \cdot \sum_{n=0}^{\infty} n! + B, \tag{1.7}$$

where $A, B \in \mathbb{Z}$, and the values of A and B are specified as follows.

Indeed, let the polynomial $p(x)$ have degree m. Let us represent it in the form

$$p(x) = a_m(x+1)(x+2)\cdots(x+m)$$
$$+ a_{m-1}(x+1)(x+2)\cdots(x+m-1) + \cdots$$
$$+ a_2(x+1)(x+2) + a_1(x+1) + a_0,$$

where $a_i \in \mathbb{Z}$, $i = 0, 1, \ldots, m$.

Then,

$$\sum_{n=0}^{\infty} p(n) \cdot n! = \sum_{n=0}^{\infty} (a_m \cdot (n+m)! + \cdots + a_1(n+1)! + a_0 n!)$$

$$= a_m \left(\sum_{n=0}^{\infty} n! - \sum_{n=0}^{m-1} n! \right) + \cdots$$

$$+ a_1 \left(\sum_{n=0}^{\infty} n! - 1 \right) + \cdots + a_0 \cdot n!$$

$$= (a_0 + \cdots + a_m) \sum_{n=0}^{\infty} n! - \left(a_m \sum_{n=0}^{m-1} n! + \cdots + a_2 \cdot 1! + a_1 \right)$$

$$= A \sum_{n=0}^{\infty} n! + B,$$

where

$$A = a_0 + \cdots + a_m \in \mathbb{Z}, \quad B = - \left(a_m \sum_{n=0}^{m-1} n! + \cdots + a_1 \right) \in \mathbb{Z}. \tag{1.8}$$

The equalities (1.7) and (1.8) serve as a source of polyadic formulae, for example,

$$\sum_{n=0}^{\infty} n \cdot n! = -1, \quad \sum_{n=0}^{\infty} (n^2 + 1) \cdot n! = 1.$$

We have already noted that any positive integer M admits a unique representation in the form

$$M = \sum_{n=1}^{N} a_n \cdot n!, \quad a_n \in \{0, 1, \ldots, n\},$$

the so-called polyadic (or factorial) representation. The equality $\sum_{n=0}^{\infty} n \cdot n! = -1$, combined with the previous one, allows us to represent any negative integer $-M - 1$ in the form

$$-M - 1 = \sum_{n=0}^{N} (n - a_n)n! + \sum_{n=N+1}^{\infty} n \cdot n!.$$

1.2. Almost Polyadic Numbers

If we consider a direct product of the fields $\mathbb{Q}_p$ over all but a finite number of primes p, we get the so-called ring of *almost polyadic numbers*. Of course, this ring depends on which primes are excluded from consideration. For example, a series

$$\sum_{n=0}^{\infty} (\gamma)_n z^n,$$

where

$$\gamma = \frac{a}{b}, \quad a \in \mathbb{Z}, \quad b \in \mathbb{N}, \quad (\gamma)_0 = 1, \quad (\gamma)_n = \gamma(\gamma+1) \cdots (\gamma+n-1),$$

converges in all fields of p-adic numbers, except for those fields for which p divides the denominator b of the fraction γ. So, it represents an almost polyadic number.

1.3.   Algebraic Extension of Polyadic Numbers

Let $\mathbb{K}$ be an algebraic number field of a finite degree κ over the field $\mathbb{Q}$. Let v be a valuation of $\mathbb{K}$. Denote by $\mathbb{K}_v$ the completion of $\mathbb{K}$ with respect to v. Any valuation v of $\mathbb{K}$ extends some valuation of $\mathbb{Q}$. If v extends the p-adic valuation (denoted by $v \,|\, p$), then $\mathbb{K}_v$ is an algebraic extension of $\mathbb{Q}_p$ of a finite degree κ_v, and for any prime p we have

$$\sum_{v \,|\, p} \kappa_v = \kappa,$$

where the summation on the left-hand side of this equation runs over all valuations $v \,|\, p$. Thus, we can consider the direct product of fields $\mathbb{K}_v$ over all primes p and all continuations $v \,|\, p$ of the p-adic valuation of the field $\mathbb{Q}$.

1.4.   Polyadic Liouville Numbers

The results in this section are published in Chirskii (2022a). We call a polyadic number λ a *polyadic Liouville number* (or a Liouville polyadic number) if, for any numbers n and P, there exists an integer A such that for all primes p satisfying the inequality $p \leq P$, the inequality $|\lambda - A|_p < |A|^{-n}$ holds. More precisely, we should write $|\lambda^{(p)} - A|_p < |A|^{-n}$; however, we will assume that when considering the field of p-adic numbers under the symbol λ, the sum of $\lambda^{(p)}$ of this series is implied.

Theorem 1.2. *A polyadic Liouville number is a transcendental element of any field $\mathbb{Q}_p$. In other words, a polyadic Liouville number is a globally transcendental number.*

Proof of Theorem 1.2.   To prove this theorem, assume that a polynomial $P(x)$ of degree d other than the identical zero has integer coefficients. Suppose that in some field $\mathbb{Q}_p$, the equality $P(\lambda) = 0$ is satisfied. Let A be an arbitrary natural number. Then, either $P(A) = 0$, or, with some constant C_1, there is an inequality

$$|\lambda - A|_p > C_1 |A|^{-d}.$$

Indeed, considering the decomposition of the polynomial $P(x)$ by the powers of the variable $x - A$, we obtain

$$P(x) = P(A) + P'(A)(x - A) + \cdots + \frac{P^{(d)}(A)}{d!}(x - A)^d.$$

All numbers

$$P'(A), \ldots, \frac{P^{(d)}(A)}{d!}$$

are integers, so their p-adic valuations do not exceed 1. By assumption, when substituting λ instead of x, we get the equality

$$0 = P(\lambda) = P(A) + P'(A)(\lambda - A) + \cdots + \frac{P^{(d)}(A)}{d!}(-A)^d. \quad (1.9)$$

Equality (1.9) can be rewritten as

$$P(A) = D(\lambda - A), \quad (1.10)$$

where D is some integer p-adic number. If the inequality $P(A) \neq 0$ holds, then it is obvious that with some constant C_0, the inequalities

$$0 < |P(A)| < C_0|A|^d$$

are fulfilled. For a non-zero rational number B, the inequality $|B|_p \geq |B|^{-1}$ holds, so

$$|P(A)|_p > \frac{1}{C_0|A|^d} = C_1|A|^{-d}.$$

According to (1.10),

$$|P(A)|_p \leq |\lambda - A|_p.$$

Thus,

$$|\lambda - A|_p > C_1|A|^{-d},$$

as stated. By assumption, $\lambda-$ is a polyadic Liouville number, and for any numbers n and P, there exists an integer A such that for all primes p satisfying the inequality $p \leq P$ the inequality $|\lambda - A|_p < |A|^{-n}$ holds. Since the numbers n and P are arbitrary, for any considered prime number p, there is an infinite set of pairs of numbers $n, A, n > d$ for which this inequality holds. Increasing the number n, we get that there is an infinite set of numbers A such that $|\lambda - A|_p < C_1|A|^{-d}$; therefore, there exists an integer A such that, simultaneously with this inequality, the inequality $P(A) \neq 0$ holds. As proved above, this means that in the field $\mathbb{Q}_p$, the inequality $P(\lambda) \neq 0$ holds. In view of the arbitrariness of the polynomial $P(x)$, the sum of the series λ is transcendental in the field $\mathbb{Q}_p$. Recall that the number p is an arbitrary prime number. The global transcendence of the polyadic Liouville number is proved. $\square$

We formulate and prove a generalisation of this theorem.

Theorem 1.3. *Let p be a prime number. Let $\alpha_i \in \mathbb{Z}_p$, $A_i \in \mathbb{Z}$, $i = 1, \ldots, m$. Let $P(x_1, \ldots, x_m)$ be a non-zero polynomial with integer coefficients whose degrees over the variables x_i, $i = 1, \ldots, m$ are equal to d_i. Then, if*

$$P(\alpha_1, \ldots, \alpha_m) = 0, \tag{1.11}$$

either the inequality

$$\max_{i=1,\ldots,m} |\alpha_i - A_i|_p > C_2|A_1|^{-d_1} \cdots |A_m|^{-d_m} \tag{1.12}$$

holds for some constant C_2, or the following equality is fulfilled:

$$P(A_1, \ldots, A_m) = 0. \tag{1.13}$$

Corollary. *If for any non-zero polynomial $P(x_1, \ldots, x_m)$ with integer coefficients whose powers in the variables x_i, $i = 1, \ldots, m$ are equal to d_i and any constant C_2 there exist $A_i \in \mathbb{Z}$, $i = 1, \ldots, m$ such that*

$$|\alpha_i - A_i|_p < C_2|A_1|^{-d_1} \cdots |A_m|^{-d_m}, \quad i = 1, \ldots, m \tag{1.14}$$

and

$$P(A_1, \ldots, A_m) \neq 0, \tag{1.15}$$

then $\alpha_1, \ldots, \alpha_m$ are algebraically independent elements of $\mathbb{Z}_p$.

Proof of Theorem 1.3. It follows from (1.11) that the polynomial $P(x_1, \ldots, x_m)$ can be represented in the form

$$P(x_1, \ldots, x_m) = \sum B_{l_1, \ldots, l_m} (x_1 - \alpha_1)^{l_1} \cdots (x_m - \alpha_m)^{l_m}, \quad (1.16)$$

where summation on the right-hand side of equality (1.16) is performed over all sets of non-negative integers $l_1, \ldots, l_m$, satisfying the conditions $l_i \leq d_i$, $i = 1, \ldots, m$, coefficients $B_{l_1, \ldots, l_m} \in \mathbb{Z}_p$, and $B_{0, \ldots, 0} = 0$. Equality (1.16) for $x_i = A_i$, $i = 1, \ldots, m$ gives

$$P(A_1, \ldots, A_m) = \sum B_{l_1, \ldots, l_m} (A_1 - \alpha_1)^{l_1} \cdots (A_m - \alpha_m)^{l_m},$$

which, due to the equality of $B_{0, \ldots, 0} = 0$, gives

$$P(A_1, \ldots, A_m) = \sum_{i=1}^{m} D_i (A_i - \alpha_i), \quad (1.17)$$

where $D_i \in \mathbb{Z}_p$, $i = 1, \ldots, m$. It follows from (1.17) that

$$|P(A_1, \ldots, A_m)|_p \leq \max_{i=1,\ldots,m} |\alpha_i - A_i|_p \max_{i=1,\ldots,m} |D_i|_p$$

$$\leq \max_{i=1,\ldots,m} |\alpha_i - A_i|_p. \quad (1.18)$$

On the other hand, if $P(A_1, \ldots, A_m) \neq 0$, then the following inequality is satisfied:

$$|P(A_1, \ldots, A_m)| \leq C_3 |A_1|^{d_1} \cdots |A_m|^{d_m}. \quad (1.19)$$

By the property of the p-adic valuation, if an integer M is different from zero, then $|M|_p \geq |M|^{-1}$, so (1.11), (1.12), (1.18), and (1.19) imply that (1.13) holds. Theorem 1.3 is proved. $\square$

Let's prove the corollary. Inequalities (1.18) and (1.19), valid for any non-zero polynomial $P(x_1, \ldots, x_m)$ with integer coefficients, mean that relations (1.14) and (1.15) are not fulfilled; therefore, for any non-zero polynomial with integer coefficients, equality (1.11) does not hold. This proves the corollary of Theorem 1.3. $\square$

For all $i = 1, \ldots, m$, we denote

$$\alpha_i = \sum_{n=0}^{\infty} a_{i,n}, \quad a_{i,n} \in \mathbb{N}, \tag{1.20}$$

$$A_{i,N} = \sum_{n=0}^{N} a_{i,n}, \quad r_{i,N} = \sum_{n=N+1}^{\infty} a_{i,N}. \tag{1.21}$$

Theorem 1.4. *Let, for a prime number p and $N \geq N_p$, where N_p is a positive integer depending on p, the inequalities*

$$|r_{i,N}|_p < \left(\max_{i=1,\ldots,m} A_{i,N} \right)^{-\gamma_p(N)}, \quad i = 1, \ldots, m, \tag{1.22}$$

hold. Here, $\gamma_p(N) \to +\infty$ as $N \to +\infty$. In addition, let

$$\lim_{N \to +\infty} \frac{\ln |r_{i,N}|_p}{\ln |r_{i+1,N}|_p} = 0, \quad i = 1, \ldots, m-1. \tag{1.23}$$

Then, series (1.20) converge to the algebraically independent elements of the field $\mathbb{Q}_p$.

Corollary. *If the conditions of Theorem 1.4 are satisfied for any prime number p, then series (1.20) are globally algebraically independent polyadic numbers.*

Proof of Theorem 1.4. Let p be a prime number. For $N \geq N_p$, consider the numbers $A_{i,N}$, $i = 1, \ldots, m$.

It follows from (1.21) and (1.22) that series (1.20) converge, and for any constant C_2 and any non-negative integers d_i, $i = 1, \ldots, m$, the following inequalities are fulfilled:

$$|\alpha_i - A_{i,N}|_p < C_2 |A_{1,N}|^{-d_1} \cdots |A_{m,N}|^{-d_m}, \quad i = 1, \ldots, m.$$

If for some N the inequality $P(A_{1,N}, \ldots, A_{m,N}) \neq 0$ holds, then, according to the corollary of Theorem 1.3, series (1.20) converge to algebraically independent elements from $\mathbb{Z}_p$. Otherwise, if for all $N \geq N_p$ we have equality

$$P(A_{1,N}, \ldots, A_{m,N}) = 0, \tag{1.24}$$

then, when passing in this equality (1.24) to the limit at $N \to +\infty$, we get

$$P(\alpha_1, \ldots, \alpha_m) = 0.$$

As in (1.16), we have an equality that is valid for all $N \geq N_p$:

$$P(x_1, \ldots, x_m) = \sum B_{l_1, \ldots, l_m} (x_1 - \alpha_1)^{l_1} \cdots (x_m - \alpha_m)^{l_m}.$$

From this, taking into account (1.21) and (1.24), the substitution $x_i = A_{i,N}$ gives

$$0 = \sum B_{l_1, \ldots, l_m} (A_{1,N} - \alpha_1)^{l_1} \cdots (A_{m,N} - \alpha_m)^{l_m}, \tag{1.25}$$

where the summation on the right-hand side of equality (1.25) is performed over all sets of non-negative integers $l_1, \ldots, l_m$, satisfying the conditions $l_i \leq d_i$, $i = 1, \ldots, m$, and coefficients $B_{l_1, \ldots, l_m} \in \mathbb{Z}_p$, where $B_{0, \ldots, 0} = 0$. In addition, since the considered polynomial is non-zero, some of the coefficients $B_{l_1, \ldots, l_m}$ are non-zero. We rewrite equality (1.25) in the form

$$0 = \sum B_{l_1, \ldots, l_m} (-r_{1,N})^{l_1} \cdots (-r_{m,N})^{l_m}. \tag{1.26}$$

Let us consider two terms distinct from zero on the right-hand side of equality (1.26). Let them be

$$B_{l_1, \ldots, l_m} (-r_{1,N})^{l_1} \cdots (-r_{m,N})^{l_m} \tag{1.27}$$

and

$$B_{k_1, \ldots, k_m} (-r_{1,N})^{k_1} \cdots (-r_{m,N})^{k_m}. \tag{1.28}$$

Moreover, the sets of numbers $(l_1, \ldots, l_m)$ and $(k_1, \ldots, k_m)$ are different. Then, from (1.27) and (1.28), it follows that

$$\ln \frac{|B_{l_1, \ldots, l_m} (-r_{1,N})^{l_1} \cdots (-r_{m,N})^{l_m}|_p}{|B_{k_1, \ldots, k_m} (-r_{1,N})^{k_1} \cdots (-r_{m,N})^{k_m}|_p}$$

$$= \ln \frac{|B_{l_1, \ldots, l_m}|_p}{|B_{k_1, \ldots, k_m}|_p} + \sum_{i=1}^{m} (l_i - k_i) \ln |r_{i,N}|_p. \tag{1.29}$$

Among the numbers $l_i - k_i$, there are some that are different from zero. Let s be the largest of the numbers i satisfying such conditions, i.e., let $l_s - k_s \neq 0$, but their differences with numbers greater than s (if any) be equal to zero. From condition (1.22), it follows that $|r_{i,N}|_p \to 0$, $\ln |r_{i,N}|_p \to -\infty$, $N \to +\infty$, $i = 1, \ldots, m$.

Combining (1.29) and (1.23), we conclude that $(l_s - k_s) \ln |r_{s,N}|_p$ is the leading term on the right-hand side of (1.29). Then, depending on the sign of the number $l_s - k_s \neq 0$, the right-hand side of formula (1.29) tends to either $+\infty$ or $-\infty$. This means that the ratio of p-adic valuations of quantities (1.27) and (1.28) tends either to infinity or to zero when $N \to +\infty$. Since there are finitely many non-zero coefficients $B_{l_1, \ldots, l_m}$ with a sufficiently large N, one of the corresponding terms on the right-hand side of equality (1.26) has a larger p-adic valuation than the rest of the terms. So, the p-adic valuation of the right-hand side of this equality for the considered N is equal to the p-adic valuation of this term and is different from 0, which contradicts equality (1.26). Theorem 1.4 is proven.   □

The corollary to Theorem 1.4 does not require a separate proof.

The following theorem gives an example of applying Theorem 1.4 to an explicitly defined set of polyadic numbers. Consider the arbitrary positive integers

$$n_{1,0} < n_{2,0} < \cdots < n_{m,0}. \tag{1.30}$$

Given a non-negative integer N, consider positive integers

$$n_{1,N} < n_{2,N} < \cdots < n_{m,N}. \tag{1.31}$$

Let

$$M_{N+1} = \min \frac{\ln p}{p - 1}, \tag{1.32}$$

where the minimum on the right-hand side of (1.32) is taken over all primes, satisfying the inequality $p \le n_{m,N}$. Assume that $\gamma(N) \to +\infty$ as $N \to +\infty$. Suppose that $n_{1,N+1}$ satisfies the inequality

$$n_{1,N+1} M_{N+1} - \ln n_{1,N+1} > \gamma(N) \ln(2(n_{m,N}!)), \tag{1.33}$$

while the positive integers $n_{i+1,N+1}$, $i = 1, \ldots, m - 1$, satisfy the conditions

$$n_{i,N+1} < \phi_i(N) n_{i+1,N+1}, \tag{1.34}$$

where

$$\phi_i(N) \to 0, N \to +\infty, \quad i = 1, \ldots, m - 1. \tag{1.35}$$

Let

$$\alpha_i = \sum_{k=0}^{\infty} (n_{i,k})!, \tag{1.36}$$

$$A_{i,N} = \sum_{k=0}^{N} (n_{i,k})!, \tag{1.37}$$

$$r_{i,N} = \sum_{k=N+1}^{\infty} (n_{i,k})!. \tag{1.38}$$

Theorem 1.5. *If the positive integers $n_{i,k}$, $i = 1, \ldots, m$, $k = 1, 2, \ldots$ satisfy conditions (1.30)–(1.35), then for any prime number p, series (1.36) converge to the algebraically independent elements of $\mathbb{Z}_p$.*

In other words, the polyadic numbers (1.36) are globally algebraically independent.

Proof of Theorem 1.5. From (1.31) and (1.36)–(1.38), the equations

$$|r_{i,N}|_p = |(n_{i,N+1}!)|_p, \quad i = 1, \ldots, m, \tag{1.39}$$

follow. Condition (1.34) implies that

$$\max_{i=1,\ldots,m} |r_{i,N}|_p = |(n_{1,N+1}!)|_p, \tag{1.40}$$

and we use the formula for the p-adic valuation of the factorial

$$|(n_{1,N+1}!)|_p = \exp\left(-\frac{\ln p}{p-1}(n_{1,N+1} - S_{n_{1,N+1}})\right), \tag{1.41}$$

where $S_{n_{1,N+1}}$ denotes the sum of digits in the p-adic expansion of the number $n_{1,N+1}$. Obviously,

$$S_{n_{1,N+1}} \leq (p-1)\log_p n_{1,N+1} = (p-1)\frac{\ln n_{1,N+1}}{\ln p}. \tag{1.42}$$

In view of (1.41) and (1.42),

$$\ln |(n_{1,N+1}!)|_p \le -\frac{\ln p}{p-1} n_{1,N+1} + \ln n_{1,N+1}. \qquad (1.43)$$

For a prime number p, we choose the number N_p as the smallest positive integer N for which the inequality $n_{m,N} \ge p$ holds. Then, it follows from equality (1.32) that for $N \ge N_p$,

$$M_{N+1} \le \frac{\ln p}{p-1}. \qquad (1.44)$$

From the inequalities (1.44) and (1.33) follows

$$-\frac{\ln p}{p-1} n_{1,N+1} + \ln n_{1,N+1} \le -M_{N+1} n_{1,N+1} + \ln n_{1,N+1}$$

$$< -\gamma(N) \ln(2(n_{m,N}!)). \qquad (1.45)$$

Using (1.43) and (1.45), we get

$$|(n_{1,N+1}!)|_p \le \exp(-\gamma(N) \ln(2(n_{m,N}!))). \qquad (1.46)$$

For all $i = 1, \ldots, m-1$, according to (1.31),

$$A_{i,N} \le A_{m,N} = \sum_{k=0}^{N} (n_{m,k})!$$

$$= (n_{m,N})! \left(1 + \frac{(n_{m,0})!}{(n_{m,N})!} + \cdots + \frac{(n_{m,N-1})!}{(n_{m,N})!} \right)$$

$$< (n_{m,N})! \left(! + (N-1) \frac{(n_{m,N-1})!}{(n_{m,N})!} \right) < 2(n_{m,N})!. \qquad (1.47)$$

It follows from (1.40), (1.46), and (1.47) that for $N \ge N_p$, inequalities (1.22) are fulfilled. Relations (1.23) immediately follow from conditions (1.35) and equalities (1.39) and (1.40). The application of Theorem 1.4 completes the proof of Theorem 1.5.   □

Let us consider examples of the above situation. Let $k \ge 2$ be a positive integer. Let

$$\Phi(k, 1) = k, \quad \Phi(k, 2) = k^k,$$

and for a positive integer m, let $\Phi(k, m)$ denote the result of k raised to the power k successively m times. In other words,

$$\Phi(k, m+1) = k^{\Phi(k,m)}. \tag{1.48}$$

Let

$$n_m = \Phi(k, m), \tag{1.49}$$

and let

$$\alpha = \sum_{m=0}^{\infty} (n_m)!. \tag{1.50}$$

Theorem 1.6. *For any positive integer $k \geq 2$ and for any prime number p, series* (1.50) *converges to a transcendental element of the ring $\mathbb{Z}_p$.*

Proof of Theorem 1.6.   Define

$$A_N = \sum_{m=0}^{N} (n_m)!, \tag{1.51}$$

$$r_N = \sum_{m=N+1}^{\infty} (n_m)!. \tag{1.52}$$

Let us prove that for $N > N_0$, it holds that

$$|r_N|_p < |A_N|^{-\varphi(N)}, \tag{1.53}$$

where $\varphi(N)$ is a function tending to $+\infty$ as $N \to \infty$.
    Definitions (1.49) and (1.52) imply that

$$|r_N|_p = |\Phi(k, N+1)!|_p.$$

Additionally, combining (1.48), (1.49), and (1.51) obviously yields the inequality

$$|A_N| < 2\Phi(k, N)!. \tag{1.54}$$

It follows from (1.48) that

$$\Phi(k, N + 1) \geq 2^{\Phi(k,N)}, \qquad (1.55)$$

which holds as an equality for $k = 2$. We use the well-known relation

$$|M!|_p = p^{-\frac{M - S_M}{p-1}},$$

where M is an arbitrary positive integer and S_M is the sum of digits in the p-adic expansion of M. It follows that

$$|M!|_p \leq p^{-\frac{M}{p-1}} p^{\frac{(p-1)\log_p M}{p-1}} = p^{-\frac{M}{p-1}} M. \qquad (1.56)$$

Thus, fixing k, we obtain

$$|r_N|_p = |\Phi(k, N+1)!|_p = |\Phi(k, N+1)!|_p \leq p^{-\frac{\Phi(k,N+1)}{p-1}} \Phi(k, N+1).$$

$$(1.57)$$

To prove inequality (1.54), in view of (1.56) and (1.57), it suffices to prove that

$$p^{-\frac{\Phi(k,N+1)}{p-1}} \Phi(k, N+1) < (2(\Phi(k, N)!)^{-\varphi(N)}. \qquad (1.58)$$

Inequality (1.58) is equivalent to

$$\Phi(k, N+1)\frac{\ln p}{p-1} - \ln \Phi(k, N+1) > \varphi(N) \ln(2\Phi(k, N)!). \qquad (1.59)$$

Using the obvious inequality $\ln a! < a \ln a$, we conclude that (1.59) follows from the inequality

$$\Phi(k, N+1)\frac{\ln p}{p-1} - \ln \Phi(k, N+1) > 2\varphi(N)\Phi(k, N) \ln(2\Phi(k, N)).$$

Consider the function

$$x\frac{\ln p}{p-1} - x,$$

which increases for $x > \frac{p-1}{\ln p}$.

Therefore, if

$$\Phi(k, N+1) > \frac{p-1}{\ln p}, \tag{1.60}$$

then, according to (1.55), inequality (1.59) follows from

$$2^{\Phi(k,N)} \frac{\ln p}{p-1} - \Phi(k, N) \ln 2 > 2\varphi(N)\Phi(k, N) \ln(2\Phi(k, N)). \tag{1.61}$$

Since

$$\frac{2^{\Phi(k,N)} \frac{\ln p}{p-1} - \Phi(k, N) \ln 2}{2\Phi(k, N) \ln(2\Phi(k, N))} \to +\infty$$

as $N \to \infty$, inequalities (1.60) and (1.61) hold for $N > N_0$, if the function $\varphi(N)$ is defined, for example, as

$$\frac{2^{\Phi(k,N)} \frac{\ln p}{p-1} - \Phi(k, N) \ln 2}{4\Phi(k, N) \ln(2\Phi(k, N))}.$$

It remains to apply Theorem 1.2, which is recalled as follows:

Any polyadic Liouville number is transcendental in any field $\mathbb{Q}_p$. In other words, any polyadic Liouville number is globally transcendental.

$\square$

Let m be a positive integer. Define

$$\Psi(N) = \Phi\left(N, \left[\frac{N}{m}\right]\right),$$

where the symbol $\left[\frac{N}{m}\right] = l$ denotes the integer part of the number $\frac{N}{m}$ and $\Phi(N, k)$ is defined in (1.48). For any positive integer N, let $N = lm + i$ and, for each $i = 0, 1, \ldots, m-1$, let

$$\alpha_i = \sum_{k=0}^{\infty} \Psi(km + i)!. \tag{1.62}$$

Additionally, let

$$A_{i,N} = \sum_{k=0}^{l} \Psi(km + i)! \tag{1.63}$$

and

$$r_{i,N} = \sum_{k=l+1}^{\infty} \Psi(km + i)!. \qquad (1.64)$$

Theorem 1.7. *The polyadic numbers α_i, $i = 0, 1, \ldots, m-1$, defined by (1.62), are globally algebraically independent.*

Proof of Theorem 1.7.   The proof of this theorem relies heavily on Theorem 1.4. Let us prove that the conditions of this theorem hold. As in the proof of Theorem 1.6, in analogy to (1.54), we have the estimate

$$A_{i,N} = \sum_{k=0}^{\left[\frac{N}{m}\right]} \Psi(km + i)! < 2\Psi(lm + i)!. \qquad (1.65)$$

In the same manner, in analogy to (1.55),

$$|r_{i,N}|_p = |\Psi((l+1)m + i)!|_p. \qquad (1.66)$$

We check that conditions (1.22) and (1.23) from Theorem 1.4 hold for numbers (1.62)–(1.64). To verify condition (1.22), it suffices to prove that, for a sufficiently large l,

$$|r_{0,lm}|_p < (A_{m-1,lm+m-1})^{-\gamma_p(l)}, \qquad (1.67)$$

where $\gamma_p(l) \to +\infty$ as $l \to +\infty$.
   Taking into account (1.64) and (1.65), we have

$$|r_{0,lm}|_p = |\Psi((l+1)m)!|_p, \qquad (1.68)$$

$$A_{m-1,lm+m-1} < 2\Psi(lm + m - 1)!. \qquad (1.69)$$

Therefore, inequality (1.67) follows from

$$|\Psi((l+1)m)!|_p < (2\Psi(lm + m - 1)!)^{-\gamma_p(l)}.$$

The last inequality follows from the inequality

$$\exp\left(-\frac{\ln p}{p-1}\Psi((l+1)m) + \ln \Psi((l+1)m)\right)$$
$$< \exp(-\gamma_p(l)\ln(2\Psi(lm + m - 1)!)). \qquad (1.70)$$

In turn, inequality (1.70) follows from the rough estimate

$$\frac{\ln p}{2(p-1)} \Psi((l+1)m) > \gamma_p(l)(2\Psi(lm+m-1))\ln(2\Psi(lm+m-1)).$$

$$(1.71)$$

By definition,

$$\Psi(N) = \Phi\left(N, \left[\frac{N}{m}\right]\right),$$

so

$$\Psi((l+1)m = \Phi((l+1)m, l+1),$$
$$\Psi(lm+m-1) = \Phi(lm+m-1, l),$$

and the definition of the function $\Phi(k,m)$ immediately implies that

$$\frac{2\Psi(lm+m-1))\ln(2\Psi(lm+m-1)}{\Phi((l+1)m, l+1)} \to 0$$

as $l \to +\infty$. Therefore, condition (1.71) obviously holds with the corresponding $\gamma_p(l)$.

The verification of the asymptotic equalities (1.23) reduces to checking the relations

$$\lim_{l \to +\infty} \frac{\ln |\Psi(lm+i)!|_p}{\ln |\Psi(lm+i+1)!|_p}, \quad i = 0, 1, \ldots, m-1,$$

which obviously hold. The theorem is proved. □

Chapter 2

F-Series*

2.1. Definition

In the introduction, we defined the Siegel E- and G-functions, and by analogy with these, we define the notion of the F-series. We say that a series

$$\sum_{n=0}^{\infty} c_n \cdot n! \cdot z^n$$

belongs to the class $F(\mathbb{K}, c_1, c_2, c_3, q)$ if the following conditions hold:

1. $c_n \in \mathbb{K}$, $n = 0, 1, \ldots$, $\mathbb{K}$ is an algebraic number field of a finite degree over the field $\mathbb{Q}$.
2. There exist constants γ, C such that

$$\overline{|c_n|} \le \gamma e^{C_1 n}, \quad n = 0, 1, 2, \ldots.$$

   Here, $\overline{|a|}$ denotes the height of the algebraic number a, i.e., the largest of the absolute values of the number itself and its conjugates in $\mathbb{K}$.
3. There exists a sequence of positive integers $d_n = q^n d_{0,n}$ such that $d_n c_k \in \mathbb{Z}_{\mathbb{K}}$, $k = 0, 1, \ldots n$, and $n = 0, 1, 2 \ldots$ (here, $\mathbb{Z}_{\mathbb{K}}$ denotes

---

*The results of this chapter are published in the work of Chirskii (2019).

the ring of integers of the field $\mathbb{K}$). In this case, the number $d_{0,n}$ is divisible only by prime factors not exceeding $c_2 n$, and

$$\operatorname{ord}_p d_{0,n} = \vartheta_p(d_{0,n}) \leq c_3 \left(\log_p n + \frac{n}{p^2} \right), \quad n = 0, 1, 2, \ldots.$$

2.2. Basic Properties

We need the following inequality, which is often used hereafter. If A is an integer from the field $\mathbb{K}$, $A \neq 0$, then for any $v \in V_0$, the following inequality holds:

$$|A|_v \geq \frac{1}{\overline{|A|}}. \tag{2.1}$$

Indeed, for any $\tilde{v} \in V_0$, the inequality $|A|_{\tilde{v}} \leq 1$ holds. Hence, for any $v \in V_0$, we have

$$\prod_{\tilde{v} \in V_0} |A|_{\tilde{v}} \leq |A|_v, \tag{2.2}$$

where the product on the left-hand side is taken over all valuations of $\tilde{v} \in V_0$. On the other hand, by the product formula,

$$\prod_{\tilde{v} \in V_0} |A|_{\tilde{v}} = \frac{1}{\prod_{v \in V_\infty} |A|_v}. \tag{2.3}$$

In addition,

$$\prod_{v \in V_\infty} |A|_v = \prod_{i=1}^{\varkappa} |A^{(i)}|^{\frac{\varkappa_v}{\varkappa}} \leq \prod_{i=1}^{\varkappa} \overline{|A|}^{\frac{\varkappa_v}{\varkappa}} \leq \overline{|A|}. \tag{2.4}$$

In this formula, $A^{(i)}$ denotes numbers algebraically conjugate to A from the field $\mathbb{K}$, and the symbol $\overline{|A|}$ denotes the greatest absolute value of these numbers. From (2.2)–(2.4), (2.1) immediately follows.

It is known that the series $\sum_{n=0}^{\infty} a_n$, $a_n \in \mathbb{K}_v$ converges in the field $\mathbb{K}_v$, $v \in V_0$, if and only if $|a_n|_v \to 0$, $n \to \infty$. In this case,

$$\left| \sum_{n=K}^{\infty} a_n \right|_v \leq \max_{n \geq K} |a_n|_v. \tag{2.5}$$

Let

$$f(z) = \sum_{n=0}^{\infty} A_n z^n, \quad A_n \in \mathbb{K}, \tag{2.6}$$

and let, for some $v \in V_0$, the set of values

$$\{|A_n|_v, \, n = 0, 1, 2, \ldots\}$$

be bounded from above. Let us denote

$$\langle f \rangle_v = \sup\{|A_n|_v, \, n = 0, 1, 2, \ldots\}. \tag{2.7}$$

Lemma 2.1. *The value defined by (2.7) has the following properties:*

1. *Let the series $f(z) = \sum_{n=0}^{\infty} A_n z^n$, $A_n \in \mathbb{K}$ and $g(z) = \sum_{n=0}^{\infty} B_n z^n$, $B_n \in \mathbb{K}$. Let $\langle f \rangle_v$ and $\langle g \rangle_v$ be determined for some valuation $v \in V_0$. Then, for the product of these series, $f(z)g(z)$, the value $\langle fg \rangle_v$ is determined, and the inequality $\langle fg \rangle_v \leq \langle f \rangle_v \langle g \rangle_v$ holds.*
2. *Let the value $\langle f \rangle_v$ be defined for series (2.6) and some valuation $v \in V_0$. Then, the value $\langle f' \rangle_v$ is defined for the formal derivative $f'(z)$ of this series, and the inequality $\langle f' \rangle_v \leq \langle f \rangle_v$ holds.*
3. *Let the value $\langle f \rangle_v$ be defined for series (2.6) and some valuation $v \in V_0$. Let $\xi \in \mathbb{K}$, $|\xi|_v \leq 1$, and let the series $f(\xi)$ converge in the field $\mathbb{K}_v$. Then, $|f(\xi)|_v \leq \langle f \rangle_v$.*

Proof of Lemma 2.1. There is an equality

$$f(z)g(z) = \sum_{k=0}^{\infty} z^k \sum_{n=0}^{k} A_n B_{k-n}.$$

For any positive integer k, we have

$$\left| \sum_{n=0}^{k} A_n B_{k-n} \right|_v \leq \sup\{|A_n|_v, \ n = 0, 1, 2, \ldots\}$$

$$\cdot \sup\{|B_n|_v, \ n = 0, 1, 2, \ldots\} = \langle f \rangle_v \langle g \rangle_v.$$

The first property is proven. Next,

$$f'(z) = \sum_{n=0}^{\infty} n A_n z^{n-1}.$$

Therefore, the inequality $|nA_n|_v \leq |A_n|_v$ entails both the existence of the value $\langle f' \rangle_v$ and the inequality $\langle f' \rangle_v \leq \langle f \rangle_v$.

Finally, from the third condition of the lemma and inequality (2.5), we obtain

$$|f(\xi)|_v = \left| \sum_{n=0}^{\infty} A_n \xi^n \right|_v \leq \max_n |A_n \xi^n|_v \leq \max_n |A_n|_v = \langle f \rangle_v. \qquad \square$$

Lemma 2.2. *Let the F-series $f(z) = \sum_{n=0}^{\infty} c_n \cdot n! \cdot z^n$ be given:*

1. *The formal derivative*

$$\sum_{n=0}^{\infty} n c_n n! z^{n-1}$$

   is an F-series.
2. *The formal integral*

$$\sum_{n=0}^{\infty} c_n n! \frac{z^{n+1}}{n+1}$$

   is an F-series.
3. *For any algebraic number μ, the series $f(\mu z)$ is an F-series.*
4. *The sum and product of a finite number of F-series is an F-series.*

Proof of Lemma 2.2. Let, for certainty, the series $f(z)$ belong to the class $F(\mathbb{K}, C_1, C_2, C_3, q)$. Let us prove the first statement of the lemma. Together with the number c_n, the number nc_n belongs to the field $\mathbb{K}$, and

$$\overline{|nc_n|} \leq n\overline{|c_n|}.$$

Since

$$\overline{|c_n|} = O(e^{C_1 n}), \quad \overline{|nc_n|} = O(e^{\check{C}_1 n}), \quad \check{C}_1 > C_1,$$

the same sequence d_n is suitable for $f(z)$ and for the series $\sum_{n=0}^{\infty} nc_n n! z^{n-1}$.

Let us prove the second statement of the lemma. Let us transform the series

$$\sum_{n=0}^{\infty} c_n n! \frac{z^{n+1}}{n+1} = \sum_{n=0}^{\infty} c_n (n+1)! \frac{z^{n+1}}{(n+1)^2}.$$

As above, together with the number c_n, the number $\frac{c_n}{(n+1)^2}$ belongs to the field $\mathbb{K}$; therefore,

$$\overline{\left| \frac{c_n}{(n+1)^2} \right|} = O(e^{C_1 n}) \quad n \to \infty.$$

Consider the least common multiple of numbers $1^2, 2^2, \ldots, (n+1)^2$ and denote it by K_{n+1}. This number is divisible only by the prime numbers p not exceeding the number $n+1$, and the degree to which these prime numbers p are included in the prime factorisation of K_{n+1} does not exceed $2 \log_p(n+1)$. Assuming, for example,

$$\check{C}_2 = \max(C_2, 1), \quad \check{C}_3 = C_3 + 2,$$

we obtain that the series $\sum_{n=0}^{\infty} c_n n! \frac{z^{n+1}}{n+1}$ belongs to the class $F(\mathbb{K}, C_1, \check{C}_2, \check{C}_3, q)$.

Let us prove the third statement of the lemma. Consider the series $f(\mu z) = \sum_{n=0}^{\infty} c_n \cdot n! \cdot (\mu z)^n$. We have $\overline{|c_n \mu^n|} = O\left(e^{C_1 n + \mu \ln n}\right)$.

Now, let $\mu = \frac{M}{Q}$, where M is an integer from the field $\mathbb{K}$ and Q is the positive integer chosen as the smallest possible number satisfying this equality. Then, $(qQ)^n d_{0,n} c_n \mu^n$ is an integer in the field $\mathbb{K}$. Thus, the series in question belongs to the class

$$F(\mathbb{K}, C_1 + \ln \overline{|\mu|}, C_2, C_3, qQ).$$

The last assertion of the lemma consists of two parts. Let the series $\sum_{n=0}^{\infty} a_n \cdot n! \cdot z^n$ belong to the class $F(\mathbb{K}, \check{C}_1, \check{C}_2, \check{C}_3, \check{q})$ and the series $\sum_{n=0}^{\infty} b_n \cdot n! \cdot z^n$ belong to the class $F(\mathbb{K}, \hat{C}_1, \hat{C}_2, \hat{C}_3, \hat{q})$.

We begin by proving that the sum of these series is an F-series:

$$\overline{|a_n + b_n|} \leq \overline{|a_n|} + \overline{|b_n|} = O(e^{C_1 n}), \quad C_1 \leq \check{C}_1 + \hat{C}_1.$$

Now, put $d_n = \check{d}_n \hat{d}_n$, where $\check{d}_n$, $\hat{d}_n$ are appropriate numbers from the definitions of the classes in question. Multiplying $a_n + b_n$ by d_n yields an integer from the field $\mathbb{K}$. Denote $\check{q}\hat{q}$, and note that the number $d_{0,n} = \check{d}_{0,n} \hat{d}_{0,n}$ is only divisible by prime numbers p not exceeding $C_2 n$, where $C_2 = \max\{\check{C}_2, \hat{C}_2\}$. In this case,

$$\vartheta_p(d_{0,n}) \leq C_3 \left(\log_p n + \frac{n}{p^2} \right),$$

where $C_3 = \check{C}_3 + \hat{C}_3$.

Consider the product of these series. There is an equality

$$\sum_{n=0}^{\infty} a_n \cdot n! \cdot z^n \cdot \sum_{n=0}^{\infty} b_n \cdot n! \cdot z^n = \sum_{n=0}^{\infty} g_n \cdot n! \cdot z^n,$$

where

$$g_n = \sum_{k+l=n} a_k b_l \frac{k! l!}{n!}.$$

Thus, $\overline{|g_n|} = O\left(e^{C_1 n}\right)$ for $C_1 \leq \check{C}_1 + \hat{C}_1$. $\qquad\square$

Lemma 2.3. *The least common multiple* γ_n *of the numbers* $\frac{n!}{k!(n-k)!}$, $k = 0, 1, \ldots, n$ *does not exceed* $e^{\tau n}$, *with some positive constant* τ.

Proof of Lemma 2.3. The number $\frac{n!}{k!(n-k)!}$ is only divisible by prime numbers p not exceeding n:

$$\vartheta_p \left(\frac{n!}{k!(n-k)!} \right) = \vartheta_p(n!) - \vartheta_p(k!) - \vartheta_p((n-k)!)$$

$$= \frac{n - S_n}{p - 1} - \frac{k - S_k}{p - 1} - \frac{(n-k) - S_{n-k}}{p - 1} = \frac{S_{n-k} + S_k - S_n}{p - 1}$$

$$\leq \frac{(p-1)(([\log_p (n-k)] + 1) + ([\log_p k] + 1)) - [\log_p n]}{p - 1}$$

$$\leq 2(\log_p n + 1).$$

Consequently, the least common multiple considered in the lemma does not exceed the number $\prod_{p \leq n} p^{\vartheta_p \left(\frac{n!}{k!(n-k)!} \right)} \leq \prod_{p \leq n} p^{2(\log_p n + 1)} = \prod_{p \leq n} p^2 n^2 \leq \prod_{p \leq n} n^4 \leq n^{4\gamma \frac{n}{\ln n}} \leq e^{\tau n}$ (here, we use P. L. Chebyshev's estimate for the number $\pi(n)$ of prime numbers p not exceeding n: $\pi(n) \leq \gamma \frac{n}{\ln n}$). The lemma is proved. □

Assume that $d_n = \check{d}_n \hat{d}_n \gamma_n$, where $\check{d}_n$, $\hat{d}_n$ are appropriate numbers from the definitions of the classes in question. The numbers $d_n g_k$, $k = 0, 1, \ldots, n$ are integers from the field $\mathbb{K}$. Note that the number $d_{0,n} = \check{d}_{0,n} \hat{d}_{0,n} \gamma_n$ is divisible only by prime numbers p not exceeding $C_2 n$, where $C_2 = \max\{\check{C}_2, \hat{C}_2, 1\}$.

From the proof of Lemma 2.3, we obtain $\vartheta_p(d_{0,n}) \leq C_3(\log_p n + \frac{n}{p^2})$, where $C_3 = \check{C}_3 + \hat{C}_3 + 2$. Lemma 2.2 is proved. □

This lemma means that the F-series as well as the E- and G-functions form a ring of power series closed with respect to formal differentiation, integration, and the replacement of the variable z by μz.

Lemma 2.4. *Let the series $f(z)$ belong to the class $F(\mathbb{K}, C_1, C_2, C_3, q)$. Then, for any prime number p that is relatively prime with the number q and any valuation v of the field $\mathbb{K}$ extending the p-adic valuation and any number $\delta > 0$, the considered series converges in the field $\mathbb{K}_v$ if*

$$|z|_v < p^{\frac{\varkappa_v}{\varkappa} \left(\frac{1}{p-1} - \frac{C_3}{p^2} - \delta \right)}.$$

Proof of Lemma 2.4. A necessary and sufficient condition for the convergence of series in the field $\mathbb{K}_v$ is the asymptotic condition $(|c_n n! z^n|_v \to 0$, $n \to \infty$. Using the definition of the class $F(\mathbb{K}, C_1, C_2, C_3, q)$, we get $q^n d_{0,n} c_n \in \mathbb{Z}_{\mathbb{K}}$, where $\mathbb{Z}_{\mathbb{K}}$ denotes the ring of integers of the field $\mathbb{K}$. The number $d_{0,n}$ is only divisible by the prime numbers

$$p \leq C_2 n \ \text{ and } \ \vartheta_p(d_{0,n}) \leq C_3 \left(\log_p n + \frac{n}{p^2} \right).$$

Consequently, for any prime number p that is relatively prime with the number q and any valuation v of the field $\mathbb{K}$ that continues the p-adic normalisation, the following inequality is satisfied:

$$|c_n|_v \leq p^{\frac{\varkappa_v C_3 \left(\log_p n + \frac{n}{p^2} \right)}{\varkappa}}.$$

Since $|n!|_v = p^{-\frac{\varkappa_v}{\varkappa} \cdot \frac{n - S_n}{p-1}}$, where S_n is the sum of digits in the p-adic expansion of n and $S_n \leq (p-1)([\log_p n] + 1)$, we obtain that if $|z|_v < p^{\frac{\varkappa_v}{\varkappa} \left(\frac{1}{p-1} - \frac{C_3}{p^2} - \delta \right)}$, then

$$|c_n n! z^n|_v \leq p^{-\frac{\varkappa_v}{\varkappa} \cdot \frac{n - S_n}{p-1} + n \frac{\varkappa_v}{\varkappa} \left(\frac{1}{p-1} - \frac{C_3}{p^2} - \delta \right) + \frac{\varkappa_v C_3 \left(\log_p n + \frac{n}{p^2} \right)}{\varkappa}}.$$

Consider the exponent

$$-\frac{\varkappa_v}{\varkappa} \cdot \frac{n - S_n}{p-1} + n \frac{\varkappa_v}{\varkappa} \left(\frac{1}{p-1} - \frac{C_3}{p^2} - \delta \right) + \frac{\varkappa_v C_3 \left(\log_p n + \frac{n}{p^2} \right)}{\varkappa}$$

$$\leq \frac{\varkappa_v}{\varkappa} (-\delta n + (C_3 + p) \log_p n).$$

This inequality shows that $|c_n n! z^n|_v \to 0$, $n \to \infty$. $\square$

Corollary 2.1. *Let $\xi = \frac{a}{b}$, $a \in \mathbb{Z}_{\mathbb{K}}$, $b \in \mathbb{N}$, let b be the smallest number that satisfies this condition. Let p be a prime number that is relatively prime with number q and satisfies the inequality $\frac{1}{p-1} - \frac{C_3}{p^2} - \delta > 0$. Then, for any valuation v of the field $\mathbb{K}$ extending the p-adic valuation and for any series $f(z)$ from the class $F(\mathbb{K}, C_1, C_2, C_3, q)$, the series $f(\xi) = \sum_{n=0}^{\infty} c_n \cdot n! \cdot \xi^n$ converges in field $\mathbb{K}_v$.*

Proof of Corollary 2.1. By convention, $|c_n n! \cdot \xi^n|_v \leq |c_n n!|_v$ and

$$||c_n n!|_v \leq p^{-\frac{\varkappa_v}{\varkappa} \cdot \frac{n - S_n}{p-1} + \frac{\varkappa_v C_3 \left(\log_p n + \frac{n}{p^2} \right)}{\varkappa}}.$$

The right-hand side of this inequality tends to zero at $n \to \infty$. The corollary is proved. ☐

Corollary 2.2. *For any series $f(z)$ of class $F(\mathbb{K}, C_1, C_2, C_3, q)$ and any valuation v of field $\mathbb{K}$ extending the p-adic valuation, the value $\langle f \rangle_v$ is defined for a prime number p satisfying the conditions of Corollary 2.1, and the equality $\langle f \rangle_v = \max_n |c_n n!|_v$ holds.*

Proof of Corollary 2.2. The series $f(1) = \sum_{n=0}^{\infty} c_n \cdot n!$ converges, so among the numbers $|c_n n!|_v$, there exists the largest one. ☐

Chapter 3

Generalisation of the Siegel–Shidlovskii Method for *F*-Series[*]

3.1. Part of the Siegel–Shidlovskii Method Common to *E*-Functions and *F*-Series

The *modified Siegel–Shidlovskii method* in the theory of transcendental numbers is used to investigate the arithmetic nature of the values of *F*-series at algebraic points. It is a product formula that can be applied to elements of direct products of the fields $\mathbb{K}_v$.

The Siegel–Shidlovskii method consists of two main parts. The first applies to solutions of an arbitrary system of differential equations of the form

$$y_i' = Q_{i,0}(z) + \sum_{j=1}^{m} Q_{i,j}(z)y_j, \quad i = 1, \dots, m. \tag{3.1}$$

It allows us to construct a set of linearly independent linear forms in the functions composed of the solution to system (3.1) having a high order of zero at $z = 0$. The lemmas constituting this part of the method are general for both *E*-functions and *F*-series and are presented below.

---

[*]The results of this chapter are published in Chirskii (2019).

Let

$$f_i(z) = \sum_{n=0}^{\infty} b_{i,n} z^n \in \mathbb{K}[[z]], \quad i = 1, \ldots, m, \ \ m \geq 2, \tag{3.2}$$

be a formal power series, $N \in \mathbb{N}$,

$$P_i(z) \in \mathbb{K}[z], \quad \deg P_i(z) \leq N. \tag{3.3}$$

Consider the linear form

$$R(z) = \sum_{i=1}^{m} P_i(z) f_i(z) = \sum_{n=0}^{\infty} r_n z^n. \tag{3.4}$$

If we consider $P_i(z) \in \mathbb{K}[z]$ to be polynomials with undetermined coefficients and require that the equations

$$r_n = 0, \quad n = 0, 1, \ldots, m(N+1) - 2$$

hold, then we obtain a system of $m(N+1) - 1$ linear homogeneous equations with respect to $m(N+1)$ unknown coefficients of these polynomials, which has a non-trivial solution. Hence, the polynomials $P_i(z)$ can be chosen so that the inequality

$$\mathrm{ord}_{z=0} R(z) \geq m(N+1) - 1$$

is fulfilled. On the other hand, if series (3.2) are not all identically zero, then for any linear form (3.4) satisfying conditions (3.3), the order $\mathrm{ord}_{z=0} R(z)$ is bounded from above by the value depending on N. Thus, the lemma is valid.

Lemma 3.1 (Lemma 4 in Shidlovskii (1989)). *Let $N \in \mathbb{N}$,*

$$f_i(z) \in \mathbb{K}[[z]], \quad i = 1, \ldots, m, \ \ m \geq 2.$$

Then, there exists a number $\theta(N) \in \mathbb{N}$ such that for any polynomial $P_i(z)$, satisfying conditions (3.3), either the linear form (3.4) is identically zero or the inequality

$$\mathrm{ord}_{z=0} R(z) \leq \theta(N)$$

holds.

This lemma can be considerably refined if series (3.2) constitute a formal solution to the system of linear differential equations

$$y_i' = \sum_{j=1}^{m} Q_{i,j}(z) y_j, \quad i = 1, \ldots, m. \tag{3.5}$$

The coefficients $Q_{i,j}(z)$ of this system belong to the field of rational functions $\mathbb{K}(z)$. Let $T = T(z) \in \mathbb{K}[z]$ be a primitive polynomial that is the least common denominator of all $Q_{i,j}(z)$, $T(z) Q_{i,j}(z) \in \mathbb{Z}_K[z]$.

Let the linear form R from variables $y_1, \ldots, y_m$ have the form

$$R = \sum_{i=1}^{m} P_i y_i, \quad P_i \in \mathbb{K}[z], \quad i = 1, \ldots, m. \tag{3.6}$$

Let the series $y_1, \ldots, y_m \in \mathbb{K}[[z]]$ constitute a formal solution to system (3.5). Substituting them into the linear form (3.6), we obtain a formal power series $R = R(z)$. The formal derivative of this series is a linear form in series $y_1, \ldots, y_m$ and their formal derivatives $y_1', \ldots, y_m'$. Let's replace $y_1', \ldots, y_m'$ with the right-hand sides of equations (3.5). Multiply the resulting linear form in $y_1, \ldots, y_m$ with coefficients from $\mathbb{K}(z)$ by the polynomial T. As a result, we obtain a linear form in these variables with coefficients from $\mathbb{K}[z]$. Let us compare system (3.5) with a linear differential operator,

$$D = \frac{\partial}{\partial z} + \sum_{k=1}^{m} \left(\sum_{i=1}^{m} Q_{k,j}(z) y_j \right) \frac{\partial}{\partial y_k}.$$

Then, the procedure described above can be written as follows:

$$TDR = TR'. \tag{3.7}$$

Further, for the arbitrary form R considered above, we use the notation

$$R = R_1, \quad R_k = TDR_{k-1}, \quad k = 2, 3, \ldots. \tag{3.8}$$

From (3.7) and (3.8),

$$R_k = TR_{k-1}', \quad k = 2, 3, \ldots. \tag{3.9}$$

According to the above, R_k is a linear form in variables $y_1, \ldots, y_m$ with coefficients from $\mathbb{K}[z]$.

We use the notation

$$R_k = \sum_{i=1}^{m} P_{k,i} y_i, \quad P_{k,i} \in \mathbb{K}[z], \quad k = 1, 2, \ldots, \quad i = 1, \ldots, m. \quad (3.10)$$

From (3.9) and (3.10), it follows that the equations

$$P_{k,i} = T \left(P'_{k-1,i} + \sum_{j=1}^{m} P_{k-1,j} Q_{j,i} \right), \quad k = 2, 3, \ldots, \quad i = 1, \ldots, m$$

$$(3.11)$$

hold.

Lemma 3.2 (Lemma 6 in Shidlovskii (1989)). *The rank of the set of linear forms $R_1, R_2, \ldots$ over the field $\mathbb{K}(z)$ is l if and only if the forms $R_1, \ldots, R_l$ are linearly independent over $\mathbb{K}(z)$, but the forms $R_1, \ldots, R_{l+1}$ are already linearly dependent over $\mathbb{K}(z)$.*

Let us denote

$$\Delta(z) = \Delta = |P_{k,i}|_{k,i=1,\ldots,m}. \quad (3.12)$$

The algebraic complements of the elements $P_{k,i}$ of this determinant are denoted by $\Delta_{k,i}$. For any $j = 1, \ldots, m$, the following equality holds:

$$\Delta(z) y_j = \sum_{k=1}^{m} \Delta_{k,j} R_k. \quad (3.13)$$

If the linear form R_1 is converted into the zero series by substituting some non-trivial solution of system (3.5) for the variables $y_1, \ldots, y_m$, then from equality (3.9), we obtain that all the forms R_k, $k = 2, 3, \ldots$ also convert into the zero series, and consequently, $\Delta(z)$ is identically 0 in z. This implies the linear dependence of $R_1, \ldots, R_m$ over the field $\mathbb{K}(z)$.

Lemma 3.3 (Lemma 7 in Shidlovskii (1989)). *Let the rank of the system of linear forms $R_1, \ldots, R_m$ over the field $\mathbb{K}(z)$ be l, $1 \leq l < m$. Then, there exist $m - l$ linearly independent formal solutions of system (3.5) such that the forms $R_1, \ldots$ turn into zero series when each of these $m-l$ solutions is substituted for the variables $y_1, \ldots, y_m$.*

The main lemma of the Siegel–Shidlovskii method in the theory of transcendental numbers, established by Shidlovskii, reads as follows.

Lemma 3.4 (Lemma 8 in Shidlovskii (1989)). *Let the set of formal power series $f_1(z), \ldots, f_m(z)$ be a formal solution of system (3.5) and linearly independent over the field $\mathbb{K}(z)$. Let the degrees of polynomials $P_i \in \mathbb{K}[z]$, $i = 1, \ldots, m$, not exceed the number N. Let the rank of the system of linear forms $R_1, \ldots, R_m$ over the field $\mathbb{K}(z)$ be l, $1 \leq l < m$. Then, there exists a number r_0 such that*

$$\mathrm{ord}_{z=0} R_1(z) \leq lN + r_0. \tag{3.14}$$

Denote

$$n_0 = 2(r_0 - m + 1). \tag{3.15}$$

In this formulation, the constant r_0 is not effective. Bertrand *et al.* (2004) established an effective upper bound on n_0. Some additional definitions will be required to formulate this result.

Denote by $\overline{\mathbb{Q}}$ the algebraic closure of the field $\mathbb{Q}$ in the field $\mathbb{C}$, i.e., the field of algebraic numbers, and by $\overline{\mathbb{Z}}$ the ring of integer algebraic numbers. The height $H(\alpha)$ of an algebraic number α is the greatest number $\mathrm{den}(\alpha)$, $\mathrm{size}(\alpha)$. The number $\mathrm{den}(\alpha)$, the denominator of the algebraic number, is the smallest positive integer d for which $d\alpha \in \overline{\mathbb{Z}}$. The number $\mathrm{size}(\alpha)$ is the largest of the Archimedean absolute values of the number α. The height of a set of polynomials with algebraic coefficients is the largest of the set of numbers consisting of the lowest common denominator of all these coefficients and the heights of these coefficients.

The height of the set of rational functions $\{Q_1, \ldots, Q_k\}$ is the height of the set of polynomials $\{T, TQ_1, \ldots, TQ_k\}$, where $T = T(z)$ is a monic polynomial of minimal degree, such that all $TQ_i \in \overline{\mathbb{Q}}[z]$. The height $H(D)$ of system (3.1) is the height of the set of its coefficients, and the degree $\deg(D)$ is the highest degree of the polynomials $T(z), T(z)Q_{i,j}(z)$, $i, j = 1, \ldots m$. Let $\mathrm{Disc}(\mathbb{K})$ denote the discriminant of the field $\mathbb{K}$.

Lemma 3.5 (Theorem 1.2 in Bertrand *et al.* (2004)). *For any system of equations (3.1), there exist positive constants C, c, effectively expressed in terms of the values $\deg(D)$, m, $\varkappa$, $\mathrm{Disc}(\mathbb{K})$*

such that

$$n_0 \le C(H(D))^c. \tag{3.16}$$

Note that it is possible to take

$$c = \log_2 C = (2\varkappa(\deg(D) + 1)m)^{(2(\deg(D)+1)m)^{8m}}. \tag{3.17}$$

Thus, (3.16) and (3.17) give an effective upper estimate for the number n_0.

Lemma 3.6 (Lemma 9 in Shidlovskii (1989)). *Let the conditions of the previous lemma be satisfied and the linear form R_1, when the series f_i are substituted into it for the variables y_i, satisfy the condition*

$$\mathrm{ord}_{z=0} R_1(z) \ge m(N+1) - s(N) - 1, \quad 0 \le s(N) < \left[\frac{N}{2}\right]. \tag{3.18}$$

Then, if $N \ge n_0$, the linear forms $R_1, \ldots, R_m$ are linearly independent over $\mathbb{K}(z)$, and determinant (3.12) of these forms is equal to

$$\Delta(z) = z^{mN - s(N) - \rho} \Delta_1(z), \tag{3.19}$$

where

$$\rho = \min_{1 \le i \le m} \mathrm{ord}_{z=0} f_i(z),$$

$$\sigma = \max_{1 \le i \le m, 1 \le k \le m} (\deg T, \deg T Q_{k,i}), \tag{3.20}$$

$$t(N) = \sigma \frac{m(m-1)}{2} + s(N) + \rho, \tag{3.21}$$

$$\Delta_1(z) \in \mathbb{K}[z], \quad \deg \Delta_1(z) \le t(N). \tag{3.22}$$

Lemma 3.7 (Siegel's lemma, Lemma 10 in Shidlovskii (1989)). *Let the set of formal power series $f_1(z), \ldots, f_m(z)$ be a formal solution of system (3.1) and linearly independent over the field $\mathbb{K}(z)$. Let $N \ge n_0$, and let conditions (3.18) be satisfied. Let ξ be an integer from the field $\mathbb{K}, \xi T(\xi) \ne 0$. Then, the matrix*

$$\|P_{k,i}(\xi)\|_{1 \le i \le m, 1 \le k \le m+t(N)}$$

has a rank m.

The above-formulated lemmas of the first part of the Siegel–Shidlovskii method are common to both E-functions and F-Series. Their proofs (in the referenced book) apply to the case of formal power series as well, and all formulae (3.13)–(3.22) remain valid.

The second part of the modification of this method applies only to F-series.

The second part of the Siegel–Shidlovskii method considers approximating numerical forms in the values of the considered set of functions. It is in this part of the method, modified for F-series, that the product formula is essentially used.

3.2. Construction of the First Approximating Form

Lemma 3.8. *Let the series*

$$f_i(z) = \sum_{n=0}^{\infty} c_{i,n} \cdot n! \cdot z^n, \quad i = 1, \dots, m,$$

belong to the class $F(\mathbb{K}, C_1, C_2, C_3, q)$*. Let* $\xi = \frac{a}{b}$*,* $a \in \mathbb{Z}_\mathbb{K}$*,* $b \in \mathbb{N}$*, be the smallest value that satisfies this condition. Then, for any positive integer* N*, there exist polynomials*

$$P_i(z) = \sum_{n=0}^{N} B_{i,n} \cdot n! \cdot z^n, \quad i = 1, \dots, m, \tag{3.23}$$

such that their coefficients $B_{i,n}$*,* $i = 1, \dots, m$*,* $n = 0, 1, \dots N$ *are integers from field* $\mathbb{K}$*, not all equal to zero, that satisfy the inequality*

$$\overline{|B_{i,n}|} \leq e^{C_4 N \sqrt{\ln N}}, \tag{3.24}$$

and the linear form

$$R(z) = \sum_{i=1}^{m} P_i(z) f_i(z) = \sum_{n=0}^{\infty} r_n \cdot n! \cdot z^n \tag{3.25}$$

has at point $z = 0$ *the order of zero*

$$\mathrm{ord}_{z=0} R(z) \geq k(N) + 1,$$

where

$$k(N) = m(N+1) - \left[\frac{N}{\sqrt{\ln N}}\right]. \qquad (3.26)$$

Let, for any prime number p that is relatively prime with the number bq and satisfies the inequalities

$$p \le k(N), \quad \frac{1}{p-1} - \frac{C_3}{p^2} > 0, \quad p\left(\frac{\ln p - 1}{\ln p}\right) > 2C_3$$

and any valuation v of field $\mathbb{K}$ that extends the p-adic valuation, the following inequalities be true:

$$\langle R \rangle_v \le \left| \left(m(N+1) - \left[\frac{N}{\sqrt{\ln N}}\right] \right)! \right|_v e^{\frac{\varkappa_v}{\varkappa}\left(C_5 log_p n + C_3 \frac{n}{p^2}\right)}, \qquad (3.27)$$

$$|R(\xi)|_v \le \left| \left(m(N+1) - \left[\frac{N}{\sqrt{\ln N}}\right] \right)! \right|_v e^{\frac{\varkappa_v}{\varkappa}\left(C_5 log_p n + C_3 \frac{n}{p^2}\right)}. \qquad (3.28)$$

Proof of Lemma 3.8. Consider the equality

$$P_i(z)f_i(z) = \sum_{k=0}^{N} B_{i,k} \cdot k! \cdot z^k \sum_{l=0}^{\infty} c_{i,l} \cdot l! \cdot z^l$$

$$= \sum_{n=0}^{\infty} b_{i,n} \cdot n! \cdot z^n, \quad i = 1, \ldots, m,$$

where for any n,

$$b_{i,n} = \frac{1}{n!} \sum_{k+l=n} B_{i,k} \cdot k! \cdot c_{i,l} \cdot l!. \qquad (3.29)$$

The condition of the order value $\mathrm{ord}_{z=0}R(z)$ means that the equalities hold:

$$\sum_{i=1}^{m} b_{i,n} = 0, \quad n = 0, 1, \ldots, m(N+1) - \left[\frac{N}{\sqrt{\ln N}}\right]. \qquad (3.30)$$

Formulae (3.29) and (3.30) give a system of equations:

$$\sum_{i=1}^{m} \frac{1}{n!} \sum_{k+l=n} B_{i,k} \cdot k! \cdot c_{i,l} \cdot l! = 0,$$

$$n = 0, 1, \ldots, m(N+1) - \left[\frac{N}{\sqrt{\ln N}}\right]. \tag{3.31}$$

Here, the internal summation is performed on the non-negative integers k and l, $k+l=n$.

Now, we give the upper estimate for the value $d_{0,n}$ from the definition of the class $F(\mathbb{K}, C_1, C_2, C_3, q)$:

$$d_{0,n} = \prod_{p \leq C_2 n} p^{\vartheta_p(d_{0,n})} \leq \prod_{p \leq C_2 n} p^{C_3 \left(\log_p n + \frac{n}{p^2}\right)}. \tag{3.32}$$

Then, using the estimate $\pi(C_2 n) \leq \gamma \frac{C_2 n}{\ln C_2 n}$ for the number of prime numbers $p \leq C_2 n$, we obtain

$$\prod_{p \leq C_2 n} p^{C_3 \log_p n} \leq e^{C_3 \ln n \cdot \gamma \frac{C_2 n}{\ln C_2 n}} \leq e^{C_6 n}. \tag{3.33}$$

Since the series $\sum_p \frac{C_3}{p^2}$ converges (summation is performed on all prime numbers p),

$$\prod_{p \leq C_2 n} p^{C_3 \frac{n}{p^2}} \leq e^{C_7 n}. \tag{3.34}$$

Therefore, from (3.32)–(3.34), we deduce

$$d_{0,n} \leq e^{C_8 n}. \tag{3.35}$$

Multiply both parts of each equation (3.31) by the least common multiple $\gamma_{k(N)}$ of the numbers,

$$\frac{k(N)!}{l!(k(N)-l)!}, \quad l = 0, 1, \ldots k(N),$$

and by the number $q^{k(N)}d_{0,k(N)}$. As a result, we obtain a system of linear equations with respect to the quantities $B_{i,k}$:

$$\gamma_{k(N)}q^{k(N)}d_{0,k(N)}\sum_{i=1}^{m}\frac{1}{n!}\sum_{k+l=n}B_{i,k}\cdot k!\cdot c_{i,l}\cdot l! = 0,$$

$$n = 0,\ldots,m(N+1) - \left[\frac{N}{\sqrt{\ln N}}\right]. \tag{3.36}$$

The coefficients of this system are integers from the field $\mathbb{K}$. In view of the definition of class $F(\mathbb{K},C_1,C_2,C_3,q)$, inequality (3.21), and Lemma 3.3, we obtain that the Archimedean absolute values of these coefficients and their conjugate numbers do not exceed the value

$$e^{C_1 k(N)}q^{k(N)}e^{C_8 k(N)}e^{\tau k(N)} \le e^{C_9 k(N)}. \tag{3.37}$$

Next, we use *Siegel's well-known lemma.*

Lemma 3.9 (Siegel, 1929; Shidlovskii, 1989). *Let*

$$a_{i,j} \in \mathbb{Z}_{\mathbb{K}}, \quad \overline{|a_{i,j}|} \le A, \quad i = 1,\ldots,P, \quad j = 1,\ldots Q, \quad P < Q.$$

Then, there exists a non-trivial solution $(x_1,\ldots,x_Q)$ of the system

$$\sum_{j=1}^{Q}a_{i,j}x_j = 0, \quad i = 1,\ldots,P$$

such that $x_j \in \mathbb{Z}_{\mathbb{K}}$, $j = 1,\ldots,Q$. In addition, the following inequalities are fulfilled:

$$\overline{|x_j|} \le C(CQA)^{\frac{P}{Q-P}}, = 1,\ldots,Q.$$

The constant C in this inequality is

$$C = 2(\operatorname{Disc}\mathbb{K})^{\frac{1}{2\varkappa}},$$

where $\operatorname{Disc}\mathbb{K}$ denotes the discriminant of the field $\mathbb{K}$ and $\varkappa$ is the degree of this field over the field $\mathbb{Q}$ of rational numbers.

In the system of equations under consideration (3.36), the number of unknowns is equal to $Q = m(N+1)$, and the number of equations is

$$P = m(N + 1) - \left[\frac{N}{\sqrt{\ln N}} \right] + 1.$$

Thus,

$$Q - P = \left[\frac{N}{\sqrt{\ln N}} \right] - 1.$$

Therefore,

$$\frac{P}{Q - P} = \frac{m(N + 1) - \left[\frac{N}{\sqrt{\ln N}} \right] + 1}{\left[\frac{N}{\sqrt{\ln N}} \right] - 1} \leq \frac{m(N + 1) - \frac{N}{\sqrt{\ln N}} + 2}{\frac{N}{\sqrt{\ln N}} - 2}$$

$$\leq C_{10} \sqrt{\ln N}. \qquad (3.38)$$

Using Siegel's lemma and estimates (3.37) and (3.38), we obtain, for $i = 1, \ldots m$, $n = 0, 1, \ldots N$,

$$\overline{|B_{i,n}|} \leq C \left(Cm(N + 1)e^{C_9 \left(m(N+1) - \left[\frac{N}{\sqrt{\ln N}} \right] \right)} \right)^{C_{10} \sqrt{\ln N}} \leq e^{C_4 N \sqrt{\ln N}}.$$

So, (3.24) is proved. □

Consider the value of

$$r_n = \sum_{i=1}^{m} \frac{1}{n!} \sum_{k+l=n} B_{i,k} \cdot k! \cdot c_{i,l} \cdot l!$$

$$= \sum_{i=1}^{m} \sum_{k+l=n} B_{i,k} \cdot c_{i,l} \frac{k! \cdot l!}{n!}.$$

Here, the internal summation is performed on the non-negative integers k and l, $k + l = n$.

By the definition of the class $F(\mathbb{K}, C_1, C_2, C_3, q)$, if p is a prime number relatively prime with number q and satisfies the inequalities

$C_3 \leq p \leq k(N)$ and if v is a valuation of the field $\mathbb{K}$, extending the p-adic valuation, then

$$|r_n|_v = \max_{k+l=n} \left\{ |B_{i,k}|_v, |c_{i,l}|_v, \left| \frac{k! \cdot l!}{n!} \right|_v \right\},$$

where the maximum is taken over all $i = 1, \ldots, m$ and all the non-negative integers k and l, $k + l = n$. By the definition of the class $F(\mathbb{K}, C_1, C_2, C_3, q)$, for any valuation $v \in V_0$ extending the p-adic valuation such that $p > C_2 n$, the inequality $|c_{i,l}|_v \leq 1$ is satisfied, and if $C_3 \leq p \leq C_2 n$, $(p, q) = 1$, then

$$|c_{i,l}|_v \leq e^{\frac{\varkappa_v}{\varkappa} C_3 \left(\log_p n + \frac{n}{p^2} \right)}.$$

For any valuation $v \in V_0$ extending the p-adic valuation such that $p > n$, we have $\left| \frac{k! \cdot l!}{n!} \right|_v = 1$, and if $p \leq n$, then

$$\left| \frac{k! \cdot l!}{n!} \right|_v \leq e^{\frac{\varkappa_v}{\varkappa} \frac{(S_k + S_l - S_n)}{(p-1)}} \leq e^{C_{11} \log_p n}.$$

Thus, for any valuation $v \in V_0$ extending the p-adic valuation such that $C_3 \leq p \leq C_2 n$, $(p, q) = 1$, the following inequality holds:

$$|r_n|_v \leq e^{C_{10} \log_p n + C_{11} \frac{n}{p^2}}. \tag{3.39}$$

If $p > C_2 n$, then $|r_n|_v \leq 1$.

Let us evaluate from above $\langle R \rangle_v$ if p is a prime number that is relatively prime with the number q and satisfies the inequalities $C_3 \leq p \leq k(N)$, and if v is a valuation of the field $\mathbb{K}$ extending the p-adic valuation:

$$\langle R \rangle_v \leq \max_{n \geq k(N)+1} \{ |r_n|_v |n!|_v \}.$$

Since for $|r_n|_v$ estimate (3.39) is true,

$$\langle R \rangle_v \leq \max_{n \geq k(N)+1} \left\{ e^{\frac{\varkappa_v}{\varkappa} \left(C_{11} \log_p n + C_{12} \frac{n}{p^2} \right)} \cdot e^{-\frac{\varkappa_v}{\varkappa} \frac{(n - S_n)}{(p-1)}} \right\}$$

$$\leq \max_{n \geq k(N)+1} \left\{ e^{\frac{\varkappa_v}{\varkappa} \left(C_{13} \log_p n + C_{12} \frac{n}{p^2} - \frac{n}{p-1} \right)} \right\}. \tag{3.40}$$

The function $C_{13} \log_p x + C_{12} \frac{x}{p^2} - \frac{x}{p-1}$ decreases for

$$x > \frac{C_{13}}{\ln p} \left(\frac{p^2(p-1)}{p^2 - C_{12}(p-1)} \right).$$

Thus, if

$$k(N) > \frac{C_{13}}{\ln p} \left(\frac{p^2(p-1)}{p^2 - C_{12}(p-1)} \right), \tag{3.41}$$

then the maximum on the right-hand side of inequality (3.40) is reached at $n = k(N)$. The right-hand side of inequality (3.41) is an increasing function of p under the condition

$$p \frac{\ln p - 1}{\ln p} > 2C_{12}. \tag{3.42}$$

Indeed, the derivative of this function on the variable p is

$$C_{13} \frac{\left((3p^2 - 2p) \ln p(p^2 - C_{12}(p-1)) - (p^3 - p^2) \left(\frac{(p^2 - C_{12}(p-1))}{p} - \ln p(2p - C_{12}) \right) \right)}{(\ln p(p^2 - C_{12}(p-1)))^2}.$$

After simplification, the function in parentheses in the numerator of the fraction takes the form

$$p^4(\ln p - 1) - 2C_{12}p^3 \ln p + (C_{12} + 1)p^3 + 4C_{12}p^2 \ln p - 2C_{12}p^2$$
$$- 2C_{12}p \ln p + C_{12}p,$$

and this value is positive under condition (3.42). Since in evaluating $\langle R \rangle_v$ we consider only prime numbers $p \le k(N)$, inequality (3.39) is satisfied for all such prime numbers satisfying inequality (3.40) if $k(N)$ satisfies the inequality

$$k(N) > \frac{C_{12}}{\ln k(N)} \left(\frac{(k(N))^2(k(N) - 1)}{(k(N))^2 - C_{11}(k(N) - 1)} \right). \tag{3.43}$$

Inequality (3.43) holds for $N > C_{14}$. So, it is proved that for $N > C_{14}$,

$$\langle R \rangle_v \le e^{\frac{\varkappa_v}{\varkappa} \left(C_{13} \log_p k(N) + C_{12} \frac{k(N)}{p^2} - \frac{k(N)}{p-1} \right)},$$

which, given (3.26), gives the desired estimate (3.27). Estimate (3.28) follows from (3.27) and the condition that the numbers p and bq are relatively prime. Lemma 3.8 is proved. □

3.3.   Estimates for Approximating Forms

Lemma 3.10. *Let* $N \geq N_0$, *and let the* F-*series* $f_1(z) \equiv 1$, $f_2(z), \ldots, f_m(z)$, *belong to the class* $F(\mathbb{K}, C_1, C_2, C_3, q)$, *constitute a solution of system* (3.5), *and be linearly independent over the field* $\mathbb{K}(z)$. *Let*

$$R_1(z) = R(z) = \sum_{i=1}^{m} P_i(z) f_i(z)$$

be of form (3.25) *constructed in Lemma 3.8, and the forms*

$$R_k(z), \quad k = 2, 3, \ldots, \quad R_k(z) = \sum_{i=1}^{m} P_{k,i}(z) f_i(z)$$

be defined by equations (3.9). *Let* ξ *belong to the field* $\mathbb{K}$ *and be a regular point of system* (3.5):

$$\xi = \frac{a}{b}, \quad a \in \mathbb{Z}_{\mathbb{K}}, \quad b \in \mathbb{N}, \quad \xi T(\xi) \neq 0.$$

Then,

$$\overline{|P_{k,i}(\xi)|} \leq \prod_{n=0}^{k-1} (n\sigma + m + N) e^{N \ln N + C_4 N \sqrt{\ln N} + C_{15} N + C_{16} k},$$

$$i = 1, \ldots, m, \quad k = 1, 2, \ldots, \tag{3.44}$$

where σ *is defined in* (3.20). *For the maximum of the absolute values of the coefficients of a polynomial* $P(z)$ *and their algebraic conjugates, the right-hand side of inequality* (3.44) *is also an upper estimate. Moreover, if* p *is a prime number relatively prime with* q *and satisfies the inequalities*

$$C_3 \leq p \leq k(N),$$

if

$$p \frac{\ln p - 1}{\ln p} > 2 C_{12}$$

holds, and if v — is a valuation of the field $\mathbb{K}$, *extending the p-adic valuation, then*

$$\langle R_k \rangle_v \leq \left| \left(m(N+1) - \left[\frac{N}{\sqrt{\ln N}} \right] \right)! \right|_v e^{\frac{x_v}{x} \left(C_5 log_p n + C_3 \frac{n}{p^2} \right)}, \quad (3.45)$$

and if the numbers p and bq are relatively prime, then

$$|R_k(\xi)|_v \leq \left| \left(m(N+1) - \left[\frac{N}{\sqrt{\ln N}} \right] \right)! \right|_v e^{\frac{x_v}{x} \left(C_5 log_p n + C_3 \frac{n}{p^2} \right)}. \quad (3.46)$$

Proof of Lemma 3.10. For the power series

$$\varphi(z) = \sum_{n=0}^{\infty} a_n z^n, \quad \vartheta(z) = \sum_{n=0}^{\infty} b_n z^n,$$

the symbol $\varphi(z) \ll \vartheta(z)$ denotes that for all n, the inequality $|a_n| \leq b_n$ is satisfied. Recall that

$$\sigma = \max_{1 \leq i \leq m, 1 \leq k \leq m} (\deg T, \deg T Q_{k,i}).$$

Moreover, let $\check{C}$ denote the maximum absolute values of the coefficients of the polynomials $T(z)$, $T(z)Q_{k,i}(z)$ and their algebraic conjugates in $\mathbb{K}$. Then,

$$T(z) \ll \check{C}(1+z)^{\sigma}, \quad T(z)Q_{k,i}(z) \ll \check{C}(1+z)^{\sigma}. \quad (3.47)$$

By induction, we prove the following statement:

$$P_{k+1,i}(z) \ll \check{C}^k (1+z)^{k\sigma+N} \prod_{n=0}^{k-1} (n\sigma + m + N) e^{N \ln N + C_4 N \sqrt{\ln N}},$$

$$i = 1, \ldots m, \quad k = 0, 1, \ldots. \quad (3.48)$$

If the upper limit of the product is less than or equal to zero, then the whole product equals 1. It follows from inequality (3.9) that

$$P_{1,i}(z) \ll (1+z)^N e^{N \ln N + C_4 N \sqrt{\ln N}}, \quad i = 1, \ldots, m. \quad (3.49)$$

So, the proved statement is true for $k = 0$.

According to (3.11),

$$P_{k+1,i}(z) = T\left(P'_{k,i} + \sum_{j=1}^{m} P_{k,j}Q_{j,i}\right), \quad k = 1, 2, 3, \ldots, \quad i = 1, \ldots, m.$$

$$(3.50)$$

From (3.23), (3.47), and (3.50) follows the inequality

$$\deg P_{k+1,i}(z) \le \sigma + \max_{i=1,\ldots,m} \deg P_{k,i}(z) \le k\sigma + N,$$

$$k = 1, 2, 3, \ldots, \quad i = 1, \ldots, m. \tag{3.51}$$

Therefore, from (3.47) and (3.51) and the inductive assumption

$$P_{k,i}(z) \ll \check{C}^{k-1}(1+z)^{(k-1)\sigma+N} \prod_{n=0}^{k-2}(n\sigma + m + N)e^{N\ln N + C_4 N\sqrt{\ln N}},$$

$$i = 1, \ldots, m, \quad k = 0, 1, \ldots, \tag{3.52}$$

it follows that

$$P'_{k,i}(z) \ll ((k-1)\sigma + N)\check{C}^{k-1}(1+z)^{(k-1)\sigma+N}$$

$$\times \prod_{n=0}^{k-2}(n\sigma + m + N)e^{N\ln N + C_4 N\sqrt{\ln N}},$$

$$i = 1, \ldots, m, \quad k = 0, 1, \ldots. \tag{3.53}$$

In addition, in view of (3.47) and (3.52),

$$T(z)\sum_{j=1}^{m} P_{k,j}(z)Q_{j,i}(z) \ll \check{C}(1+z)^{\sigma}\check{C}^{k-1}(1+z)^{(k-1)\sigma+N}$$

$$\cdot \prod_{n=0}^{k-2}(n\sigma + m + N)e^{N\ln N + C_4 N\sqrt{\ln N}},$$

$$k = 0, 1, \ldots. \tag{3.54}$$

From (3.49), (3.50), (3.53), and (3.54), the proved relation (3.48) follows immediately, and inequalities (3.44) and the corresponding upper-bound estimate for the maximum absolute values of the coefficients of a polynomial $P(z)$ and their algebraic conjugates follow in turn.

Let us proceed to the proof of formulae (3.45) and (3.46).

Prove that $\langle R_k \rangle_v \leq \langle R_1 \rangle_v$. Indeed, by construction,

$$R_k = TR'_{k-1}, \quad k = 2, 3, \ldots .$$

Use Lemma 2.1 and obtain

$$\langle R_k \rangle_v \leq \langle T \rangle_v \langle R'_{k-1} \rangle_v \leq \langle R'_{k-1} \rangle_v \leq \langle R_{k-1} \rangle_v, \quad k = 2, 3, \ldots . \quad (3.55)$$

From (3.55), we immediately obtain

$$\langle R_k \rangle_v \leq \langle R_1 \rangle_v, \quad k = 2, 3, \ldots .$$

According to (3.27), it follows from (3.55) that

$$\langle R_k \rangle_v \leq \left| \left(m(N+1) - \left[\frac{N}{\sqrt{\ln N}} \right] \right)! \right|_v e^{\frac{\varkappa_v}{\varkappa} \left(C_5 log_p n + C_3 \frac{n}{p^2} \right)}.$$

By Lemma 3.8, in view of (3.28),

$$|R_k(\xi)|_v \leq \left| \left(m(N+1) - \left[\frac{N}{\sqrt{\ln N}} \right] \right)! \right|_v e^{\frac{\varkappa_v}{\varkappa} \left(C_5 log_p n + C_3 \frac{n}{p^2} \right)}. \quad \square$$

Lemma 3.11. *Let $N \geq N_0$, and let F-series $f_1(z) \equiv 1$, $f_2(z), \ldots,$ $f_m(z)$ belong to the class $F(\mathbb{K}, C_1, C_2, C_3, q)$, constitute a solution of system (3.5), and be linearly independent over the field $\mathbb{K}(z)$.*

Let $\xi \in K$ be a regular point of system (3.5),

$$\xi = \frac{a}{b}, \quad a \in \mathbb{Z}_\mathbb{K}, \quad b \in \mathbb{N}, \quad \xi T(\xi) \neq 0.$$

Then, there exists a set of m linearly independent forms

$$L_k(y_1, \ldots, y_m) = \sum_{i=1}^m h_{k,i} y_i, \quad k = 1, \ldots, m, \quad (3.56)$$

where $h_{k,i} \in \mathbb{Z}_\mathbb{K}$, $k, i = 1, \ldots, m$, and

$$\overline{|h_{k,i}|} \leq e^{N \ln N + C_{17} N \sqrt{\ln N}}. \quad (3.57)$$

Moreover, if p is a prime number that is relatively prime with the number q and satisfies the inequalities

$$C_3 \leq p \leq k(N)$$

and

$$p\frac{\ln p - 1}{\ln p} > 2C_{12},$$

if v is a valuation of the field $\mathbb{K}$ extending the p-adic valuation, and if the numbers p and bq are relatively prime, then

$$|L_k(f_1(\xi), \ldots, f_1(\xi))|_v$$

$$\leq \left| \left(m(N+1) - \left[\frac{N}{\sqrt{\ln N}}\right] \right)! \right|_v e^{\frac{\varkappa_v}{\varkappa} \left(C_5 \log_p n + C_{18}\frac{n}{p^2} \right)}. \qquad (3.58)$$

Proof of Lemma 3.11. Let N be any number satisfying the conditions of the lemma.

Let $R_1(z)$ be of form (3.25) constructed in Lemma 3.8, and let the forms $R_k(z)$, $k = 2, 3, \ldots$ be defined by equations (3.9). By Siegel's lemma (Lemma 3.6), the matrix

$$\|P_{k,i}(\xi)\|_{1 \leq i \leq m, 1 \leq k \leq m + t(N)}$$

has rank m over the field $\mathbb{K}$. This means that among its rows with numbers $1, 2, \ldots, m + t(N)$, there are m linearly independent forms over the field $\mathbb{K}$. Let these lines have numbers $s_1 < \cdots < s_m$. Consider the linear forms

$$\sum_{i=1}^{m} P_{k,i}(\xi)y_i, \quad k = s_1, \ldots, s_m.$$

From the inequality

$$\deg P_{l,i}(z) \leq (l-1)\sigma + N, \quad l = 1, 2, 3, \ldots, \quad i = 1, \ldots, m, \qquad (3.59)$$

it follows from (3.59) that the degrees of polynomials $P_{l,i}(z)$, $l = s_1, \ldots, s_m$ do not exceed the number $(m + t(N) - 1)\sigma + N$. Therefore, the numbers

$$b^{(m+t(N)-1)\sigma+N} P_{l,i}(\xi) \in \mathbb{Z}_{\mathbb{K}}.$$

Let us denote

$$L_k(y_1, \ldots, y_m) = \sum_{i=1}^{m} b^{(m+t(N)-1)\sigma+N} P_{s_k,i}(\xi) y_i = \sum_{i=1}^{m} h_{k,i} y_i,$$

$$k = 1, \ldots, m.$$

For the coefficients $h_{k,i}$ of these forms, in view of (3.44), the following inequality is true:

$$\overline{|P_{l,i}(\xi)|} \le \prod_{n=0}^{l-1} (n\sigma + m + N) e^{N \ln N + C_4 N \sqrt{\ln N} + C_{15} N + C_{16} l},$$

$$i = 1, \ldots, m, \quad l = 1, 2, \ldots.$$

According to formula (3.21), given that $\rho = 0$, since $f_0(z) = 1$, we obtain

$$t(N) = \sigma \frac{m(m-1)}{2} + s(N).$$

In order to fulfil the condition of Lemma 3.5,

$$\mathrm{ord}_{z=0} R_1(z) \ge m(N+1) - s(N) - 1,$$

let's require that

$$k(N) + 1 = m(N+1) - \left[\frac{N}{\sqrt{\ln N}} \right] + 1 \ge m(N+1) - s(N) - 1$$

or

$$s(N) + 2 \ge \left[\frac{N}{\sqrt{\ln N}} \right].$$

Put $s(N) = \left[\frac{N}{\sqrt{\ln N}} \right] - 2.$

Then, we get

$$t(N) = \sigma \frac{m(m-1)}{2} + s(N) = \left[\frac{N}{\sqrt{\ln N}}\right] - 2 + \sigma \frac{m(m-1)}{2}$$

$$\leq \frac{N}{\sqrt{\ln N}} + C_{17}.$$

Thus,

$$\overline{|h_{k,i}|} = \overline{|b^{(m+t-1)\sigma+N} P_{k,i}(\mathfrak{z})|}$$

$$\leq b^{(m+t(N)-1)\sigma+N} \prod_{n=0}^{s_k-1} (n\sigma + m + N)$$

$$\times e^{N\ln N + C_4 N\sqrt{\ln N} + C_{15}N + C_{16}s_k},$$

$$i = 1, \ldots, m, \quad k = 1, \ldots, m.$$

Note that $l \leq s_m \leq m + t(N)$. We have the estimate

$$\prod_{n=0}^{s_m-1} (n\sigma + m + N) \leq ((m + t(N) - 1)\sigma + m + N)^{m+t(N)}$$

$$\leq \left(N + C_{18}\frac{N}{\sqrt{\ln N}}\right)^{\frac{N}{\sqrt{\ln N}}+C_{17}} \leq e^{C_{19}N\sqrt{\ln N}}.$$

Therefore,

$$\overline{|h_{k,i}|} \leq e^{N\ln N + C_{20}N\sqrt{\ln N}},$$

and inequality (3.57) is proved.
Further, for all $k = 1, \ldots, m$,

$$L_k(f_1(\xi), \ldots, f_m(\xi)) = \sum_{i=1}^{m} b^{(m+t(N)-1)\sigma+N} P_{s_k,i}(\xi) y_i$$

$$= b^{(m+t(N)-1)\sigma+N} R_{s_k}(\xi).$$

Since $b^{(m+t(N)-1)\sigma+N}$ is a positive integer, in view of (3.46), we get

$$|L_k(f_1(\xi), \ldots, f_m(\xi))|_v$$
$$\leq \left|\left(m(N+1) - \left[\frac{N}{\sqrt{\ln N}}\right]\right)!\right|_v e^{\frac{\varkappa_v}{\varkappa}\left(C_5 \log_p n + C_{18}\frac{n}{p^2}\right)}.$$

So, (3.58) holds, and Lemma 3.11 is fully proved.   □

3.4.   Infinite Linear and Algebraic Independence of *F*-Series Values

Further calculations are cumbersome, but they are necessary for the exact formulation of the results. Let $\xi \in \mathbb{K}$ be a regular point of system (3.5),

$$\xi = \frac{a}{b}, \quad a \in \mathbb{Z}_\mathbb{K}, \quad b \in \mathbb{N}, \quad \xi T(\xi) \neq 0,$$
$$h(\xi) = \ln(1 + size(\xi)) + \ln b. \tag{3.60}$$

(In other words, $h(\xi) = \ln H(\xi)$.) In addition, let us put

$$C_4 = C_1 + \ln q + \frac{5}{4}C_2 C_3 + 2C_3 + 5, \quad C_5 = m^2 C_4 + 2. \tag{3.61}$$

We denote

$$N_0 = \max\left(n_0, \left(\ln 2 + \frac{\mathrm{Disc}(\mathbb{K})}{\varkappa}\right)^2, \exp(4(2(m+3) + C_4(m-1))^2),\right.$$

$$\left.\exp(h(\mathfrak{z})^2 + \deg(D)(h(\mathfrak{z}) + 2) + 2\ln H(D))\right), \tag{3.62}$$

$$H_0 = \exp\left(N_0 \ln N_0 \left(1 - \frac{m+3+(m-1)C_5}{(\ln N_0)^{\frac{1}{2}}}\right)\right). \tag{3.63}$$

These formulae show that the value H_0 is explicitly expressed through n_0, the parameters of the series in question and the system of differential equations (3.5) they satisfy, the logarithmic height of the algebraic number ξ, the degree, and the discriminant of the field $\mathbb{K}$.

Let us also denote, for $x > 3$,

$$l(x) = \exp((\ln x)^{\frac{1}{2}}), \quad u(x) = m(x+1) - [x(\ln x)^{-\frac{1}{2}}] \qquad (3.64)$$

(the symbol $[a]$ denotes the integer part of the number a) and

$$P_l(x) = l\left(\frac{x}{\ln x}\right), \quad P_u(x) = u\left(\frac{x}{\ln x}\left(1 + 2\frac{m+3+(m-1)C_5}{(\ln x/\ln\ln x)^{1/2}}\right)\right).$$
$$(3.65)$$

Theorem 3.1 (Theorem 1.1 in Bertrand *et al.* (2004)). *Let the F-series $f_1(z) \equiv 1, f_2(z), \ldots, f_m(z)$ form a solution of system (3.5) and be linearly independent over the field $\mathbb{K}(z)$. Let $\xi \in \mathbb{K}$ be a regular point of system (3.5). Let*

$$L(y_1, \ldots, y_m) = h_1 y_1 + \cdots + h_m y_m \qquad (3.66)$$

be a non-zero linear form, $h_i \in \mathbb{Z}_\mathbb{K}$, $i = 1, \ldots, m$. Let

$$H(L) = \max H(h_i).$$

Then, for any $H \geq \max(H_0, H(L))$, there exists a prime number p satisfying the inequalities

$$P_l(\ln H) \leq p \leq P_u(\ln H) \qquad (3.67)$$

and a valuation v of the field $\mathbb{K}$, extending the p-adic valuation of the field $\mathbb{Q}$ such that

$$L(\xi) = L(f_1(\xi), \ldots, f_m(\xi)) \neq 0. \qquad (3.68)$$

Moreover, in the field $\mathbb{K}_v$, the following inequality is true:

$$|L(\xi)|_v \geq H^{-m - \frac{m+3+2mC_{20}}{\sqrt{\ln\ln H}}}. \qquad (3.69)$$

Proof of Theorem 3.1. Since the forms

$$L_k(y_1, \ldots, y_m) = \sum_{i=1}^{m} h_{k,i} y_i, \quad k = 1, \ldots, m, \qquad (3.70)$$

are linearly independent, from them we can choose m linearly independent forms

$$L(y_1, \ldots, y_m) = h_1 y_1 + \cdots + h_m y_m.$$

By renumbering forms (3.70) if necessary, we assume that the forms $L(y_1, \ldots, y_m), L_2(y_1, \ldots, y_m), \ldots, L_m(y_1, \ldots, y_m)$ are linearly independent. Then, the determinant Δ of these forms is a non-zero element of the field $\mathbb{K}$.

Multiplying the ith column of the determinant by $f_i(\xi)$, $i = 2, \ldots, m$, and adding these columns to the first one, we obtain that the initial determinant is equal to the determinant whose first column consists of the linear forms

$$L(\xi) = L(f_1(\xi), \ldots, f_m(\xi)),$$
$$L_2(\xi) = L_2(f_1(\xi), \ldots, f_m(\xi)),$$
$$\vdots$$
$$L_m(\xi) = L_m(f_1(\xi), \ldots, f_m(\xi)).$$

In each field $\mathbb{K}_v$, where the valuation v satisfies the conditions of the theorem, we have the equality

$$\Delta = L(\xi)\Delta_1 + \sum_{i=2}^{m} L_i(\xi)\Delta_i, \qquad (3.71)$$

where Δ_i denotes the algebraic complement of the ith element of the first column of the determinant. Equality (3.71) will enable us to obtain the proof of the theorem using the product formula.

The set of non-Archimedean valuations V_0 of the field $\mathbb{K}$ for each positive integer N will be divided into three parts:

$$V_0 = V_1 \cup V_2 \cup V_3.$$

The set V_1 consists of the valuations v of the field $\mathbb{K}$ extending the p-adic valuations of the field $\mathbb{Q}$ for prime numbers p satisfying the inequalities $l(N) < p \le u(N)$. The set V_2 consists of the valuations v of the field $\mathbb{K}$ extending the p-adic valuations of the field $\mathbb{Q}$ for prime numbers p satisfying the inequalities $p \le l(N)$. The set V_3 consists of the valuations v of the field $\mathbb{K}$ extending the p-adic valuations of the field $\mathbb{Q}$ for prime numbers p satisfying the inequalities $u(N) \leq p$. The symbols $\prod_i(\ )$ denote the products over all $\in V_i$, the symbols $\prod_{i,j}(\ )$ denote the products of all $v \in V_i \cup V_j$, and so on.

Then, we assume that a positive integer N satisfies a set of conditions that will have the form $N \geq N_i$, $i = 1, 2, \ldots$. These numbers N_i, $i = 1, 2, \ldots$ represent the boundaries for which certain inequalities are satisfied. Of course, for these numbers, it is possible to give some explicit (very cumbersome) estimates from above, expressed through the constants

$$C_i, i = 1, 2, \ldots, q, m, \deg(D), H(D), H(\xi)$$

and through the degree and discriminant of the field $\mathbb{K}$; however, for further presentation, it is essential only to be able to give such estimates. For example, if $N \geq N_1$, let the following inequality be true:

$$N \geq 2m \ln N + (m+2)(\ln N)^{\frac{1}{2}}.$$

The existence of N_1 and the possibility of finding for this number an estimate from above are obvious. Let the following inequalities also hold (where C is a constant from Siegel's Lemma 3.9):

$$N \geq e^{(4(2(m+3)+C_4(m-1)))^2}, \quad N > (\ln C)^2.$$

For the given ξ, let for $N \geq N_2$ the inequality

$$\ln N > h(\xi)(\ln N)^{\frac{1}{2}} + \deg_1(A)(h(\xi)+2) + \ln H_1(A)$$

hold. Denote

$$\varepsilon(x) = (\ln x)^{-\frac{1}{2}}\varepsilon_1(x) = (m+3+(m-1)C_5)(\ln x)^{-\frac{1}{2}}. \qquad (3.72)$$

Let, for $N \geq N_3$,

$$(\ln N)^{\frac{1}{2}} > \ln bq.$$

Consider that N_0 is the smallest positive integer N for which all the above inequalities are satisfied, and put

$$H_0 = e^{N_0 \ln N_0(1-\varepsilon_1(N_0))}.$$

For any $H \geq \max\{H(L), H_0\}$, we set the positive integer N to satisfy the inequalities

$$(N-1)\ln(N-1)(1-\varepsilon_1(N-1)) \leq \ln H < N \ln N(1-\varepsilon_1(N)). \tag{3.73}$$

By the definition of the numbers N_0, H_0, the number N satisfies the inequality $N \geq N_0 \geq n_0$.

This allows us to use Lemma 3.6.

Since Δ is a non-zero integer from the field $\mathbb{K}$, the inequality

$$\prod_{2,3} |\Delta|_v \leq 1$$

holds, and the product formula

$$\prod_{1,2,3,\infty} |\Delta|_v = 1$$

yields the inequality

$$\prod_1 |\Delta|_v \geq |\Delta|^{-1} \geq \left(m! H e^{\left((m-1)\left(N \ln N + C_{20} N (\ln N)^{\frac{1}{2}}\right)\right)} \right)^{-1}. \tag{3.74}$$

From Lemma 3.11 (3.58), we derive the estimate

$$\prod_1 \left| \sum_{i=2}^{m} L_i(\xi) \Delta_i \right|_v \leq \prod_1 \max_{i=2,3,\ldots,m} |L_i(\xi)|_v \leq \prod_1 |u(N)!|_v$$

$$\cdot \prod_{1,2} p^{\left(\frac{\varkappa_v}{\varkappa}\left((C_3+2)\log_p u(N) + C_3 u(N) p^{-2}\right)\right)}. \tag{3.75}$$

Let us prove the inequality

$$\prod_{1,2} p^{\left(\frac{\varkappa_v}{\varkappa}\left((C_3+2)\log_p u(N) + C_3 u(N) p^{-2}\right)\right)}$$

$$= \prod_{p \leq u(N)} p^{(C_3+2)\log_p u(N) + C_3 u(N) p^{-2}}$$

$$= e^{\left(\sum_{p \leq u(N)} \ln p (C_3+2)\log_p u(N) + C_3 u(N) p^{-2}\right)}$$

$$\leq e^{\left(\sum_{p \leq u(N)} (\ln u(N)(C_3+2) + C_3 u(N) p^{-2} \ln p)\right)}$$

$$\leq e^{(C_{21}) \ln u(N) \cdot \frac{u(N)}{\ln u(N)} + C_3 u(N) \sum_{p \leq u(N)} \frac{\ln p}{p^2}} \leq e^{C_{22} N}$$

since, according to (3.64),

$$u(x) = m(x+1) - [x(\ln x)^{-\frac{1}{2}}],$$

and because the series $\sum_{p=1}^{\infty} \frac{\ln p}{p^2}$ converges.

For numbers N satisfying the above conditions, we get

$$\prod_2 |u(N)!|_v = \prod_{p \leq l(N)} |u(N)!|_p = e^{-\sum_{p \leq l(N)} \ln p \cdot \frac{u(N) - S_p(u(N))}{p-1}}$$

$$\geq e^{-\sum_{p \leq l(N)} \frac{u(N) \ln p}{p-1}}.$$

Consider

$$\sum_{p \leq l(N)} \frac{u(N) \ln p}{p-1}.$$

The sum

$$\sum_{p \leq l(N)} \frac{\ln p}{p-1}$$

differs from the sum

$$\sum_{p \leq l(N)} \frac{\ln p}{p}$$

by the partial sum of the convergent series $\sum_p \frac{\ln p}{p(p-1)}$. Therefore,

$$\sum_{p \leq l(N)} \frac{u(N) \ln p}{p-1} = u(N) \left(\sum_{p \leq l(N)} \frac{\ln p}{p} + C_{23} \right).$$

According to a known estimate

$$\sum_{p \leq x} \frac{\ln p}{p} = \ln x + O(1), \quad x \to \infty,$$

we get

$$\sum_{p \leq l(N)} \frac{\ln p}{p} + C_{23} = \ln l(N) + C_{24}.$$

Since, according to (3.64),

$$l(x) = \exp\left((\ln x)^{\frac{1}{2}}\right), \quad u(x) = m(x+1) - \left[x(\ln x)^{-\frac{1}{2}}\right],$$

we obtain

$$u(N)\left(\sum_{p \leq l(N)} \frac{\ln p}{p} + C_{23}\right) \leq C_{25} N (\ln N)^{\frac{1}{2}}.$$

Thus,

$$\prod_2 |u(N)!|_v \geq e^{-C_{25} N (\ln N)^{\frac{1}{2}}}.$$

Since

$$\prod_{1,2} |u(N)!|_v = (u(N)!)^{-1},$$

we get

$$\prod_1 |u(N)!|_v \leq e^{-u(N)\ln u(N) + C_{25} N (\ln N)^{\frac{1}{2}}} \leq e^{-mN \ln N + C_{26} N (\ln N)^{\frac{1}{2}}}.$$

$$(3.76)$$

Thus, (3.74)–(3.76) give

$$\prod_1 \left|\sum_{i=2}^m L_i(\xi)\Delta_i\right|_v \leq e^{-mN \ln N + (m+2)N \ln N \varepsilon(N)}. \qquad (3.77)$$

If $|L(\xi)|_v = 0$ for all $v \in V_1$, then it follows from (3.71), (3.73), and (3.77) that

$$-\ln m! - \ln H - (m-1)N \ln N - (m-1)C_{20} N (\ln N)^{\frac{1}{2}}$$
$$\leq -mN \ln N + (m+2)N \ln N \varepsilon(N).$$

This means that

$$N \ln N (1 - \varepsilon_1(N)) - \ln H \leq 0.$$

This inequality contradicts the choice (3.72) of the number N. Thus, it is proved that $|L(\xi)|_v \neq 0$ for at least one $v \in V_1$. Let us proceed to obtain estimate (3.69). We have proved that

$$|L(\xi)|_v \neq 0.$$

Equality (3.71) and inequality (3.77) then give

$$|L(\xi)|_v |\Delta_1|_v = |\Delta|_v \geq \prod_1 |\Delta|_v \geq |\Delta|^{-1}$$

$$\geq \left(m! H e^{\left((m-1)\left(N \ln N + C_{20} N (\ln N)^{\frac{1}{2}}\right)\right)} \right)^{-1}. \tag{3.78}$$

It follows from (3.72) that the inequality

$$(N-1)\ln(N-1)\left(1 - \frac{C_{27}}{\sqrt{\ln N - 1}}\right) \leq \ln H < N \ln N \left(1 - \frac{C_{27}}{\sqrt{\ln N}}\right) \tag{3.79}$$

holds. It follows from inequality (3.79) that

$$\left(N \ln N - \ln N + N \ln\left(1 - \frac{1}{N}\right) - \ln\left(1 - \frac{1}{N}\right)\right)\left(1 - \frac{C_{27}}{\sqrt{\ln N - 1}}\right)$$

$$\leq \ln H < N \ln N \left(1 - \frac{C_{27}}{\sqrt{\ln N}}\right). \tag{3.80}$$

The left-hand side of inequality (3.80) at $N \geq N_4$ can be estimated from below:

$$N \ln N - \frac{C_{27} N \ln N}{\sqrt{\ln N - 1}} - \ln N + \frac{C_{27} \ln N}{\sqrt{\ln N - 1}} + N \ln\left(1 - \frac{1}{N}\right)$$

$$- N \ln\left(1 - \frac{1}{N}\right)\frac{C_{27}}{\sqrt{\ln N - 1}} - \ln\left(1 - \frac{1}{N}\right)$$

$$+ \ln\left(1 - \frac{1}{N}\right)\frac{C_{27}}{\sqrt{\ln N - 1}} > N \ln N - \frac{C_{28} N \ln N}{\sqrt{\ln N - 1}},$$

replacing C_{27} with a larger constant C_{28}. Thus,

$$N \ln N \left(1 - \frac{C_{28}}{\sqrt{\ln N}}\right) < \ln H < N \ln N \left(1 - \frac{C_{27}}{\sqrt{\ln N}}\right). \tag{3.81}$$

Inequalities (3.81) are equivalent to the inequalities

$$\frac{\ln H}{\left(1 - \frac{C_{27}}{\sqrt{\ln N}}\right)} < N \ln N < \frac{\ln H}{\left(1 - \frac{C_{28}}{\sqrt{\ln N}}\right)}, \qquad (3.82)$$

from which we get

$$\ln \ln H - \ln \ln N - \ln \left(1 - \frac{C_{28}}{\sqrt{\ln N}}\right)$$

$$< \ln N < \ln \ln H - \ln \ln N - \ln \left(1 - \frac{C_{27}}{\sqrt{\ln N}}\right). \qquad (3.83)$$

It follows from these inequalities that for any $\delta_1, \delta_2, 0 < \delta_1 < \delta_2 < 1$ at $N \geq N_5$, the inequalities

$$\ln \ln H - (1 - \delta_1) \ln \ln N < \ln N < \ln \ln H - (1 - \delta_2) \ln \ln N \quad (3.84)$$

hold. Let us rewrite these inequalities as

$$\ln N \left(1 + \frac{(1 - \delta_2) \ln \ln N}{\ln N}\right) < \ln \ln H < \ln N \left(1 + \frac{(1 - \delta_1) \ln \ln N}{\ln N}\right)$$

or

$$\frac{\ln \ln H}{\left(1 + \frac{(1-\delta_1) \ln \ln N}{\ln N}\right)} < \ln N < \frac{\ln \ln H}{\left(1 + \frac{(1-\delta_2) \ln \ln N}{\ln N}\right)}. \qquad (3.85)$$

From (3.82) and (3.85), it follows that

$$\frac{\ln H}{\ln \ln H} \left(\frac{\left(1 + \frac{(1-\delta_2) \ln \ln N}{\ln N}\right)}{1 - \frac{C_{27}}{\sqrt{\ln N}}}\right) < N < \frac{\ln H}{\ln \ln H} \left(\frac{\left(1 + \frac{(1-\delta_1) \ln \ln N}{\ln N}\right)}{1 - \frac{C_{27}}{\sqrt{\ln N}}}\right).$$

$$(3.86)$$

Consider an expression of the form

$$\frac{\left(1 + \frac{(1-\delta) \ln \ln N}{\ln N}\right)}{1 - \frac{C}{\sqrt{\ln N}}} = 1 + \frac{C}{\sqrt{\ln N}} + \frac{(1 - \delta) \ln \ln N}{\ln N} + \frac{C^2}{\ln N}$$

$$+ \frac{(1 - \delta) \ln \ln N}{(\ln N)^{\frac{3}{2}}} + \cdots. \qquad (3.87)$$

The terms on the right-hand side of equality (3.87), starting from the third, are those of a geometric progression with denominators $\frac{1}{\sqrt{\ln N}}$ and $\frac{C}{\sqrt{\ln N}}$, correspondingly. When $N \geq N_6$, both these numbers are less than 1. Therefore, for the right-hand side of (3.87), there are upper and lower estimates of the form $1 + \frac{\tilde{C}}{\sqrt{\ln N}}$. (Naturally, for the upper estimate, we have $\tilde{C} > C$, and vice versa for the lower estimate.) Thus, from (3.86), we obtain

$$\frac{\ln H}{\ln \ln H}\left(1 + \frac{C_{29}}{\sqrt{\ln N}}\right) < N < \frac{\ln H}{\ln \ln H}\left(1 + \frac{C_{30}}{\sqrt{\ln N}}\right). \qquad (3.88)$$

From inequalities (3.88), we obtain

$$\ln \ln H - \ln \ln \ln H + \ln\left(1 + \frac{C_{29}}{\sqrt{\ln N}}\right)$$

$$< \ln N < \ln \ln H - \ln \ln \ln H + \ln\left(1 + \frac{C_{30}}{\sqrt{\ln N}}\right);$$

therefore,

$$\ln \ln H \left(1 - \frac{(1 - \delta_3)\ln \ln \ln H}{\ln \ln H}\right)$$

$$< \ln N < \ln \ln H\left(1 - \frac{(1 - \delta_4)\ln \ln \ln H}{\ln \ln H}\right). \qquad (3.89)$$

From (3.88) and (3.89), the inequality

$$N < \frac{\ln H}{\ln \ln H}\left(1 + \frac{C_{30}}{\sqrt{\ln N}}\right)$$

$$< \frac{\ln H}{\ln \ln H}\left(1 + \frac{C_{30}}{\sqrt{\ln \ln H \left(1 - \frac{(1-\delta_3)\ln \ln \ln H}{\ln \ln H}\right)}}\right)$$

$$< \frac{\ln H}{\ln \ln H}\left(1 + \frac{C_{31}}{\sqrt{\ln \ln H}}\right) \qquad (3.90)$$

follows. From (3.82) and (3.89), it follows that

$$N \ln N < \cfrac{\ln H}{\left(1 - \cfrac{C_{28}}{\sqrt{\ln \ln H \left(1 - \frac{(1-\delta_3) \ln \ln \ln H}{\ln \ln H}\right)}}\right)} < \ln H \left(1 + \frac{C_{32}}{\sqrt{\ln \ln H}}\right),$$

(3.91)

$$N\sqrt{\ln N} < \frac{\ln H}{\ln \ln H} \left(1 + \frac{C_{31}}{\sqrt{\ln \ln H}}\right) \sqrt{\ln \ln H \left(1 - \frac{(1-\delta_4) \ln \ln \ln H}{\ln \ln H}\right)}$$

$$< \frac{\ln H}{\sqrt{\ln \ln H}} \left(1 + \frac{C_{33}}{\sqrt{\ln \ln H}}\right).$$

(3.92)

From (3.91) and (3.92), we obtain

$$N \ln N + C_{20} N (\ln N)^{\frac{1}{2}} < \ln H \left(1 + \frac{C_{32}}{\sqrt{\ln \ln H}}\right)$$

$$+ C_5 \frac{\ln H}{\sqrt{\ln \ln H}} \left(1 + \frac{C_{33}}{\sqrt{\ln \ln H}}\right)$$

$$< \ln H + \frac{C_{34} \ln H}{\sqrt{\ln \ln H}}.$$

(3.93)

Finally, (3.77) and (3.93) give

$$|L(\xi)|_v \geq \left(m! H e^{\left((m-1)\left(N \ln N + C_{20} N(\ln N)^{\frac{1}{2}}\right)\right)}\right)^{-1}$$

$$\geq \left(m! H e^{\left((m-1)\left(\ln H + \frac{C_{34} \ln H}{\sqrt{\ln \ln H}}\right)\right)}\right)^{-1} \geq H^{-m - \frac{C_{35}}{\sqrt{\ln \ln H}}}.$$
$\qquad\square$

Theorem 3.2. *Let the F-series $f_1(z) \equiv 1$, $f_2(z), \ldots, f_m(z)$ consti-tute a solution to system (3.5) and be linearly independent over the field $\mathbb{K}(z)$. Let $\xi \in \mathbb{K}$ be a regular point of system (3.5). Let*

$$L(y_1, \ldots, y_m) = h_1 y_1 + \cdots + h_m y_m$$

(3.94)

be a non-zero linear form, $h_i \in \mathbb{Z}_\mathbb{K}$, $i = 1, \ldots, m$. Then, there exists an infinite set of prime numbers p and valuations v of the field $\mathbb{K}$, *extending the p-adic valuation of the field* $\mathbb{Q}$, *such that in the field* $\mathbb{K}_v$,

$$L(f_1(\xi), \ldots, f_m(\xi)) \neq 0.$$

In other words, the series $f_1(\xi), \ldots, f_m(\xi)$ *are infinitely linearly independent.*

Corollary. *There are no linear global relations between* $f_1(\xi), \ldots,$ $f_m(\xi)$.

Proof of Theorem 3.2. By choosing the number H_1 to satisfy the inequality $H \geq \max(H_0, H(L))$ (where L is defined in (3.94)), we find by Theorem 3.1 a prime number p_1 satisfying the inequalities $P_l(\ln H_1) \leq p_1 \leq P_u(\ln H_1)$ and the valuation v of the field $\mathbb{K}$ extending the p-adic valuation of the field $\mathbb{Q}$, such that the inequality

$$L(\xi) = L(f_1(\xi), \ldots, f_m(\xi)) \neq 0$$

holds. Determine the number H_2 from the condition

$$P_u(\ln H_1) < P_l(\ln H_2).$$

Find the prime number p_2 that satisfies the inequalities

$$P_l(\ln H_2) \leq p_2 \leq P_u(\ln H_2)$$

and the valuation v of the field $\mathbb{K}$ extending the p-adic valuation of the field $\mathbb{Q}$, such that the inequality

$$L(\xi) = L(f_1(\xi), \ldots, f_m(\xi)) \neq 0$$

is true, and determine the number H_3 from the condition $P_u(\ln H_2) < P_l(\ln H_3)$. The numbers p_1, p_2 are distinct since they belong to non-intersecting intervals. Let us continue this process and determine, given the values of $H_1, \ldots, H_k$ already found, the number H_{k+1} from the condition

$$P_u(\ln H_k) \leq P_l(\ln H_{k+1}).$$

The corresponding prime numbers $p_1, p_2, \ldots, p_{k+1}$ are different because they belong to non-intersecting intervals. $\square$

Theorem 3.3. *Let the F-series $f_1(z), f_2(z), \ldots, f_m(z)$ constitute a solution to system (3.1) and be algebraically independent over the field $\mathbb{K}(z)$. Let $\xi \in K$ be a regular point of system (3.1). Let $P(y_1, \ldots, y_m)$ be a non-zero polynomial with coefficients from $\mathbb{Z}_\mathbb{K}$.*

Then, there exists an infinite set of prime numbers p and valuations v of the field $\mathbb{K}$, extending the p-adic valuation of the field $\mathbb{Q}$, such that in the field $\mathbb{K}_v$,

$$P(f_1(\xi), \ldots, f_m(\xi)) \neq 0.$$

In other words, the series $f_1(\xi), \ldots, f_m(\xi)$ are infinitely algebraically independent.

Proof of Theorem 3.3. Let the degree of a polynomial $P(y_1, \ldots, y_m)$ be denoted by K. Consider the products of powers

$$\varphi_{k_1 \ldots k_m} = f_1^{k_1} \ldots f_m^{k_m}, \quad k_1 + \cdots + k_m \leq K, \quad k_i \geq 0, \quad i = 1, \ldots, m. \tag{3.95}$$

If the series $f_1(z), f_2(z), \ldots, f_m(z)$ satisfy system (3.1), then series (3.95) satisfy the system of linear differential equations

$$y'_{k_1 \ldots k_m} = \sum Q^*_{l_1 \ldots l_m} y_{l_1 \ldots l_m} \tag{3.96}$$

of order $\frac{(K+m)!}{K!m!}$. Indeed,

$$(f_1^{k_1} \ldots f_m^{k_m})' = \sum_{i=1}^{m} k_i f_1^{k_1} \ldots f_i^{k_i-1} \ldots f_m^{k_m} f_i' = \sum Q^*_{l_1 \ldots l_m} \varphi_{l_1 \ldots l_m}.$$

Here, the summation is performed on all possible sets of non-negative integers $l_1 \ldots l_m$ such that $l_1 + \cdots + l_m \leq K$. The coefficients $Q^*_{l_1 \ldots l_m} = Q^*_{l_1 \ldots l_m}(z)$ of these equations are linear combinations of the functions $Q_{i,j}(z)$, $i = 1, \ldots, m$, $j = 0, 1, \ldots, m$. For this system, the polynomial $T(z)$ such that $T(z)Q^*_{l_1 \ldots l_m} \in \mathbb{Z}_K[z]$ coincides with the polynomial $T(z)$ for system (3.1). The height of the system will increase to no more than $(K+m)^{K+m}$ times that of system (3.1).

Let us prove that if the series $f_1, \ldots, f_m$ belong to the class $F(\mathbb{K}, C_1, C_2, C_3, q)$, then the considered products of their powers

(3.95) belong to the class $F(\mathbb{K}, \tilde{C}_1, \tilde{C}_2, \tilde{C}_3, \tilde{q})$, where

$$\tilde{C} = C_1 + \ln 2, \quad \tilde{C}_2 = C_2, \quad \tilde{C}_3 = (C_3 + 1)K, \quad \tilde{q} = q^{1+\{\ln K\}}. \tag{3.97}$$

Let's represent the series $\varphi_{k_1 \ldots k_m}$ in the form

$$f_1^{k_1}(z) \ldots f_m^{k_m}(z) = \sum_{n=0}^{\infty} a_n n! z^n.$$

We consider the product on the left-hand side of this equality to be a product of $s \leq K$ factors, among which there may be equal factors. We denote the coefficient at z^n of the factor with the number i by $a_{i,n} n!$. Then, we have

$$a_n = \sum \frac{n_1! \cdots n_s!}{n!} a_{1,n_1} \ldots a_{s,n_s}. \tag{3.98}$$

Here, the summation is performed on all possible sets of non-negative integers $n_1 \cdots n_s$ such that $n_1 + \cdots + n_s = n$. The following estimate is true:

$$\overline{|a_{1,n_1} \ldots a_{s,n_s}|} = O(e^{C_1(n_1 + \cdots + n_s)}) = O(e^{C_1 n}), \quad n \to \infty. \tag{3.99}$$

The quantity

$$\sum \frac{n_1! \ldots n_s!}{n!}$$

can be estimated from above as follows:

$$\sum \frac{n_1! \ldots n_s!}{n!} \leq \sum 1 = \frac{(n+s-1)!}{n!(s-1)!} \leq 2^{n+s-1} \leq 2^{n+K}. \tag{3.100}$$

In this evaluation, the summation is performed on all sets of non-negative integers $n_1 \ldots n_s$ such that $n_1 + \cdots + n_s = n$. Here, we use a well-known combinatorial formula,

$$\frac{(n+s-1)!}{n!(s-1)!},$$

for the number of sets of such integers. Therefore, by (3.99) and (3.100),

$$\overline{|a_n|} = 2^{n+K} O(e^{C_1 n}) = O(e^{(C_1 + \ln 2)n}), \quad n \to \infty. \tag{3.101}$$

Let's put for each n and each set of non-negative integers $n_1 \ldots n_s$ such that $n_1 + \cdots + n_s = n$ the numbers $n_1 \ldots n_s$ in non-increasing order and denote the resulting numbers by $l_1 \ldots l_s$. Then, the following inequalities are evident:

$$n_{l_1} \leq n, \quad n_{l_2} \leq \left[\frac{n}{2}\right], \ldots, n_{l_s} \leq \left[\frac{n}{s}\right]. \tag{3.102}$$

Consider Ξ_n as the least common denominator of the numbers

$$\frac{n_1! \ldots n_s!}{n^*!}, \quad n_1 + \cdots + n_s = n^*, \quad n^* \leq n. \tag{3.103}$$

It is divisible only by the prime numbers p that do not exceed the number n. Moreover, for any such prime number p, according to (3.103) and the equality

$$\mathrm{ord}_p n! = \frac{n - S_n}{p - 1},$$

where S_n denotes the sum of digits in the p-adic expansion of the number n, we get

$$\mathrm{ord}_p\left(\frac{n_1! \ldots n_s!}{n^*!}\right) = \frac{n_1 - S_{n_1} + \cdots + n_s - S_{n_s} - n^* + S_{n^*}}{p - 1}$$

$$= \frac{S_{n^*} - S_{n_1} - \cdots - S_{n_s}}{p - 1} \geq -s \log_p n.$$

Therefore,

$$\mathrm{ord}_p \Xi_n \leq s \log_p n \leq K \log_p n. \tag{3.104}$$

It follows from (3.102) that

$$q^{n + \left[\frac{n}{2}\right] + \cdots + \left[\frac{n}{s}\right]} d_{0,n} d_{0,\left[\frac{n}{2}\right]} \cdots d_{0,\left[\frac{n}{s}\right]} \Xi_n a_n \in \mathbb{Z}_{\mathbb{K}},$$

and the number

$$d_{0,n} d_{0,\left[\frac{n}{2}\right]} \cdots d_{0,\left[\frac{n}{s}\right]} \Xi_n \tag{3.105}$$

is divisible only by prime numbers p not exceeding the number

$$\max(C_2, 1) n. \tag{3.106}$$

Moreover, the following inequalities are satisfied:

$$\text{ord}_p \left(d_{0,n} d_{0,\left[\frac{n}{2}\right]} \cdots d_{0,\left[\frac{n}{s}\right]} \Xi_n \right)$$

$$\leq C_3 \left(\log_p n + \log_p \left[\frac{n}{2}\right] + \cdots + \log_p \left[\frac{n}{s}\right] \right.$$

$$+ \frac{n}{p^2} + \frac{\left[\frac{n}{2}\right]}{p^3} + \cdots + \frac{\left[\frac{n}{s}\right]}{p^{s+1}} \right)$$

$$+ K \log_p n \leq K(c_3 + 1) \left(\log_p n + \frac{n}{p^2} \right). \qquad (3.107)$$

If we put

$$\tilde{q} = q^{1+[\ln K]},$$

then in view of the obvious inequality of

$$n(1 + [\ln K]) \geq n + \left[\frac{n}{2}\right] + \cdots + \left[\frac{n}{K}\right] \geq n + \left[\frac{n}{2}\right] + \cdots + \left[\frac{n}{s}\right],$$

the number

$$q^{n+\left[\frac{n}{2}\right]+\cdots+\left[\frac{n}{s}\right]}$$

divides the number $(\tilde{q})^n$. It follows from (3.98)–(3.107) that series (3.95) belong to the class $F(\mathbb{K}, \tilde{C}_1, \tilde{C}_2, \tilde{C}_3, \tilde{q})$ whose parameters satisfy the conditions (3.97).

Moreover, if the series $f_1(z), f_2(z), \ldots, f_m(z)$ are algebraically independent over the field $\mathbb{K}(z)$, then the series $\varphi_{k_1 \ldots k_m}$ defined by (3.95) are linearly independent over this field.

Thus, the considered series $\varphi_{k_1 \ldots k_m}$ satisfy all the conditions of Theorem 3.2 for any natural K. Applying this theorem, we obtain the statement of Theorem 3.3. $\qquad \square$

3.5. Non-Trivial Global Relations

We call a global relation

$$P(f_1(\xi), \ldots, f_m(\xi)) = 0$$

trivial if it is obtained by substituting $z = \xi$ into a polynomial relation,

$$P(z, f_1(z), \ldots, f_m(z)) = 0,$$

which is identical in z. Here, $P(z, x_1, \ldots, x_m)$ is a non-zero polynomial with coefficients that are integers from the field $\mathbb{K}$. In the other case, we call the global relation *non-trivial*.

Theorem 3.4. *Let the F-series $f_1(z) \equiv 1, f_2(z), \ldots, f_m(z)$ constitute a solution of system (3.5). Let $\xi \in \mathbb{K}$ be a regular point of system (3.5).*

Then, there are no non-trivial global relations between

$$f_1(\xi) \equiv 1, f_2(\xi), \ldots, f_m(\xi).$$

To prove the theorem, we need some lemmas.

Lemma 3.12. *Let the F-series $g(z) \equiv 1, g_2(z), \ldots, g_M(z)$ constitute a solution of the system*

$$y_i' = \sum_{j=1}^{M} \hat{Q}_{i,j}(z) y_j, \quad i = 1, \ldots, M. \tag{3.108}$$

The coefficients $\hat{Q}_{i,j}(z)$ of this system belong to the field of rational functions $\mathbb{K}(z)$. Let $\hat{T} = \hat{T}(z) \in \mathbb{K}[z]$ be a primitive polynomial that is the least common denominator of all $\hat{Q}_{i,j}(z), \hat{T}(z)\hat{Q}_{i,j}(z)(z) \in \mathbb{Z}_{\mathbb{K}}[z]$.

Let $\xi \in \mathbb{K}$ be a regular point of system (3.108).

Then, there are no non-trivial global relations between $g_1(\xi) \equiv 1$, $g_2(\xi), \ldots, g_M(\xi)$.

To prove Lemma 3.12, we need the following assertion.

Lemma 3.13. *Let the greatest number of series linearly independent over $\mathbb{K}(z)$ among the F-series $g_1(z) \equiv 1, g_2(z), \ldots, g_M(z)$ be equal to $l, 1 \le l \le M - 1$. Let $\xi \in K$. Then, these series can be renumbered so that the resulting series $\varphi_1(z), \ldots, \varphi_l(z)$ are linearly independent over $K(z)$, and the other series can be represented as linear combinations,*

$$\varphi_j(z) = B_{j,1}(z)\varphi_1(z) + \cdots + B_{j,l}(z)\varphi_l(z), \quad j = l+1, \ldots, M,$$

where $B_{j,k}(z)$, $j = l+1, \ldots, M$, $k = 1, \ldots, l$ is a rational function from $\mathbb{K}(z)$ and the point $\xi \in \mathbb{K}$ is not a pole of any of these rational functions.

The proof can be derived by induction over the number M of the series in question. In the case of $M = 2$, only the value $l = 1$ is possible. Then, the series are related, up to a constant factor, by a single linear equation

$$A_1(z) + A_2(z)g_2(z) = 0,$$

with relatively prime polynomials $A_1(z), A_2(z) \in \mathbb{K}[z]$. If $A_2(\xi) = 0$, we get $A_1(\xi) = 0$. But this is impossible since $A_1(z), A_2(z)$ are relatively prime polynomials. Thus, the equation

$$g_2(z) = -\frac{A_1(z)}{A_2(z)}$$

gives the desired representation.

Let the lemma be valid for a number of series not exceeding $M-1$ for any possible value of l, $1 \le l \le M - 2$. Let us prove that its statement remains valid for M series and for any value of l, $1 \le l \le M - 1$.

For any n, these series are related over the field $\mathbb{K}(z)$ by exactly $M - l \ge 1$ linearly independent equations. Consider any of these equations:

$$\sum_{i=1}^{M} A_i(z)g_i(z) = 0. \tag{3.109}$$

In equation (3.109), all $A_i(z)$ are relatively prime polynomials from $\mathbb{K}[z]$. Consequently, they cannot simultaneously vanish at $z = \xi$. By renumbering the series $g_1(z) \equiv 1, g_2(z), \ldots, g_M(z)$, if necessary, we can assume that it is the coefficient at $\varphi_M(z)$ that does not vanish. Then, from (3.109) follows the equality

$$\sum_{i=1}^{M-1} B_{M,i}^*(z)g_i(z) = \varphi_M(z).$$

Here, all $B^*_{M,i}(z) \in \mathbb{K}(z)$ and have no pole at $z = \xi$. If $l = M - 1$, we are through. If $l < M - 1$, then the series $\varphi_1(z), \ldots, \varphi_{M-1}(z)$ are connected over the field $\mathbb{K}(z)$ by $M - l - 1$ linearly independent equations, and by the assumption of induction, after proper renumbering, we obtain

$$\varphi_j(z) = B_{j,1}(z)\varphi_1(z) + \cdots + B_{j,l}(z)\varphi_l(z), \quad j = l+1, \ldots, M - 1,$$
$$(3.110)$$

where $B_{j,k}(z) \in \mathbb{K}(z)$, $j = l + 1, \ldots, M - 1$, $k = 1, \ldots, l$, and these rational functions have no pole at $z = \xi$. In this case, the series $\varphi_1(z), \ldots, \varphi_l(z)$ are linearly independent over the field $\mathbb{K}(z)$. It remains to substitute the right parts of the corresponding equations (3.110) into equality (3.109) instead of the series $\varphi_{l+1}(z), \ldots, \varphi_{M-1}(z)$. This will result in the following equality:

$$\varphi_M(z) = \sum_{i=1}^{l} B_{M,i}(z)\varphi_i(z).$$
$$(3.111)$$

Here, all $B_{M,i}(z) \in \mathbb{K}(z)$ and have no pole at $z = \xi$. The induction is complete, and Lemma 3.13 is proved.

Let us return to the proof of Lemma 3.12. Substituting into the system (3.108) the expressions (3.110) and (3.111) for the series $\varphi_1(z), \ldots, \varphi_M(z)$, we obtain that the series $\varphi_1(z), \ldots, \varphi_l(z)$ satisfy the system of linear differential equations

$$y_i' = \sum_{j=1}^{l} Q^*_{i,j}(z)y_j(z), \quad i = 1, \ldots, l.$$
$$(3.112)$$

Here, all $Q^*_{i,j}(z) \in \mathbb{K}(z)$ and have no pole at $z = \xi$.

Let us prove that the series $\varphi_1(z), \ldots, \varphi_l(z)$ can be chosen so that $\varphi_1(z) \equiv 1$. Indeed, otherwise, for some j, $l+1 \le j \le M$, the following equality holds:

$$1 = B_{j,1}(z)\varphi_1(z) + \cdots + B_{j,l}(z)\varphi_l(z).$$
$$(3.113)$$

If there exists k, $1 \leq k \leq l$, such that $B_{j,k}(\xi) \neq 0$, then from (3.113), we obtain

$$\varphi_k(z) = -\frac{B_{j,1}(z)}{B_{j,k}(z)}\varphi_1(z) - \cdots - \frac{B_{j,k-1}(z)}{B_{j,k}(z)}\varphi_{k-1} - \frac{1}{B_{j,k}(z)}$$

$$- \frac{B_{j,k+1}(z)}{B_{j,k}(z)}\varphi_{k+1}(z) - \cdots - \frac{B_{j,l}(z)}{B_{j,k}(z)}\varphi_l(z), \qquad (3.114)$$

and all rational functions $\frac{B_{j,i}(z)}{B_{j,k}(z)}$ have no pole at the point $z = \xi$.

Let us substitute the right-hand side of equality (3.114), instead of $\varphi_k(z)$, into all equations (3.112) and (3.113). As a result, we obtain the representations of other series $\varphi_r(z)$, $l + 1 \leq r \leq M$, $r \neq j$ as linear combinations of series $\varphi_1(z), \ldots, \varphi_{k-1}(z), \varphi_{k+1}(z), \ldots, \varphi_l(z)$ with coefficients that are rational functions from $\mathbb{K}(z)$ and have no pole at $z = \xi$. But this contradicts the assumption made above. If the equations

$$B_{j,k}(\xi) = 0. \ k = 1, \ldots, l$$

hold, then by differentiating equality (3.113) using system (3.112), we obtain

$$0 = \sum_{i=1}^{l}(B'_{j,i}(z)\varphi_i(z) + B_{j,i}(z)\varphi'_i(z))$$

$$= \sum_{i=1}^{l} B'_{j,i}(z)\varphi_i(z) + \sum_{i=1}^{l} B_{j,i}(z) \sum_{k=1}^{l} Q^*_{i,k}(z)\varphi_k(z)$$

$$= \sum_{i=1}^{l}\left(B'_{j,i}(z) + \sum_{k=1}^{l} Q^*_{i,k}(z)B_{j,k}(z)\right)\varphi_i(z). \qquad (3.115)$$

However, the series $\varphi_1(z), \ldots, \varphi_l(z)$ are linearly independent over the field $\mathbb{K}(z)$, so it follows from equality (3.115) that

$$B'_{j,i}(z) + \sum_{k=1}^{l} Q^*_{i,k}(z)B_{j,k}(z) = 0, \quad i = 1, \ldots, l. \qquad (3.116)$$

This system of linear differential equations differs from system
(3.112) only by the sign of the coefficients, and these coefficients
have no pole at $z = \xi$. The rational functions $B_{j,k}(z)$ that make up
the solution of this system are zero at the point $z = \xi$. By the Cauchy
theorem on the uniqueness of the solution of the Cauchy problem for
this system, we obtain the identities $B_{j,k}(z) \equiv 0$, $k = 1, \ldots, l$, which
mean that equality (3.113) of formal power series is impossible. Thus,
the series $\varphi_1(z), \ldots, \varphi_l(z)$ satisfy all the conditions of the corollary
of Theorem 3.2. Applying this corollary, we obtain that there is no
global linear relation connecting $\varphi_1(\xi), \ldots, \varphi_l(\xi)$ with the coefficients
from the field $\mathbb{K}$.

Any global linear relation,

$$c_1' + c_2' g_2(\xi) + \cdots + c_M' g_M(\xi) = 0,$$

with coefficients $c_1', \ldots, c_M'$ from the field $\mathbb{K}$, after the series $g_1(z) \equiv 1$,
$g_2(z), \ldots, g_M(z)$ are renumbered in the above way, will change into
a linear global relation,

$$c_1 + c_2 \varphi_2(\xi) + \cdots + c_M \varphi_M(\xi) = 0, \qquad (3.117)$$

where $c_1, \ldots, c_M$ from the field $\mathbb{K}$ are the appropriately renumbered
$c_1', \ldots, c_M'$.

From equations (3.110) and (3.111) follow the global relations

$$\varphi_j(\xi) = B_{j,1}(\xi)\varphi_1(\xi) + \cdots + B_{j,l}(\xi)\varphi_l(\xi), \quad j = l+1, \ldots, M,$$

or

$$B_{j,1}(\xi)\varphi_1(z) + \cdots + B_{j,l}(\xi)\varphi_l(\xi) - \varphi_j(\xi) = 0, \quad j = l+1, \ldots, M.$$
$$(3.118)$$

Let us prove that any linear global relation (3.117) is a linear combination with coefficients from the field $\mathbb{K}$ of relations (3.118).

Suppose, on the contrary, that the vector $(c_1, \ldots, c_M)$ is not a
linear combination with coefficients from the field $\mathbb{K}$ of the vectors
of the form

$$(B_{j,1}(\xi); \ldots; B_{j,l}(\xi); 0; \ldots; 0; -1; 0; \ldots; 0), \quad j = l+1, \ldots, M$$

(the vector with the number j has number -1 at the jth position).
Then, the vector

$$(c_1, \ldots, c_M) + \sum_{j=l+1}^{M} c_j (B_{j,1}(\xi); \ldots ; B_{j,l}(\xi); 0; \ldots ; 0; -1; 0; \ldots ; 0)$$

$$= \left(c_1 + \sum_{j=l+1}^{M} c_j B_{j,1}(\xi); \ldots ; c_l + \sum_{j=l+1}^{M} c_j B_{j,l}(\xi); 0; \ldots ; 0 \right)$$

is different from the zero vector. This means that at some value of
$k, 1 \le k \le l$, the number

$$c_k + \sum_{j=l+1}^{M} c_j B_{j,k}(\xi)$$

is different from zero. But then, the linear combination of the global
relations obtained by adding to relation (3.117) the global relations
(3.118) multiplied by the corresponding numbers c_j, $j = l+1, \ldots, M$,
will give the global relation

$$c_1 + \sum_{j=l+1}^{M} c_j B_{j,1}(\xi) + \cdots + \left(c_l + \sum_{j=l+1}^{M} c_j B_{j,l}(\xi) \right) \varphi_l(\xi) = 0,$$

whose coefficients, as shown above, include numbers different from
zero. It has been proven above that this is impossible. The lemma is
proved. □

Let us proceed to the proof of the theorem. Let there be a global
algebraic relation

$$P(f_1(\xi), \ldots, f_m(\xi)) = 0,$$

where $P(x_1, \ldots, x_m)$ is a non-zero polynomial with coefficients from
the ring $\mathbb{Z}_{\mathbb{K}}$. Let the degree of the polynomial $P(x_1, \ldots, x_m)$ in the
set of variables $x_1, \ldots, x_m$ be equal to K. Consider the products of
powers

$$f_1^{k_1}(z) \ldots f_m^{k_m}(z)$$

of the series $f_1(z), \ldots, f_m(z)$, where $k_1, \ldots, k_m$ are non-negative integers satisfying the inequality $k_1 + \cdots + k_m \le K$. Let us denote them

by $g_1(z) \equiv 1, g_2(z), \ldots, g_M(z)$, where

$$M = \frac{(K+m)!}{K!m!}.$$

By (3.96), the series $g_1(z) \equiv 1, g_2(z), \ldots, g_M(z)$ belong to the class $F(\mathbb{K}, \tilde{C}_1, \tilde{C}_2, \tilde{C}_3, \tilde{q})$, where

$$\tilde{C} = C_1 + \ln 2, \quad \tilde{C}_2 = C_2, \quad \tilde{C}_3 = (C_3 + 1)K, \quad \tilde{q} = q^{1+\{\ln K\}}.$$

Moreover, by (3.95), these series constitute a solution to the system of linear differential equations

$$y'_{k_1 \ldots k_m} = \sum Q^*_{l_1 \ldots l_m} y_{l_1 \ldots l_m}.$$

Thus, all the conditions of Lemma 3.13 are satisfied, and applying them we obtain that the algebraic global relation

$$P(f_1(\xi), \ldots, f_m(\xi)) = 0$$

considered as a linear global relation between the series

$$g_1(\xi) \equiv 1, g_2(\xi), \ldots, g_M(\xi)$$

is trivial. This means that it can be obtained by substituting $z = \xi$ into some linear relation between series $g_1(z) \equiv 1, g_2(z), \ldots, g_M(z)$ with coefficients from $\mathbb{ZK}[z]$. But this relation can be viewed as an algebraic relation between series $f_1(\xi) \equiv 1, f_2(\xi), \ldots, f_m(\xi)$ with coefficients from $\mathbb{Z}_\mathbb{K}$. Hence, the considered global algebraic relation is trivial, and the theorem is proved. $\qquad\square$

Chapter 4

Euler's Factorial Series and Direct Generalisations

The series of the form

$$E_1(z) = \sum_{n=0}^{\infty} n! z^n \qquad (4.1)$$

is usually called *Euler's factorial series* (or Euler's divergent series since, in the field $\mathbb{C}$ of complex numbers, it converges only at $z = 0$). This series, introduced by the great L. Euler in 1760, can be considered the asymptotic expansion of the integral

$$\int_0^{+\infty} \frac{e^{-w}}{1 - zw} dw.$$

Indeed, if we substitute $1 + wz + \cdots + (wz)^n + \cdots$ for $\frac{1}{1-zw}$ and then formally integrate, while keeping in mind that

$$\int_0^{+\infty} w^n e^{-w} dw = \Gamma(n + 1) = n!,$$

we simply get the series (4.1). As mentioned above, this series in the field $\mathbb{C}$ of complex numbers converges only at $z = 0$. But in any field $\mathbb{Q}_p$ of p-adic numbers, it converges at any z such that

$$|z|_p < p^{\frac{1}{p-1}},$$

and this allows us to investigate the arithmetic problems of the values of this series, in particular, the properties of the series $\sum_{n=0}^{\infty} n!$.

The series of the form

$$\sum_{n=0}^{\infty} a_n \cdot n!, \quad a_n \in \mathbb{Z}$$

represent the so-called polyadic numbers (Postnikov, 1988).

The series $\sum_{n=0}^{\infty} n!$ is related to *Kurepa's conjecture* (Kurepa, 1971; Vladimirov, 2002), which asserts that for any prime $p > 2$, the partial sum of this series, i.e., the number $\sum_{n=0}^{p-1} n!$, is not divisible by p.

Let us consider a somewhat more general series with the form

$$E_\gamma(z) = \sum_{n=0}^{\infty} (\gamma)_n z^n, \tag{4.2}$$

where $(\gamma)_n$ is the *Pochhammer symbol*, which is determined by the equalities $(\gamma)_0 = 1$ and $(\gamma)_n = \gamma(\gamma + 1) \cdots (\gamma + n - 1)$ at $n \geq 1$.

Theorem 4.1. *Let*

$$\gamma \in \mathbb{Q}, \ \gamma \neq 0, -1, -2, \ldots, \gamma = \frac{a}{b}, \ a \in \mathbb{Z}, \ b \in \mathbb{N}.$$

Let $\xi \neq 0$ be an arbitrary integer algebraic number.

Then, $E_\gamma(\xi)$ is an infinitely transcendental, almost polyadic number.

Note that the term 'almost polyadic number' here means that the series $E_\gamma(\xi)$ converges in any field $\mathbb{Q}_p$ of p-adic numbers, except for the case where p divides $b \in \mathbb{N}$.

Corollary. *The value*

$$E_1(1) = \sum_{n=0}^{\infty} n!$$

is an infinitely transcendental polyadic number.

Proof of Theorem 4.1. Under the conditions of the theorem, the series (4.2) is not an algebraic function since it converges only at

$z = 0$. This series is a formal solution of the differential equation

$$y' = \frac{\gamma z - 1}{z^2} y + \frac{1}{z^2}.$$

It only remains to prove that the series (4.2) belongs to some class $F(C_1, C_2, C_2, q)$. We recall that a series

$$\sum_{n=0}^{\infty} c_n \cdot n! \cdot z^n$$

belongs to the class $F(\mathbb{K}, c_1, c_2, c_3, q)$ if the following conditions hold:

1. $c_n \in \mathbb{K}$, $n = 0, 1, \ldots$, $\mathbb{K}$ is an algebraic number field of a finite degree over the field $\mathbb{Q}$.
2. There exist constants γ and C_1 such that

$$\overline{|c_n|} \leq \gamma e^{C_1 n}, \ n = 0, 1, 2, \ldots.$$

Here, $\overline{|a|}$ denotes the height of an algebraic number a, i.e., the largest of the absolute values of the number itself and its conjugates in $\mathbb{K}$.

3. There exists a sequence of positive integers $d_n = q^n d_{0,n}$ such that $d_n c_k \in \mathbb{Z}_{\mathbb{K}}$, for $k = 0, 1, \ldots, n$ and $n = 0, 1, 2 \ldots$.

(Here, $\mathbb{Z}_{\mathbb{K}}$ denotes the ring of integers of the field $\mathbb{K}$.) In this case, the number $d_{0,n}$ is divisible only by prime factors not exceeding $C_2 n$, and

$$\mathrm{ord}_p d_{0,n} = \vartheta_p(d_{0,n}) \leq C_3 \left(\log_p n + \frac{n}{p^2} \right), \ n = 0, 1, 2, \ldots.$$

Given that all $c_n = 1$, we get $C_1 = 0$, $\gamma = 1$, $d_n = 1$, $q = 1$, and $C_3 = 0$.

So, we are in a position to apply Theorem 3.3 and complete the proof of Theorem 4.1. □

The proof of Theorem 4.1 used a generalisation of the Siegel–Shidlovskii method. Theorem 3.3 will be applied in Chapter 6 to a wide class of hypergeometric F-series.

It is possible to use another approach and study the arithmetic properties of $E_1(z)$ by applying the Hermite–Padé approximations.

We use this approach in Chapters 7 and 8, where it enables us to prove the infinite linear independence of the values of hypergeometric series with algebraic irrational and certain transcendental parameters, respectively.

To show the possibilities of this approach, we formulate and prove some theorems from the works of Matala-aho and Zudilin (2018) and Ernwall-Hytönen *et al.* (2019), obtained using the Hermite–Padé approximations. We use the notations from these papers. There,

$$E(z) = \sum_{k=0}^{\infty} k!(-z)^k,$$

and the symbols $E_p(\xi)$ and $E_q(\xi)$ stand for the sums of this series in the fields $\mathbb{Q}p$ and $\mathbb{Q}_q$, respectively.

Theorem 4.2 (Theorem 1 in Matala-aho and Zudilin (2018)).
Given $\xi \in \mathbb{Z}\backslash\{0\}$, let $\mathcal{P}$ be a subset of prime numbers such that

$$\lim_{n\to\infty} \sup c^n \prod_{p\in\mathcal{P}} |n!|_p^2 = 0,$$

where

$$c = 4|\xi| \prod_{p\in\mathcal{P}} |\xi|_p^2.$$

Then, either there exists a prime $p \in \mathcal{P}$ for which $E_p(\xi)$ is irrational, or there are two distinct primes $p, q \in \mathcal{P}$ such that $E_p(\xi) \neq E_q(\xi)$ while $E_p(\xi)$, $E_q(\xi) \in \mathbb{Q}$.

Proof of Theorem 4.2. Assume that we can construct approximations $\frac{p_n}{q_n}$ to $E_p(\xi)$ that do not depend on p and satisfy the following condition: for any p, we have $q_n E_p(\xi) - p_n = r_{n,p}$, $n = 0, 1, 2, \ldots$; $|r_{n,p}|_p \to 0$ as $n \to \infty$. Moreover, let there exist infinitely many indices n for which $r_{n,p} \neq 0$, for at least one prime $p \in \mathcal{P}$. If we assume that for each $p \in \mathcal{P}$, we have $E_p(\xi) = \frac{a}{b} \in \mathbb{Q}$, then $q_n a - p_n b \neq 0$,

$q_n a - p_n b \in \mathbb{Z}$. So,

$$0 < |q_n a - p_n b|_p \leq 1,$$

for infinitely many n and all primes p. Therefore, using the product formula,

$$1 = |q_n a - p_n b| \prod_p |q_n a - p_n b|_p \leq |q_n a - p_n b| \prod_{pp \in \mathcal{P}} |q_n a - p_n b|_p$$

$$\leq (|a| + |b|) \max(|q_n|, |p_n|) \prod_{p \in \mathcal{P}} |b r_{n,p}|_p$$

$$\leq (|a| + |b|) \max(|q_n|, |p_n|) \prod_{p \in \mathcal{P}} |r_{n,p}|_p$$

for those n. This means that the condition

$$\lim_{n \to \infty} \sup \max(|q_n|, |p_n|) \prod_{p \in \mathcal{P}} |r_{n,p}|_p = 0 \tag{4.3}$$

contradicts the latter estimate; therefore, the relation $E_p(\xi) = \frac{a}{b}$ cannot hold for all $p \in \mathcal{P}$.

It remains to prove the existence of $\frac{p_n}{q_n}$ with the declared properties. To do this, consider the series (4.2). To preserve the notation adopted by Matala-aho and Zudilin (2018), we put $\gamma = a$ and use the classical notation for the hypergeometric series

$$\sum_{n=0}^{\infty} (a)_n z^n = {}_2F_0(a, 1|z). \qquad \square$$

Lemma 4.1 (Theorem 2 in Matala-aho and Zudilin (2018)).
For $n, \lambda \in \mathbb{Z}$, $n \geq 0$, and $\lambda \geq 0$, take

$$B_{n,\lambda}(z) = \sum_{i=0}^{n} \binom{n}{i} \frac{(-1)^i z^{n-i}}{(a)_{i+\lambda}}.$$

Then, $\deg_z B_{n,\lambda}(z) = n$, and for a polynomial $A_{n,\lambda}(z)$ of degree $\deg_z A_{n,\lambda}(z) \leq n + \lambda - 1$, we have

$$B_{n,\lambda}(z) {}_2F_0(a, 1 \mid z) - A_{n,\lambda}(z) = L_{n,\lambda}(z), \tag{4.4}$$

where $\text{ord}_{z=0}L_{n,\lambda}(z) = 2n + \lambda$. *Explicitly,*

$$L_{n,\lambda}(z) = (-1)^n z^{2n+\lambda} \sum_{k=0}^{\infty} k! \binom{n+k}{k} \binom{n+k+a+\lambda-1}{k} z^k.$$

$$(4.5)$$

Proof of Lemma 4.1. Relation (4.4) means that there is a 'gap' of length n in the power series expansion

$$B_{n,\lambda}(z)_2F_0(a, 1 \mid z) = A_{n,\lambda}(z) + L_{n,\lambda}(z).$$

Write $B_{n,\lambda}(z) = \sum_{h=0}^{n} b_h z^h$, and consider the series expansion of the product

$$B_{n,\lambda}(z)_2F_0(a, 1 \mid z) = \sum_{l=0}^{\infty} r_l z^l,$$

where $r_l = \sum_{h+k=l} b_h(a)_k$, in the explicit form,

$$r_l = \sum_{i=0}^{n} (-1)^i \binom{n}{i} \frac{(a)_{i+\lambda+m}}{(a)_{i+\lambda}} = \sum_{i=0}^{n} (-1)^i \binom{n}{i} (a+i+\lambda)_m,$$

with $m = l - n - \lambda$ for $l > n + \lambda - 1$. To verify the condition,

$$r_{n+\lambda} = r_{n+\lambda+1} = \cdots = r_{n+\lambda+n-1} = 0,$$

$$(4.6)$$

it is convenient to introduce the shift operators $N = N_a$ and $\Delta = \Delta_a = N - \text{id}$, defined on the functions $f(a)$ by the equalities

$$Nf(a) = f(a+1), \quad \Delta f(a) = f(a+1) - f(a).$$

For $n \geq 0$, the following equality holds:

$$\Delta^n(a+\lambda)_m = (-1)^n(-m)_n(a+\lambda+n)_{m-n}.$$

Therefore,

$$r_l = \sum_{i=0}^{n} (-1)^i \binom{n}{i} N^i(a+\lambda)_m = (\text{id} - N)^n(a+\lambda)_m$$

$$= (-\Delta)^n(a+\lambda)_m = (-m)_n(a+\lambda+n)_{m-n},$$

which implies (4.6) since $(-m)_n = 0$ for $m = 0, 1, \ldots, n-1$. The explicit expression for r_l also implies that

$$r_{2n+\lambda+k} = (-n-k)_n (a+\lambda+n)_k$$
$$= (-1)^n n! \binom{n+k}{k} k! \binom{n+k+a+\lambda-1}{k},$$

so (4.5) is proven. We also have

$$A_{n,\lambda}(z) = \sum_{l=0}^{n+\lambda-1} r_l z^l = \sum_{l=0}^{n+\lambda-1} z^l \sum_{i=0,i\geq n-l}^{n} (-1)^i \binom{n}{i} \frac{(a)_{i+l-n}}{(a)_{i+\lambda}}.$$

Lemma 4.1 is proved. $\qquad\square$

Now, we put $a = 1$, $\lambda = 0$ and change z to $-z$. We also write

$$Q_n(z) = (-1)^n n! B_{n,0}(-z) = \sum_{i=0}^{n} \binom{n}{i} \frac{n!}{i!} z^{n-i} = \sum_{i=0}^{n} i! \binom{n}{i}^2 z^i,$$

$$P_n(z) = (-1)^n n! A_{n,0}(-z)$$
$$= (-1)^n \sum_{l=0}^{n-1} (-z)^l \sum_{i=n-l}^{n} (-1)^{n-i} \binom{n}{i} \frac{n!(i+l-n)!}{i!}$$
$$= (-1)^n \sum_{l=0}^{n-1} (-z)^l \sum_{i=0}^{l} (-1)^i i! (l-i)! \binom{n}{i}^2,$$

and

$$R_n(z) = (-1)^n n! L_{n,0}(-z) = (n!)^2 z^{2n} \sum_{k=0}^{\infty} (-1)^k k! \binom{n+k}{k}^2 z^k.$$

In this notation (recall that $_2F_0(a, 1 \mid z) = E(z)$), equality (4.4) for $n \geq 0$ gives

$$Q_n(z)E(z) - P_n(z) = R_n(z). \tag{4.7}$$

The following equality holds:

$$Q_n(z)P_{n+1}(z) - Q_{n+1}(z)P_n(z) = (n!)^2 z^{2n}. \tag{4.8}$$

Indeed,

$$Q_n(z)P_{n+1}(z) - Q_{n+1}(z)P_n(z) = Q_{n+1}(z)R_n(z) - Q_n(z)R_{n+1}(z).$$

The degree of the left-hand side of this equality is not more than $2n$, and the order of zero of the right-hand side at $z = 0$ is not less than $2n$.

Now, we put $z = \xi \in \mathbb{Z}\backslash\{0\}$, and for $n \geq 0$,

$$p_n = P_n(\xi) \in \mathbb{Z}, \ q_n = Q_n(\xi) \in \mathbb{Z}.$$

We also define, for each $p \in \mathcal{P}$,

$$r_{n,p} = R_n(\xi) = q_n E_p(\xi) - p_n.$$

The following estimates hold:

$$|q_n| \leq |\xi|^n n! \sum_{i=0}^{n} \binom{n}{i}^2 = |\xi|^n n! \binom{2n}{n} < 4^n |\xi|^n n!,$$

$$|p_n| \leq |\xi|^{n-1} n \sum_{i=0}^{n-1} i!(n-1-i)! \binom{n}{i}^2$$

$$\leq |\xi|^n n! \sum_{i=0}^{n} \binom{n}{i}^2 < 4^n |\xi|^n n!.$$

We also have

$$|r_{n,p}|_p = |\xi|_p^{2n} |n!|_p^2 \left| \sum_{k=0}^{\infty} (-1)^k k! \binom{n+k}{k}^2 \xi^k \right|_p \leq |\xi|_p^{2n} |n!|_p^2.$$

Therefore, (4.9) means that

$$\lim_{n\to\infty} \sup 4^n |\xi|^n n! \prod_{p\in\mathcal{P}} |\xi|_p^{2n} |n!|_p^2 = 0.$$

From (4.8), we deduce that, for at least one $p \in \mathcal{P}$, either $r_{n,p} \neq 0$ or $r_{n=1,p} \neq 0$. Theorem 4.2 is proven. $\qquad\square$

Now, we formulate the result from Ernwall-Hytönen *et al.* (2019). Denote

$$\Lambda(x) = a - bx \in \mathbb{Z}[x], \ b \neq 0.$$

For a given $m \in \mathbb{Z}$, $m \geq 3$, write $\boldsymbol{a}_1, \ldots, \boldsymbol{a}_{\varphi(m)}$ for the residue classes in the reduced residue system modulo m. Let $\mathbb{P}$ denote the set of prime numbers. Denote further

$$F(z) = \sum_{n=0}^{\infty} n! z^n.$$

Let $\xi \in \mathbb{Z} \backslash \{0\}$. Let $F_p(\xi)$ be the sum of the series $F(\xi)$ in the field $\mathbb{Q}_p$.

Theorem 4.3 (Theorem 3 in Ernwall-Hytönen *et al.* (2019)).
Let $m \in \mathbb{Z}$, $m \geq 3$. Assume that

$$R = \bigcup_{j=1}^{r} (\boldsymbol{a}_j \cap \mathbb{P})$$

is any union of the primes in r residue classes in the reduced residue system modulo m, where $r > \frac{\varphi(m)}{2}$. Then, there are infinitely many primes $p \in R$ such that $\Lambda(F_p(\xi)) \neq 0$.

Theorems 4.1–4.3 do not assert that at least for one prime p, the considered value $F_p(\xi)$ is irrational in the field $\mathbb{Q}_p$. They imply only that any linear relation does not hold in infinitely many fields $\mathbb{Q}_p$. The problem of proving the irrationality of the value $F_p(\xi)$ seems to be difficult enough.

Recently, Ernwall-Hytönen *et al.* (2023) proved several theorems with estimates of linear forms, such as $\Lambda(x)$ above in $F_p(\xi)$. In these estimates, the p-adic value of ξ depends on the height of the considered form; therefore, the results do not imply the irrationality of the number $F_p(\xi)$. Still, they are of great interest since they involve various techniques used to prove the estimates. Now, we describe the results from Ernwall-Hytönen *et al.* (2023) concerning lower estimates for the form $\Lambda(F_p(\xi))$ in the field $\mathbb{Q}_p$ for some fixed prime p.

As was noted above, the series

$$E(t) = \sum_{k=}^{\infty} k! t^k, \tag{4.9}$$

which was denoted by $E_1(t)$ in (4.1); however, here, we retain the notations used by Ernwall-Hytönen *et al.* (2023). We denote by $\mathbb{C}_p$ the algebraic closure of the field $\mathbb{Q}_p$. Recall that we considered the standard p-adic valuation given by

$$|p|_p = p^{-1}.$$

The series (4.9) converges for t from the disc

$$\left\{ z \in \mathbb{C}_p, |z|_p < p^{\frac{1}{p-1}} \right\},$$

and we denote by $E_p(t)$ the sum of the series (4.9) in $\mathbb{C}_p$.

On the other hand, the formal expansion of the integral

$$\mathcal{H}(t) = \int_0^{\infty} \frac{e^{-s}}{1 - ts} ds$$

coincides with $E(t)$, as was shown in the beginning of this chapter. The integral $\mathcal{H}(t)$ is not defined on the positive real axis, but it can be analytically continued there (we denote the resulting multivalued function by $\hat{\mathcal{H}}(t)$).

The Padé polynomials for the series $E(t)$ are given above. They converge to $E(t)$ in the p-adic case. The function $\hat{\mathcal{H}}(t)$ has the same Padé polynomials. They converge to this function in the Archimedean case for appropriate values of t. A very interesting interconnection between $E(t)$ and $\mathcal{H}(t)$ is expressed with the help of the continued fraction

$$\cfrac{1}{1 - t - \cfrac{t^2}{1 - 3t - \cfrac{2^2 t^2}{1 - 5t + \cdots}}} = \cfrac{a_1}{b_1 + \cfrac{a_2}{b_2 + \cdots}}.$$

Here,

$$a_1 = 1, \ b_1 = 1 - t, \ a_k = (k-1)^2 t^2, \ b_k = 1 - (2k-1)t, \text{ and } k \geq 2.$$

Theorem 4.4 (Theorem 1.1 in Ernwall-Hytönen *et al.* (2023)).
Let $p \in \mathbb{P}$, $t \in \mathbb{C}_p$, and $|t|_p < 1$. Then, in the field $\mathbb{Q}_p$,

$$\cfrac{1}{1 - t - \cfrac{t^2}{1 - 3t - \cfrac{2^2 t^2}{1 - 5t + \cdots}}} = \sum_{k=}^{\infty} k! t^k.$$

Let $t \in \mathbb{R}$ and $t \leq 0$. Then, in the field $\mathbb{R}$,

$$\cfrac{1}{1 - t - \cfrac{t^2}{1 - 3t - \cfrac{2^2 t^2}{1 - 5t + \cdots}}} = \int_0^{\infty} \frac{e^{-s}}{1 - ts} ds.$$

Let us formulate the theorem concerning lower estimates for the form $\Lambda(F_p(\xi))$ in the field $\mathbb{Q}_p$ for some fixed prime p.

Theorem 4.5 (Theorem 1.4 in Ernwall-Hytönen *et al.* (2023)).
Let p be a prime number and H, $l \in \mathbb{Z}$, $H \geq 1$, $l \geq 1$. Let $t \in \mathbb{Z}\backslash\{0\}$, $|t|_p < 1$, and denote

$$B(l, t) = \left(\frac{l|t|}{4}\right)^{\frac{1}{4}} + \left(\frac{l|t|}{4}\right)^{\frac{-1}{4}}.$$

If

$$2(l + 1)! B(l + 1, t) \cdot |t|^{t+1} e^{2\sqrt{\frac{l+1}{|t|}}} |t|_p^{2l} < \frac{1}{H},$$

then, for all c, $d \in \mathbb{Z}$, $c \neq 0$ with $\lceil c \rceil + |d| \leq H$, the following holds:

$$|c E_p(t) - d|_p \geq |t|_p^{2l}.$$

Corollary 4.1 (Corollary 1.5 in Ernwall-Hytönen *et al.* (2023)). *Let p be a prime number and H, $l \in \mathbb{Z}$, $H \geq 1$, $l \geq 1$. Suppose that a is a positive integer such that*

$$p^a > \left(2(l + 1)! B(l + 1, p^a) e^{2\sqrt{\frac{l+1}{|t|}}}\right)^{\frac{1}{l-1}},$$

where

$$B(l, \pm p^a) = \left(\frac{lp^a}{4}\right)^{\frac{1}{4}} + \left(\frac{lp^a}{4}\right)^{\frac{-1}{4}}.$$

Then, for all $c, d \in \mathbb{Z}$, $c \neq 0$ with $\lceil c \rceil + |d| \leq H$, the following holds:

$$|cE_p(\pm p^a) - d|_p \geq \frac{1}{p^{2al}}.$$

Corollary 4.2 (Corollary 1.6 in Ernwall-Hytönen *et al.* (2023)).

Let p be a prime number, $c, d \in \mathbb{Z}$, $c \neq 0$, $H \in \mathbb{Z}$, $H \geq 1$, and $\lceil c \rceil + |d| \leq H$. Suppose that a is a positive integer such that

$$p^a > (\log H)^4, \ \log H \geq e^8.$$

Then,

$$|cE_p(\pm p^a) - d|_p > \frac{1}{H^{\frac{a}{8}\log p}}.$$

Corollary 4.3 (Corollary 1.7 in Ernwall-Hytönen *et al.* (2023)).

Let p be a prime number, $H \in \mathbb{Z}$, and $H \geq 4$. Suppose that a is a positive integer such that

$$p^a > \frac{16}{11} e^{\frac{11+6\log 4 + \log 5 + 4\sqrt{10}}{11}} \log\left(2He^{\frac{11}{16}}\right).$$

Then, for all $c, d \in \mathbb{Z}$, $c \neq 0$ with $\lceil c \rceil + |d| \leq H$, the following holds:

$$|cE_p(\pm p^a) - d|_p > \frac{1}{\left(2He^{\frac{11}{16}}\right)^{\frac{32}{11}a\log p}}.$$

Chapter 5

Arithmetic Properties of Polyadic Series with Periodic Coefficients*

5.1. Formulation of the Theorem

Let us consider a sequence of integers $\{a_n\}$ of period T.

For every $k = 0, 1, 2, \ldots$, we have the equation $a_{k+T} = a_k$, which implies that

$$\sum_{k=0}^{\infty} a_k \cdot k! = \sum_{k=0}^{T-1} a_k \sum_{s=0}^{\infty} (k + sT)!. \tag{5.1}$$

We write

$$f_0(z) = \sum_{s=0}^{\infty} (zT)^{Ts} \cdot (1)_s \left(\frac{1}{T}\right)_s \cdots \left(\frac{T-1}{T}\right)_s, \tag{5.2}$$

$$f_k(z) = \sum_{s=0}^{\infty} (zT)^{Ts} \cdot (1)_s \left(\frac{1}{T}+1\right)_s \cdots \left(\frac{k}{T}+1\right)_s \left(\frac{k+1}{T}\right)_s$$

$$\cdots \left(\frac{T-1}{T}\right)_s, \quad k = 1, \ldots, T-2, \tag{5.3}$$

$$f_{T-1}(z) = \sum_{s=0}^{\infty} (zT)^{Ts} \cdot (1)_s \left(\frac{1}{T}+1\right)_s \cdots \left(\frac{T-1}{T}+1\right)_s. \tag{5.4}$$

---

*The results of this chapter are published in Chirskii (2017).

For $k = 1, \ldots, T - 1$, we have the equations

$$(k + sT)! = 1 \cdot 2 \cdot 3 \cdot \ldots \cdot (k + sT)$$

$$= 1(1 + T) \cdots (1 + sT) \cdot 2(2 + T) \cdot \ldots \cdot (2 + sT) \cdot \ldots$$

$$\cdot\, k(k + T) \cdot \ldots \cdot (K + sT) \cdot (k + 1)(k + 1 + T) \cdot \ldots$$

$$\cdot\, (k + 1 + (s - 1)T) \cdots T(2T) \cdots (sT)$$

$$= k! \cdot s! \cdot T^{sT} \left(\frac{1}{T} + 1 \right) \cdots \left(\frac{1}{T} + s \right) \cdots \left(\frac{k}{T} + 1 \right) \cdots$$

$$\left(\frac{k}{T} + s \right) \cdot \left(\frac{k + 1}{T} \right) \cdots \left(\frac{k + 1}{T} + s - 1 \right) \cdots$$

$$\left(\frac{T - 1}{T} \right) \cdots \left(\frac{T - 1}{T} + s - 1 \right)$$

$$= k! \cdot T^{sT} \cdot (1)_s \cdot \left(\frac{1}{T} + 1 \right)_s \cdots \left(\frac{k}{T} + 1 \right)_s$$

$$\times \left(\frac{k + 1}{T} \right)_s \cdots \left(\frac{k + 1}{T} \right)_s. \tag{5.5}$$

Moreover, for $k = 0$, we have the equation

$$T^{Ts}(1)_s \left(\frac{1}{T} \right)_s \cdots \left(\frac{T - 1}{T} \right)_s$$

$$= T^s \cdot s! \cdot T^s \left(\frac{1}{T} \right) \cdot \left(\frac{1}{T} + 1 \right) \cdots \left(\frac{1}{T} + s - 1 \right) \cdots T^s \left(\frac{T - 1}{T} \right)$$

$$\cdot \left(\frac{T - 1}{T} + 1 \right) \cdots \left(\frac{T - 1}{T} + s - 1 \right)$$

$$= T \cdot 2T \ldots sT \cdot 1 \cdot (T + 1) \cdots (T + (s - 1)T) \cdots (T - 1)$$

$$\cdot (2T - 1) \cdots (sT - 1) = (Ts)!. \tag{5.6}$$

Hence, by (5.2)−(5.6), we obtain

$$\sum_{s=0}^{\infty} (k + sT)! = \sum_{s=0}^{\infty} k! \cdot T^{sT} \cdot (1)_s \cdot \left(\frac{1}{T} + 1 \right)_s \cdots \left(\frac{k}{T} + 1 \right)_s$$

$$\times \left(\frac{k + 1}{T} \right)_s \cdots \left(\frac{k + 1}{T} \right)_s$$

$$= k! \sum_{s=0}^{\infty} T^{Ts}(1)_s \cdot \left(\frac{1}{T} + 1\right)_s \cdots \left(\frac{k}{T} + 1\right)_s$$

$$\times \left(\frac{k+1}{T}\right)_s \cdots \left(\frac{k+1}{T}\right)_s = k! f_k(1). \tag{5.7}$$

By (5.2) and (5.7), we see that

$$\sum_{k=0}^{\infty} a_k k! = \sum_{k=0}^{T-1} a_k \sum_{s=0}^{\infty} (k + sT)! = \sum_{k=0}^{T-1} a_k k! f_k(1).$$

Theorem 5.1. *Let $\xi \in \mathbb{Z}$, $\xi \neq 0$, then $f_0(\xi), f_1(\xi), \ldots, f_{T-1}(\xi)$ are infinitely algebraically independent polyadic numbers.*

Corollary 5.1. *Let $\{a_n\}$ be a periodic sequence of integers with non-zero numbers among a_n. Then, the series*

$$\sum_{n=1}^{\infty} a_n \cdot n!$$

represents an infinitely transcendental polyadic number. In particular, the Euler series $\sum_{n=1}^{\infty} n!$ is an infinitely transcendental polyadic number.

5.2. Algebraic Independence of Function Series

To prove Theorem 5.1, one must first prove the algebraic independence over $\mathbb{C}(z)$ of the formal power series $f_0(z), f_1(z), \ldots, f_{T-1}(z)$. This fact follows from Theorem 5.2, formulated and proved as follows. The scheme of the proof of Theorem 5.2 coincides with that of the proofs of Salikhov's theorems (Salikhov, 1973), see also Shidlovskii (1989).

Before formulating Theorem 5.2, we need some lemmas.

For two families of series S_1 and S_2, we denote their linear equivalence over the field $\mathbb{C}(z)$ by writing $S_1 \sim S_2$.

Lemma 5.1. *The following relation of linear equivalence holds:*

$$\{f_0(z), \ldots, f_{T-1}(z)\} \sim \{f_0(z), \ldots, f_0^{(T-1)}(z)\}.$$

92   *Polyadic Transcendental Number Theory*

Proof of Lemma 5.1. We write $\delta = z\frac{d}{dz}$ and claim the validity of the equation

$$(\delta + k + 1)f_k(z) = (k+1)f_{k+1}(z), \quad k = 0, \ldots, T-2. \tag{5.8}$$

Indeed,

$$(\delta + k + 1)f_k(z) = \left(z\frac{d}{dz} + k + 1\right)\left(\sum_{s=0}^{\infty}(zT)^{Ts}(1)_s\left(\frac{1}{T}+1\right)_s\right.$$
$$\times \left.\left(\frac{k}{T}+1\right)_s\left(\frac{k+1}{T}\right)_s \cdots \left(\frac{T-1}{T}\right)_s\right)$$
$$= \sum_{s=0}^{\infty}(Ts+k+1)(zT)^{Ts}(1)_s\left(\frac{1}{T}+1\right)_s \cdots$$
$$\left(\frac{k}{T}+1\right)_s\left(\frac{k+1}{T}\right)_s \cdots \left(\frac{T-1}{T}\right)_s. \tag{5.9}$$

Further,

$$\left(\frac{k+1}{T}\right)_s \cdot (k+1+Ts)$$
$$= \left(\frac{k+1}{T}\right)\left(\frac{k+1}{T}+1\right)\cdots\left(\frac{k+1}{T}+s-1\right)\cdot T\cdot\left(\frac{k+1}{T}+s\right)$$
$$= (k+1)\left(\frac{k+1}{T}+1\right)\cdots\left(\frac{k+1}{T}+s-1\right)\cdot\left(\frac{k+1}{T}+s\right)$$
$$= (k+1)\left(\frac{k+1}{T}+1\right)\cdots\left(\frac{k+1}{T}+1+s-2\right)$$
$$\cdot\left(\frac{k+1}{T}+1+s-1\right) = (k+1)\left(\frac{k+1}{T}+1\right)_s. \tag{5.10}$$

Using (5.9), (5.10), and (5.2)–(5.4), we obtain equations (5.8), which imply the assertion of Lemma 5.1. □

We write

$$\Phi(z) = \sum_{n=0}^{\infty} n!z^n. \tag{5.11}$$

Lemma 5.2. *The following equation holds:*

$$f_0(z) = \frac{1}{T} \sum_{j=0}^{T-1} \Phi(\zeta^j z), \tag{5.12}$$

where $\zeta = \exp\left(\frac{2\pi i}{T}\right)$.

Proof of Lemma 5.2. It follows from equations (5.2) and (5.6) that

$$f_0(z) = \sum_{s=0}^{\infty} (zT)^{Ts} (1)_s \left(\frac{1}{T}\right)_s \cdots \left(\frac{T-1}{T}\right)_s = \sum_{s=0}^{\infty} z^{Ts} (Ts)!.$$

Further, using (5.11), we obtain

$$\frac{1}{T} \sum_{j=0}^{T-1} \Phi(\zeta^j z) = \frac{1}{T} \sum_{j=0}^{T-1} \sum_{r=0}^{\infty} (\zeta)^{jr} z^r \cdot r! = \frac{1}{T} \sum_{r=0}^{\infty} \left(\sum_{j=0}^{T-1} \zeta^{jr}\right) z^r \cdot r!.$$

Since $\zeta = \exp\left(\frac{2\pi i}{T}\right)$, it follows that when r is divisible by T, we have the equation

$$\sum_{j=0}^{T-1} \zeta^{jr} = \sum_{j=0}^{T-1} 1 = T.$$

When r is not divisible by T, we have the equation

$$\sum_{j=0}^{T-1} \zeta^{jr} = \frac{1 - \zeta^{rT}}{1 - \zeta^r} = 0.$$

Thus, we see that

$$\frac{1}{T} \sum_{j=0}^{T-1} \Phi(\zeta^j z) = \frac{1}{T} \sum_{s=0}^{\infty} T z^{Ts} (Ts)! = f_0(z).$$

This completes the proof of Lemma 5.2. $\qquad\square$

We introduce the function

$$\Psi(z) = z\Phi(z). \tag{5.13}$$

It follows immediately from formula (5.12) in Lemma 5.2 that

$$f_0(z) = \frac{1}{zT} \sum_{j=0}^{T-1} \zeta^{-j} \Psi(\zeta^j z). \tag{5.14}$$

Lemma 5.3. *The following relation of linear equivalence holds:*

$$\{f_0(z), \dots, f_{T-1}(z)\} \sim \{\Psi(\zeta^j z), \; j = 0, 1, \dots, T-1\}.$$

Proof of Lemma 5.3. For every $a \neq 0$, the series $\Psi(az)$ satisfies the differential equation

$$(\Psi(az))' = \frac{\Psi(az)}{az^2} - \frac{1}{z} \tag{5.15}$$

because

$$(\Psi(az))' = \left(\sum_{n=0}^{\infty} n!(az)^{n+1}\right)' = a \sum_{n=0}^{\infty} (n+1)!(az)^n$$

$$= \frac{1}{az^2} \left(\sum_{n=0}^{\infty} (n+1)!(az)^{n+2}\right)$$

$$= \frac{1}{az^2} \left(\sum_{n=1}^{\infty} n!(az)^{n+1}\right) = \frac{1}{az^2} \left(\sum_{n=0}^{\infty} n!(az)^{n+1} - az\right)$$

$$= \frac{\Psi(az)}{az^2} - \frac{1}{z}.$$

By equations (5.12) and (5.13),

$$f_0'(z) = \frac{-1}{Tz^2} \sum_{j=0}^{T-1} \zeta^{-j} \Psi(\zeta^j z) + \frac{1}{zT} \sum_{j=0}^{T-1} \zeta^{-j} \left(\frac{\Psi(\zeta^{-j} z)}{\zeta^j z^2} - \frac{1}{z}\right)$$

$$= \frac{1}{Tz^3} \sum_{j=0}^{T-1} \zeta^{-2j} \Psi(\zeta^j z) - \frac{1}{z} f_0(z) - \frac{1}{z^2 T} \sum_{j=0}^{T-1} \zeta^{-j}.$$

For every $l = 1, \ldots, T - 1$, we have the equation

$$\sum_{j=0}^{T-1} \zeta^{-lj} = \frac{1}{\zeta^{l(T-1)}} \left(1 + \zeta^l + \cdots + \zeta^{l(T-1)}\right) = \frac{1 - \zeta^{lT}}{\zeta^{l(T-1)}(1 - \zeta^l)} = 0.$$

$$(5.16)$$

Hence,

$$f_0'(z) + \frac{1}{z} f_0(z) = \frac{1}{Tz^3} \sum_{j=0}^{T-1} \zeta^{-2j} \Psi(\zeta^j z). \qquad (5.17)$$

Suppose that the following equation holds for $k = 0, \ldots, T - 3$:

$$f_0^{(k+1)}(z) + L_k \left(f_0(z), f_0'(z), \ldots, f_0^{(k)}(z)\right)$$

$$= \frac{1}{Tz^{2k+3}} \sum_{j=0}^{T-1} \zeta^{-(k+2)j} \Psi(\zeta^j z), \qquad (5.18)$$

where $L_k(f_0(z), f_0'(z), \ldots, f_0^{(k)}(z))$ stands for a linear form in the series $f_0(z), f_0'(z), \ldots, f_0^{(k)}(z)$ and the coefficients of the form are rational functions of z. It follows from (5.15)–(5.18) that

$$f_0^{(k+2)}(z) + L_k' \left(f_0(z), f_0'(z), \ldots, f_0^{(k)}(z)\right)$$

$$= \frac{-(2k+3)}{Tz^{2k+4}} \sum_{j=0}^{T-1} \zeta^{-(k+2)j} \Psi(\zeta^j z)$$

$$+ \frac{1}{Tz^{2k+3}} \sum_{j=0}^{T-1} \zeta^{-(k+2)j} \left(\frac{\Psi(\zeta^j z)}{\zeta^j z^2} - \frac{1}{z}\right)$$

$$= \frac{-(2k+3)}{z} \cdot \frac{1}{Tz^{2k+3}} \sum_{j=0}^{T-1} \zeta^{-(k+2)j} \Psi(\zeta^j z)$$

$$+ \frac{1}{Tz^{2k+5}} \sum_{j=0}^{T-1} \zeta^{-(k+3)j} \Psi(\zeta^j z) - \frac{1}{Tz^{2k+4}} \sum_{j=0}^{T-1} \zeta^{-(k+2)j}$$

$$= \frac{-(2k+3)}{z} \cdot \left(f_0^{(k+1)}(z) + L_k \left(f_0(z), f_0'(z), \ldots, f_0^{(k)}(z) \right) \right)$$

$$+ \frac{1}{Tz^{2k+5}} \sum_{j=0}^{T-1} \zeta^{-(k+3)j} \Psi(\zeta^j z).$$

The equation proved above can be represented in the form

$$f_0^{(k+2)}(z) + L_{k+1} \left(f_0(z), f_0'(z), \ldots, f_0^{(k+1)}(z) \right)$$

$$= \frac{1}{Tz^{2k+5}} \sum_{j=0}^{T-1} \zeta^{-(k+3)j} \Psi(\zeta^j z).$$

Thus, equations (5.18) hold for $k = 0, \ldots, T-2$. By Lemma 5.1, these equations, together with equation (5.14), can be represented in matrix form as

$$A\mathbf{f_0} = B\mathbf{\Psi},$$

where $\mathbf{f_0}$ stands for the column vector with the coordinates $\{f_0(z), \ldots, f_{T-1}(z)\}$ and $\mathbf{\Psi}$ the column vector with the coordinates $\{\Psi(\zeta^j z), j = 0, 1, \ldots, T-1\}$. The matrix A is triangular with a value of 1 on the main diagonal, and the determinant of the matrix B is the Vandermonde determinant with a coefficient that is a non-zero rational function of z. Hence, the transformation matrix for the passage from the system $\{f_0(z), \ldots, f_{T-1}(z)\}$ into the system $\{\Psi(\zeta^j z), j = 0, 1, \ldots, T-1\}$ is non-singular. □

Theorem 5.2. *Let $\alpha_1, \ldots, \alpha_m$ be distinct non-zero numbers. Then, the formal power series $\Psi(\alpha_1 z), \ldots, \Psi(\alpha_m z)$ are algebraically independent over the field $\mathbb{C}(z)$.*

Proof of Theorem 5.2. We carry out the proof by induction. When $m = 1$, the series $\Psi(\alpha z)$ has a zero radius of convergence in the field $\mathbb{C}$ of complex numbers, and hence it is not an algebraic function.

Moreover, the following lemma holds.

Lemma 5.4. *The equation*

$$y' = \frac{y}{\alpha z^2} - \frac{1}{z}, \quad \alpha \neq 0, \tag{5.19}$$

has no solutions that are algebraic functions of z.

Proof of Lemma 5.4. Suppose the contrary, and consider an expansion of an algebraic solution y into a series:

$$y = \sum_{k=0}^{\infty} a_k \cdot z^{r_k}, \quad r_{k+1} > r_k, \quad k = 0, 1, \ldots, \quad (5.20)$$

where $r_k \in \mathbb{Q}$ are fractions with the same denominator. Then, we have the equation

$$y' = \sum_{k=0}^{\infty} a_k \cdot r_k z^{r_k-1},$$

and equation (5.19) gives

$$\sum_{k=0}^{\infty} a_k \cdot r_k \cdot z^{r_k-1} = \sum_{k=0}^{\infty} \frac{a_k}{\alpha} z^{r_k-2} - \frac{1}{z}. \quad (5.21)$$

By considering the least degree of z in equation (5.21), we see that the equation $r_0 - 2 = -1$ must hold, whence $r_0 = 1$. Hence, $a_0 = \alpha$. The following equations hold for $k = 0, 1, 2, \ldots$:

$$r_{k+1} - 2 = r_k - 1, \quad \frac{a_{k+1}}{\alpha} = a_k \cdot r_k,$$

whence $r_{k+1} = k + 2$, $a_{k+1} = a_k \cdot (k+1) \cdot \alpha$.

Thus, the series y in (5.20) becomes

$$y = \sum_{k=0}^{\infty} k! (\alpha z)^{k+1},$$

which is impossible because for $\alpha \neq 0$, this series has a zero radius of convergence in the field $\mathbb{C}$ and hence cannot represent any algebraic function.   □

Lemma 5.5. *Let the series y_i, $i = 1, \ldots, m$, satisfy the equations*

$$y_i' = \frac{y_i}{\alpha_i z^2} - \frac{1}{z}, \quad i = 1, \ldots, m, \quad (5.22)$$

where α_i, $i = 1, \ldots, m$, are pairwise distinct non-zero numbers. Then, the formal power series $y_1(z), \ldots, y_m(z)$ and 1 are linearly independent over the field $\mathbb{C}(z)$.

Proof of Lemma 5.5. Suppose the contrary, and let the dimension of the vector space over $\mathbb{C}(z)$, spanned by these series, be equal to k, $k < m$. Since it follows from Lemma 5.4 that there are no rational functions among the series $y_1, \ldots, y_m$, then $k \geq 2$. Suppose that $1, y_1, \ldots, y_{k-1}$ are linearly independent, $1, y_1, \ldots, y_k$ are linearly dependent, and the equation

$$q_0 + q_1 y_1 + \cdots + q_k y_k = 0, \quad \text{where } q_i \in \mathbb{C}[z], \quad i = 0, 1, \ldots, k, \tag{5.23}$$

holds. We note that according to the way in which k is chosen, $q_k \neq 0$. Moreover, since there are no rational functions among $y_1, \ldots, y_k$, it follows that there is an index l, $1 \leq l \leq k-1$, such that $q_l \neq 0$.

Differentiating equation (5.23) and using system (5.22), we obtain

$$q_0' + q_1' y_1 + q_1 \left(\frac{y_1}{\alpha_1 z^2} - \frac{1}{z} \right) + q_l' y_l + q_l \left(\frac{y_l}{\alpha_l z^2} - \frac{1}{z} \right) + \cdots$$

$$+ q_k' y_k + q_k \left(\frac{y_k}{\alpha_k z^2} - \frac{1}{z} \right) = 0$$

or

$$q_0' + \frac{1}{z} (q_1 + \cdots + q_k) + \left(q_1' + \frac{q_1}{\alpha_1 z^2} \right) y_1 + \cdots + \left(q_l' + \frac{q_l}{\alpha_l z^2} \right) y_l$$

$$+ \cdots + \left(q_k' + \frac{q_k}{\alpha_k z^2} \right) y_k = 0. \tag{5.24}$$

It follows from (5.23) and (5.24) that

$$\frac{q_l' + \frac{q_l}{\alpha_l z^2}}{q_l} = \frac{q_k' + \frac{q_k}{\alpha_k z^2}}{q_k}, \tag{5.25}$$

for otherwise it would be possible to eliminate the variable y_k from these two equations and obtain a non-trivial equation connecting the series $1, y_1, \ldots, y_{k-1}$ over $\mathbb{C}(z)$. We transform equation (5.25) into the form

$$\frac{q_l'}{q_l} - \frac{q_k'}{q_k} = \frac{1}{\alpha_k z^2} - \frac{1}{\alpha_l z^2},$$

and hence

$$\alpha_l \alpha_k z^2 (q_l' q_k - q_k' q_l) = (\alpha_l - \alpha_k) q_l q_k.$$

We denote by R_l, R_k, the orders of zero at the point $z = 0$ of the polynomials q_l, q_k, respectively. The orders of zero at the point $z = 0$ of the polynomials q_l', q_k' are not less than $R_l - 1, R_k - 1$, respectively. Since $\alpha_l \neq \alpha_k$, it follows that the right-hand side of the last equation has an order of zero at the point $z = 0$, equal to $R_l + R_k$. The left-hand side of this equation has an order of zero at the point $z = 0$ not less than $R_l + R_k + 1$. This proves the lemma. $\qquad \square$

We continue the proof of Theorem 5.2 by induction. Let $m > 1$, let the formal power series

$$\Psi_1 = \Psi(\alpha_1 z), \ldots, \Psi_m = \Psi(\alpha_m z) \tag{5.26}$$

be algebraically dependent over $\mathbb{C}(z)$, and let l be such that every $l - 1$ among the series (5.26) are algebraically independent, while l of them are algebraically dependent. Since we can control the indexing of series (5.26), we may assume that the series $\Psi_1, \ldots, \Psi_{l-1}$ are algebraically independent and the series $\Psi_1, \ldots, \Psi_l$ are algebraically dependent. Consider a system of differential equations satisfied by these series,

$$w_i' = \frac{w_i}{\alpha_i z^2} - \frac{1}{z}, \quad i = 1, \ldots, l, \tag{5.27}$$

and a differential operator corresponding to system (5.27),

$$D = \frac{\partial}{\partial z} + \sum_{i=1}^{l} \left(\frac{w_i}{\alpha_i z^2} - \frac{1}{z} \right) \frac{\partial}{\partial w_i}. \tag{5.28}$$

Let $P = P(z, w_1, \ldots, w_l) \in \mathbb{C}[z, w_1, \ldots, w_l]$ be an irreducible polynomial that is not identically zero and such that

$$P(z, \Psi_1, \ldots, \Psi_l) = 0. \tag{5.29}$$

Applying Lemma 4 from Chapter 5 of the book by Shidlovskii (1989), we see that for some $Q(z) \in \mathbb{C}(z)$, we have the following equation for polynomials in $\mathbb{C}[z, w_1, \ldots, w_l]$:

$$DP = QP. \tag{5.30}$$

Let the degree of the polynomial P with respect to the family of variables $w_1, \ldots, w_l$ be equal to s. We represent P in the form

$$P = \sum_{k_1 + \cdots + k_l \leq s} P_{\overline{k}} \cdot w_1^{k_1} \ldots w_l^{k_l}, \tag{5.31}$$

where $\overline{k} = (k_1, \ldots, k_l)$, $P_{\overline{k}} \in \mathbb{C}[z]$, $k_i \in \mathbb{Z}$, $k_i \geq 0$, $i = 1, \ldots, l$. It follows from formulae (5.27)–(5.31) that

$$D \left(\sum_{k_1 + \cdots + k_l \leq s} P_{\overline{k}} \, w_1^{k_1} \ldots w_l^{k_l} \right) = Q \left(\sum_{k_1 + \cdots + k_l \leq s} P_{\overline{k}} \, w_1^{k_1} \ldots w_l^{k_l} \right),$$

whence

$$\sum_{k_1 + \cdots + k_l \leq s} \left(P_{\overline{k}}' + \sum_{i=1}^{l} P_{\overline{k}} \left(\frac{k_i w_1^{k_1} \ldots w_l^{k_l}}{\alpha_i z^2} - \frac{k_i}{z} w_1^{k_1} \ldots w_i^{k_i - 1} \ldots w_l^{k_l} \right) \right)$$

$$= \sum_{k_1 + \cdots + k_l \leq s} Q P_{\overline{k}} \, w_1^{k_1} \ldots w_l^{k_l}. \tag{5.32}$$

Consider an arbitrary non-zero term of the polynomial P with the largest degree s with respect to the family of variables $w_1, \ldots, w_l$. Let it be of the form

$$P_{\overline{r}} w_1^{r_1} \ldots w_l^{r_l}.$$

It follows from (5.32) that $P_{\overline{r}}$ satisfies the differential equation

$$y' = \left(Q - \sum_{i=1}^{l} \frac{r_i}{\alpha_i z^2} \right) y. \tag{5.33}$$

The corresponding equation in which r_i is replaced by t_i holds for every coefficient $P_{\overline{t}}$ with index $\overline{t} = (t_1, \ldots, t_l)$, satisfying the condition $t_1 + \cdots + t_l = s$.

Let v be chosen in such a way that $r_v > 0$, and let $P_{\overline{k}}$ be the coefficient of P with index $\overline{k} = (k_1, \ldots, k_l)$, for which the equations $k_i = r_i$, $i \neq v$, and $k_v = r_v - 1$ hold. It follows from formula (5.32) that

$$P_{\overline{k}}' = \left(Q + \sum_{i=1}^{l} \left(-\frac{k_i}{\alpha_i z^2} \right) \right) P_{\overline{k}} + \sum_{j=1}^{l} (k_j + 1) \cdot \left(\frac{1}{z} \right) \cdot P_{\overline{k}_i}, \quad (5.34)$$

where

$$\overline{k}_j = (k_1, \ldots, k_{j-1}, k_j + 1, k_{j+1}, \ldots, k_l). \quad (5.35)$$

Consider the series

$$\Phi = \Phi(z) = -\sum_{j=1}^{l} (k_j + 1) P_{\overline{k}_j} \Psi_j \quad (5.36)$$

and evaluate its derivative using (5.15), (5.26), (5.33), and (5.36) and the fact that for $P_{\overline{k}_j}$ the sum of subscripts is equal to s:

$$\Phi' = -\sum_{j=1}^{l} (k_j + 1)(P_{\overline{k}_j}' \Psi_j + P_{\overline{k}_j} \Psi_j')$$

$$= \sum_{j=1}^{l} (k_j + 1) \left(Q + \sum_{i=1}^{l} k_i \left(-\frac{1}{\alpha_i z^2} \right) \right) P_{\overline{k}_j} \Psi_j$$

$$+ \sum_{j=1}^{l} (k_j + 1) \cdot \frac{1}{z} \cdot P_{\overline{k}_j}$$

$$= \left(Q + \sum_{i=1}^{l} k_i \left(-\frac{1}{\alpha_i z^2} \right) \right) \Phi + \sum_{j=1}^{l} P_{\overline{k}_j} \cdot (k_j + 1) \cdot \frac{1}{z}. \quad (5.37)$$

It follows from equations (5.34) and (5.37) that

$$P_{\overline{k}}' - \Phi' = \left(Q + \sum_{i=1}^{l} \left(-\frac{k_i}{\alpha_i z^2} \right) \right) (P_{\overline{k}} - \Phi). \quad (5.38)$$

We write $Y_{\overline{k}} = P_{\overline{k}} - \Phi$. Let

$$w = \frac{Y_{\overline{k}}}{P_{\overline{k}_v}},$$

where $P_{\overline{k}_v} = P_{\overline{r}} \neq 0$ by the definition of $\overline{k}_v$. Then, by (5.33) and (5.38), the following equations hold:

$$\frac{w'}{w} = \frac{Y'_{\overline{k}}}{Y_{\overline{k}}} - \frac{P'_{\overline{k}_v}}{P_{\overline{k}_v}} = \left(Q - \sum_{i=1}^{l} \frac{k_i}{\alpha_i z^2} \right) - \left(Q - \sum_{\substack{i=1 \\ i \neq v}}^{l} \left(-\frac{k_i}{\alpha_i z^2} \right) \right) + \frac{k_v + 1}{\alpha_v z^2}$$

$$= \frac{1}{\alpha_v z^2}.$$

Hence,

$$w' = \frac{w}{\alpha_v z^2}. \tag{5.39}$$

Consider the series

$$\Omega = \Psi_v - \frac{1}{k_v + 1} w = \Psi_v - \frac{1}{k_v + 1} \left(\frac{P_{\overline{k}} - \Phi}{P_{\overline{r}}} \right)$$

$$= \frac{1}{(k_v + 1)P_{\overline{r}}} (P_{\overline{r}} \Psi_v (k_v + 1) - P_{\overline{k}} + \Phi$$

$$= \frac{1}{(k_v + 1)P_{\overline{r}}} (P_{\overline{r}} \Psi_v (k_v + 1) - P_{\overline{k}} + \sum_{i=1}^{l} (k_i + 1) P_{\overline{k}_i} \cdot \Psi_i$$

$$= -\frac{1}{(k_v + 1)P_{\overline{r}}} \left(P_{\overline{k}} + \sum_{\substack{i=1 \\ i \neq v}}^{l} (k_i + 1) P_{\overline{k}_i} \Psi_i \right), \tag{5.40}$$

which takes into account that $\overline{r} = \overline{k}_v$ by the definition of $\overline{k}$ and equation (5.35).

Since

$$\Psi'_v = \frac{\Psi_v}{\alpha_v z^2} - \frac{1}{z}, \frac{w'}{k_v + 1} = \frac{w}{\alpha_v (k_v + 1) z^2},$$

by (5.15) and (5.39), we obtain

$$\Omega' = \left(\Psi_v - \frac{w}{k_v + 1}\right)' = \frac{\Psi_v}{\alpha_v z^2} - \frac{1}{z} - \frac{1}{(k_v + 1)}\frac{w}{\alpha_v z^2}$$

$$= \frac{1}{\alpha_v z^2}\left(\Psi_v - \frac{w}{k_v + 1} - \frac{1}{z}\right),$$

that is, Ω is a solution to the equation satisfied by Ψ_v. However, equation (5.40) implies that the series

$$\Omega, 1, \Psi_1, \ldots, \Psi_{v-1}, \Psi_{v+1}, \ldots, \Psi_l$$

are linearly dependent over $\mathbb{C}(z)$, which contradicts Lemma 5.5. This completes the proof of Theorem 5.2.   □

5.3.   Completion of the Proof of Theorem 5.1

The series $f_0(z), \ldots, f_{T-1}(z)$ defined by (5.2)–(5.4) belong to the class $F(\mathbb{Q}, c_1, c_2, c_3, d)$, where $c_1 = T \ln 2$, $c_2 = c_3 = 0$, and $d = 1$. They constitute a solution of the system of linear differential equations,

$$y_{k'} = (k + 1)\frac{y_{k+1} - y_k}{z}, k = 0, \ldots, T - 2,$$

$$y_{T-1'} = \frac{T}{z^{T+1}(T - 1)!}(y_0 - 1) - \frac{T}{z} \cdot y_{T-1}.$$

Now, we can apply Theorem 3.3. Theorem 5.1 is proved.   □

Chapter 6

Hypergeometric *F*-Series*

6.1. Hypergeometric Gevrey Series

The *Pohhammer symbol* is defined by the equations

$$(a)_0 = 1, \quad (a)_n = a(a+1)\cdots(a+n-1), \quad n \geq 1. \tag{6.1}$$

For the set of real numbers

$$S = \{\alpha_1, \ldots, \alpha_r; \beta_1, \ldots, \beta_s\}, \tag{6.2}$$

with respect to which we assume that the numbers $\beta_1, \ldots, \beta_s$ are not negative integers, it is customary to denote by

$$_rF_s\left(\begin{matrix} \alpha_1 \cdots \alpha_r \\ \beta_1 \cdots \beta_s \end{matrix}; z\right) = \sum_{n=0}^{\infty} \frac{(\alpha_1)_n \cdots (\alpha_r)_n}{(\beta_1)_n \cdots (\beta_s)_n n!} z^n \tag{6.3}$$

the so-called *generalised hypergeometric series*.
We shall consider a somewhat more general series of the form

$$\sum_{n=0}^{\infty} \frac{(\alpha_1)_n \cdots (\alpha_r)_n}{(\beta_1)_n \cdots (\beta_s)_n} z^n. \tag{6.4}$$

---

*The results of this chapter are published in Chirskii (2022a).

Series (6.3) and (6.4) belong to the so-called Gevrey series of orders $s + 1 - r$ and $s - r$, respectively.

If $r - s < 0$, then the series under consideration (6.4) represents an entire function (series (6.3) represents an entire function if $r - s \leq 0$). These cases include the exponential function, the Bessel functions, the Kummer functions, and many other functions that are important in mathematics. If the numbers $\alpha_1, \ldots, \alpha_r; \beta_1, \ldots, \beta_s$ are rational, then the functions belong to the class of Siegel E-functions, and the well-known Siegel–Shidlovskii method in the theory of transcendental numbers (Shidlovskii, 1989) can be applied. The works of Salikhov (1989, 1990, 1991), in which a near-complete solution to the problem was obtained, are devoted to the application of this method to hypergeometric E-functions. The case in which there are algebraic irrational numbers among the numbers $\alpha_1, \ldots, \alpha_r; \beta_1, \ldots, \beta_s$ was considered in the works of V. G. Sprindzhuk (1968), A. I. Galochkin (1984), A. N. Korobov (1981), P. L. Ivankov (1993), etc.

For $r - s = 0$, series (6.4) has a finite a radius of convergence (series (6.3) has a finite radius of convergence for $r - s = -1$). These functions include the logarithmic function, the Gauss hypergeometric function, many algebraic functions, and the incomplete elliptic integrals. Under the condition of rationality of the numbers $\alpha_1, \ldots, \alpha_r; \beta_1, \ldots, \beta_s$, these functions belong to the class of Siegel G-functions, and the papers of Galochkin (1974), Chudnovsky (1985), Bombieri (1981), André (1996), and Hata (1992) were devoted to their study.

If $r - s > 0$, then series (6.4), which differs from a polynomial, has a zero radius of convergence in the field $\mathbb{C}$ (series (6.3) has a zero radius of convergence if $r - s \geq 0$). In this case, the series can be treated as a formal one. Some series of this kind are asymptotic expansions that are of interest for mathematics and applications. In Chapter 4, we studied the arithmetic properties of the Euler series

$$\sum_{n=0}^{\infty} (-1)^n n! z^n.$$

6.2. Hypergeometric *F*-Series: Infinite Algebraic Independence of Their Values

Let $t = r - s > 0$. Consider the series

$$f(z) = \sum_{n=0}^{\infty} \frac{(\alpha_1)_n \cdots (\alpha_r)_n}{(\beta_1)_n \cdots (\beta_s)_n} (zt)^{tn}. \tag{6.5}$$

For the families of real numbers $\bar{a} = (a_1, \ldots, a_m)$ and $\bar{b} = (b_1, \ldots, b_m)$, we use the notation

$$\bar{a} \approx \bar{b} \tag{6.6}$$

if there is a permutation $i_1, \ldots, i_m$ of the numbers $1, \ldots, m$ such that $b_j - a_{i_j} \in \mathbb{Z}$, $j = 1, \ldots, m$. We use the notation $\bar{a} + c$ for the family $(a_1 + c, \ldots, a_m + c)$.

Theorem 6.1. *Let* $t = r - s = 2k$ *and let set* (6.2) *of rational parameters* $S = (\alpha_1, \ldots, \alpha_r; \beta_1, \ldots, \beta_s)$ *satisfy the following conditions:*

$$\alpha_i \notin \mathbb{Z}, \quad \beta_j \notin \mathbb{Z}, \quad \alpha_i - \beta_j \notin \mathbb{Z}, \quad i = 1, \ldots, t + s, \quad j = 1, \ldots, s.$$

For all common divisors d *of numbers* t *and* s, *either* $\bar{\alpha} + \frac{1}{d} \approx \bar{\alpha}$ *fails to hold or* $\bar{\beta} + \frac{1}{d} \approx \bar{\beta}$ *fails to hold.*

Let none of the following conditions hold:

1. *If* $s = 0$, *then there are* $x_0, \ldots, x_{k-1} \in \mathbb{C}$ *such that*

$$\bar{\alpha} + x_0 \approx \left(0, -\frac{1}{2}, x_1, -x_1, \ldots, x_{k-1}, -x_{k-1}\right).$$

2. *If* $s > 0$, $s = 2q$, *then there are* $x_0, \ldots, x_{k+s-1} \in \mathbb{C}$ *such that*

$$\bar{\alpha} + x_0 \approx \left(0, -\frac{1}{2}, x_1, -x_1, \ldots, x_{k+q-1}, -x_{k+q-1}\right),$$

$$\bar{\beta} + x_0 \approx (x_{k+q}, -x_{k+q}, \ldots, x_{k+s-1}, -x_{k+s-1}),$$

or

$$\bar{\alpha} + x_0 \approx (x_1, -x_1, \ldots, x_{k+q}, -x_{k+q}),$$

$$\bar{\beta} + x_0 \approx \left(0, -\frac{1}{2}, x_{k+q+1}, -x_{k+q+1}, \ldots, x_{k+s-1}, -x_{k+s-1}\right).$$

3. *If $s > 0$, $s = 2q + 1$, then there are $x_0, \ldots, x_{k+l-1} \in \mathbb{C}$ such that*

$$\bar{\alpha} + x_0 \approx (0, x_1, -x_1, \ldots, x_{k+q-1}, -x_{k+q-1}),$$

$$\bar{\beta} + x_0 \approx \left(-\frac{1}{2}, x_{k+q}, -x_{+q}, \ldots, x_{k+s-1}, -x_{k+s-1}\right).$$

Let $\mathbb{K}$ be an algebraic number field of finite degree $\varkappa$ over the field $\mathbb{Q}$ of rational numbers. Let $f'(z), f''(z), \ldots, f^{(r-1)}(z)$ be the formal derivatives of series (6.5). Let $\gamma \in \mathbb{K}$, $\gamma \neq 0$.

Then, the series $f(\gamma), f'(\gamma), f''(\gamma), \ldots, f^{(r-1)}(\gamma)$ are infinitely algebraically independent.

Remark. Since $\gamma \in \mathbb{K}$, there is a least positive integer b such that $b\gamma \in \mathbb{Z}_{\mathbb{K}}$ ($\mathbb{Z}_{\mathbb{K}}$ is the ring of integers of the field $\mathbb{K}$). Write

$$a_i = \frac{a_i}{A_i}, \quad a_i \in \mathbb{Z}, \quad A_i \in \mathbb{N}, \quad (a_i, A_i) = 1, \quad i = 1, \ldots, r, \quad q = \prod_{i=1}^{r} A_i^2.$$

Note that the infinite product of fields under consideration consists of the fields $\mathbb{K}_v$, where the valuation v of the field $\mathbb{K}$ extends the p-adic valuation of the field $\mathbb{Q}$, for all primes p except for those that divide the number bq.

Proof of Theorem 6.1. For convenience, we recall the definition of the F-series. The series

$$f(z) = \sum_{n=0}^{\infty} c_n \cdot n! \cdot z^n$$

belongs to the class $F(\mathbb{K}, C_1, C_2, C_3, q)$ if the following are true.

1. All coefficients c_n belong to some algebraic number field $\mathbb{K}$ of finite degree $\varkappa$ over the field $\mathbb{Q}$ of rational numbers.

2. The maxima of absolute values of numbers algebraically conjugate to the number c_n are $O(C_1^n)$, $n \to \infty$, with some real constant $C_1 > 1$.
3. There is a sequence of positive integers d_n such that, for $k = 0, 1, \ldots, n$, the numbers $d_n c_k$ belong to the ring of integers $\mathbb{Z}_{\mathbb{K}}$ of the field $\mathbb{K}$ and $d_n = q^n d_{0,n}$, $q \in \mathbb{N}$, the numbers $d_{0,n}$ are divisible only by primes p not exceeding $C_2 n$, and for all such primes p, the following inequality holds:

$$\vartheta_p(d_{0,n}) \leq C_3 \left(\log_p n + \frac{n}{p^2} \right).$$

(The symbol $\vartheta_p(a)$ indicates the power with which the prime p enters the factorisation of the integer a.) □

This definition shows that the F-series are related to the so-called Gevrey series.

Lemma 6.1. *If all numbers* $\alpha_1, \ldots, \alpha_r; \beta_1, \ldots, \beta_s$ *are rational, then series* (6.5) *is an F-series.*

Proof of Lemma 6.1. The first condition of the definition of F-series is satisfied. Consider the coefficient of series (6.5) at the term with $(nt)! z^{nt}$. Let a, b be rational numbers, while the number b is neither zero nor a negative integer. Consider the quantity (see (6.1))

$$\frac{(a)_n}{(b)_n} = \prod_{k=1}^{n} \frac{1 + \frac{a-1}{k}}{1 + \frac{b-1}{k}} = \prod_{k=1}^{n_0} \frac{1 + \frac{a-1}{k}}{1 + \frac{b-1}{k}}$$

$$\times \exp \left(\sum_{k=n_0+1}^{n} \left(\ln \left(1 + \frac{a-1}{k} \right) \right) - \left(\ln \left(1 + \frac{b-1}{k} \right) \right) \right),$$

where the number n_0 is chosen in such a way that $n_0 \geq \max(|a-1|, |b-1|)$. Let us use the Taylor formula and the equation

$$\sum_{k=1}^{n} \frac{1}{k} = \ln n + C + o \left(\frac{1}{n} \right), \quad n \to +\infty,$$

(the letter C stands for the Euler constant). As a result, we see that there is a positive number $\widetilde{c}_0$ for which

$$\frac{(a)_n}{(b)_n} = O(n^{\widetilde{c}_0}), \quad n \to +\infty. \tag{6.7}$$

The following inequality is obvious:

$$\frac{(tn)!}{(n!)^t} \leq t^n. \tag{6.8}$$

It follows from (6.7) and (6.8) that there is a positive integer C_1 for which

$$\left| \frac{(\alpha_1)_n \cdots (\alpha_r)_n t^n}{(\beta_1)_n \cdots (\beta_s)_n (tn)!} \right| \leq \exp(C_1 n), \quad n = 0, 1, 2, \ldots.$$

This inequality means that the other condition in the definition of F-series also holds.

Let us consider the least common denominator of the numbers

$$\frac{(a)_k}{(b)_k}, \quad k = 1, \ldots, n.$$

Let $a = \frac{\alpha}{A}$, $b = \frac{\beta}{B}$, $\alpha, \beta \in \mathbb{Z}$, $A, B \in \mathbb{N}$, $(\alpha, A) = (\beta, B) = 1$. Write

$$\frac{M_k}{N_k} = \frac{A^{2k}(a)_k}{B^k(b)_k} = \frac{A^k \alpha(\alpha + A) \cdots (\alpha + (k-1)A)}{\beta(\beta + B) \cdots (\beta + (k-1)B)}. \tag{6.9}$$

For every prime divisor p of the number N_k, the condition $(p, B) = 1$ holds. For every $l = 1, 2, \ldots$, if the variable x ranges over p^l consecutive integers, then exactly one of the corresponding numbers $\beta + Bx$ is divisible by p^l. Therefore, among the numbers $\beta(\beta + B) \cdots (\beta + (k-1)B)$ that are factors of the number N_k, at least $\left[\frac{k}{p^l}\right]$ and at most $1 + \left[\frac{k}{p^l}\right]$ numbers are divisible by p^l. Under the assumption that $p^l > |\beta| + (k-1)B$, none of these numbers is divisible by the number p^l.

Let N_k be divisible by p^ρ and not divisible by $p^{\rho+1}$. Then, the following inequalities hold:

$$\sum_{l=1}^{l_k} \left[\frac{k}{p^l}\right] \leq \rho \leq \sum_{l=1}^{l_k} \left(1 + \left[\frac{k}{p^l}\right]\right),$$

where we have used the notation

$$l_k = \left[\frac{\ln|\beta| + (k-1)B}{\ln p} \right].$$

Hence,

$$\left[\frac{k}{p} \right] \le \rho \le \left[\frac{k}{p} \right] + \tilde{c}_1 \frac{k}{p^2} + \tilde{c}_2 \frac{\ln(k+1)}{\ln p} \tag{6.10}$$

for $p \le |\beta| + (k-1)B$.

If a prime number p satisfies the condition $(p, A) = 1$, then, for the maximum degree τ of the number p for which the number M_k is divisible by p^τ, an inequality similar to the inequality on the left in (6.10) holds:

$$\left[\frac{k}{p} \right] \le \tau. \tag{6.11}$$

Inequality (6.11) remains valid for the primes p dividing the number A since M_k contains the factor A^k. Let r_p stand for the exponent with which the prime number p enters the prime factorisation of the exact denominator of the fraction in (6.9). It follows from inequalities (6.10) and (6.11) that

$$r_p \le \tilde{c}_1 \frac{n}{p^2} + \tilde{c}_2 \frac{\ln n}{\ln p}, \quad k = 0, 1, \dots, n. \tag{6.12}$$

Write

$$b_j = \frac{\beta_j}{B_j}, \quad \beta_j \in \mathbb{Z}, \quad B_j \in \mathbb{N}, \quad (\beta_j, B_j) = 1, \quad j = 1, \dots, s.$$

Every prime p entering the least common denominator of the fractions

$$\frac{(\alpha_1)_k \cdots (\alpha_r)_k t^{tk}}{(\beta_1)_k \cdots (\beta_s)_k (tk)!}, \quad k = 0, 1, \dots, n, \tag{6.13}$$

satisfies the inequality

$$p \le \max \left(\max_{J=1,\dots,s} (|\beta_j| + (n-1)B_j), tn \right). \tag{6.14}$$

Inequality (6.14) means that, for some positive constant C_2, the following inequality holds:

$$p \le C_2 n. \tag{6.15}$$

For every prime p satisfying the inequality (6.14), the degree of the power of p entering the expansion of the least common denominator of the fractions in (6.13) does not exceed the sum of r terms, each of which is bounded above by the right-hand side of inequality (6.12). Therefore, this degree is not higher than

$$C_3 \left(\frac{n}{p^2} + \frac{\ln n}{\ln p} \right) \tag{6.16}$$

for a corresponding constant C_3.

Write

$$a_i = \frac{\alpha_i}{A_i}, \quad \alpha_i \in \mathbb{Z}, \quad A_i \in \mathbb{N}, \quad (\alpha_i, A_i) = 1, \quad i = 1, \dots, r.$$

For the number q in the definition of the F-series, we can take the number

$$q = \prod_{i=1}^{r} A_i^2.$$

This completes the proof of Lemma 6.1. □

Lemma 6.2. *Series* (6.5) *is a formal solution of the differential equation*

$$\left\{ \prod_{j=1}^{s} \left(z \frac{d}{dz} + t(\beta_j - 1) \right) - z^t \prod_{i=1}^{r} \left(z \frac{d}{dz} + t\alpha_i \right) \right\} y = \prod_{j=1}^{s} t(\beta_j - 1). \tag{6.17}$$

Proof of Lemma 6.2. Indeed, on the one hand,

$$\prod_{j=1}^{s}\left(z\frac{d}{dz}+t\left(\beta_j-1\right)\right)\sum_{n=0}^{\infty}\frac{(\alpha_1)_n\cdots(\alpha_r)_n}{(\beta_1)_n\cdots(\beta_s)_n}(zt)^{tn}$$

$$=\sum_{n=0}^{\infty}\frac{(\alpha_1)_n\cdots(\alpha_r)_n}{(\beta_1)_n\cdots(\beta_s)_n}\left(\prod_{j=1}^{s}\left(tn+t(\beta_j-1)\right)\right)(zt)^{tn}$$

$$=\prod_{j=1}^{s}t\left(\beta_j-1\right)+t^s\sum_{n=1}^{\infty}\frac{(\alpha_1)_n\cdots(\alpha_r)_n}{(\beta_1)_{n-1}\cdots(\beta_s)_{n-1}}(zt)^{tn}. \qquad (6.18)$$

On the other hand,

$$z^t\prod_{i=1}^{r}\left(z\frac{d}{dz}+t\alpha_i\right)\frac{(\alpha_1)_n\cdots(\alpha_r)_n}{(\beta_1)_n\cdots(\beta_s)_n}(zt)^{tn}$$

$$=z^t t^r\sum_{n=0}^{\infty}\frac{(\alpha_1)_{n+1}\cdots(\alpha_r)_{n+1}}{(\beta_1)_n\cdots(\beta_s)_n}(zt)^{tn}$$

$$=t^s\sum_{n=0}^{\infty}\frac{(\alpha_1)_{n+1}\cdots(\alpha_r)_{n+1}}{(\beta_1)_n\cdots(\beta_s)_n}(zt)^{t(n+1)}$$

$$=t^s\sum_{n=1}^{\infty}\frac{(\alpha_1)_n\cdots(\alpha_r)_n}{(\beta_1)_{n-1}\cdots(\beta_s)_{n-1}}(zt)^{tn}. \qquad (6.19)$$

Equations (6.18) and (6.19) imply (6.17). This completes the proof of Lemma 6.2. □

Series of the form (6.5) if $l < m$ give the entire function

$$\sum_{n=0}^{\infty}\frac{(a_1)_n\cdots(a_l)_n}{(b_1)_n\cdots(b_m)_n}\left(\frac{z}{m-l}\right)^{(m-l)n},$$

satisfying differential equations similar to equation (6.17). After changing $z=\frac{1}{u}$ to the differential operator $z\frac{d}{dz}$, the corresponding differential operator becomes $-u\frac{d}{du}$, and equation (6.17) passes to the equation

$$\left\{ \prod_{j=1}^{s} \left(-u\frac{d}{du} + t(\beta_j - 1) \right) - u^{-t} \prod_{i=1}^{r} \left(-u\frac{d}{du} + t\alpha_i \right) \right\} \tilde{y}$$

$$= \prod_{j=1}^{s} t(\beta_j - 1), \tag{6.20}$$

where the symbol $\tilde{y}$ denotes the result of substituting $z = \frac{1}{u}$ into the formal power series $y \in \mathbb{R}[[z]]$. We restrict ourselves to the case of $t = 2k$, $k \in \mathbb{Z}$. In this case, one can readily transform equation (6.20) into the form

$$\left\{ \prod_{i=1}^{r} \left(u\frac{d}{du} + t((1 - \alpha_i) - 1) \right) - u^{t} \prod_{j=1}^{s} \left(u\frac{d}{du} + t(1 - \beta_j) \right) \right\} \tilde{y}$$

$$= (-1)^{r+1} u^{t} \prod_{j=1}^{s} t(\beta_j - 1). \tag{6.21}$$

Note that the homogeneous equation

$$\left\{ \prod_{j=1}^{s} \left(z\frac{d}{dz} + t(\beta_j - 1) \right) - z^{t} \prod_{i=1}^{r} \left(z\frac{d}{dz} + t\alpha_i \right) \right\} y = 0, \tag{6.22}$$

corresponding to equation (6.17), passes under the change $z = \frac{1}{u}$ to the homogeneous equation corresponding to equation (6.20) and has the form

$$\left\{ \prod_{i=1}^{r} \left(u\frac{d}{du} + t((1 - \alpha_i) - 1) \right) - u^{t} \prod_{j=1}^{s} \left(u\frac{d}{du} + t(1 - \beta_j) \right) \right\} \tilde{y} = 0. \tag{6.23}$$

Using the notations $r = t + l$, $s = l$, $\lambda_i = 1 - \alpha_i$, $\mu_j = 1 - \beta_j$, it coincides with the equation

$$\left\{ \prod_{i=1}^{t+l} \left(u\frac{d}{du} + t(\lambda_i - 1) \right) - u^{t} \prod_{j=1}^{l} \left(u\frac{d}{du} + t\mu_j \right) \right\} \tilde{y} = 0, \tag{6.24}$$

which was studied in the paper by Salikhov (1991). Let us introduce the definitions and notations necessary to apply the results of this paper.

Definition. A differential equation

$$y^{(m)} + r_{m-1}y^{(m-1)} + \cdots + r_0 y = 0, \quad r_i \in \mathbb{C}(z)$$

is said to be reducible over $\mathbb{C}(z)$ if there is a non-trivial solution $y(z)$ of this equation such that the functions $y(z), y'(z), \ldots, y^{(m-1)}(z)$ are algebraically dependent over $\mathbb{C}(z)$ and irreducible over $\mathbb{C}(z)$ otherwise.

Write $\bar{\mu} = (\mu_1, \ldots, \mu_l)$, $\bar{\lambda} = (\lambda_1, \ldots, \lambda_{t+l})$. (In the following, we use the notation in (6.6).)

Lemma 6.3 (Part of Theorem 2 from Salikhov (1991)). *Let $t = 2k$, and let the set of parameters $S = (\mu_1, \ldots, \mu_l; \lambda_1, \ldots, \lambda_{t+l})$ satisfy the following conditions:*

$$\mu_i \neq 0, -1, -2, \ldots, \quad \lambda_j \neq 0, -1, -2, \ldots,$$

$$\mu_i - \lambda_j \notin \mathbb{Z}, \quad i = 1, \ldots, l, \; j = 1, \ldots, t+l.$$

For all common divisors d of the numbers t, l, either $\bar{\mu} + \frac{1}{d} \approx \bar{\mu}$ fails to hold or $\bar{\lambda} + \frac{1}{d} \approx \bar{\lambda}$ fails to hold.

None of the following conditions holds:

1. *If $l = 0$, then there are $x_0, \ldots, x_{k-1} \in \mathbb{C}$ such that*

$$\bar{\lambda} + x_0 \approx \left(0, \frac{1}{2}, x_1, -x_1, \ldots, x_{k-1}, -x_{k-1}\right).$$

2. *If $l > 0$, $l = 2q$, then there are $x_0, \ldots, x_{k+l-1} \in \mathbb{C}$ such that either*

$$\bar{\lambda} + x_0 \approx \left(0, \frac{1}{2}, x_1, -x_1, \ldots, x_{k+q-1}, -x_{k+q-1}\right),$$

$$\bar{\mu} + x_0 \approx (x_{k+q}, -x_{+q}, \ldots, x_{k+l-1}, -x_{k+l-1})$$

or

$$\bar{\lambda} + x_0 \approx (x_1, -x_1, \ldots, x_{k+q}, -x_{k+q}),$$

$$\bar{\mu} + x_0 \approx \left(0, \frac{1}{2}, x_{k+q+1}, -x_{k+q+1}, \ldots, x_{k+l-1}, -x_{k+l-1}\right).$$

3. *If $l > 0$, $l = 2q + 1$, then there are $x_0, \ldots, x_{k+l-1} \in \mathbb{C}$ such that*

$$\bar{\lambda} + x_0 \approx (0, x_1, -x_1, \ldots, x_{k+q-1}, -x_{k+q-1}),$$

$$\bar{\mu} + x_0 \approx \left(\frac{1}{2}, x_{k+q}, -x_{+q}, \ldots, x_{k+l-1}, -x_{k+l-1}\right).$$

Then, equation (6.24) is irreducible.

Lemma 6.4. *Let $\alpha_1, \ldots, \alpha_r; \beta_1, \ldots, \beta_s$, $r - s = 2k \in \mathbb{N}$ satisfy the conditions of Theorem 6.1. (In other words, these numbers are such that the numbers $\lambda_i = 1 - \alpha_i$, $\mu_j = 1 - \beta_j$ satisfy the conditions of Lemma 6.3.) Then, for every solution y of equation (6.23), the series $y, y', \ldots, y^{(r-1)}$ were algebraically independent $\mathbb{C}(z)$.*

Proof of Lemma 6.4. If there were a solution y to this equation such that the series $y, y', \ldots, y^{(r-1)}$ were algebraically dependent, then the following equation would hold:

$$P(z, y, y', \ldots, y^{(r-1)}) = 0,$$

with some polynomial $P(z, x_1, \ldots, x_m)$, which differs from the zero polynomial. After the change $z = \frac{1}{u}$, this equation would be transformed into the equation

$$\tilde{P}(z, \tilde{y}, \tilde{y}', \ldots, \tilde{y}^{(r-1)}) = 0,$$

which contradicts Lemma 6.3.   □

The singular points of equation (6.23) are the points $u = 0$, $u = \infty$. Since the numbers $\alpha_1, \ldots, \alpha_r; \beta_1, \ldots, \beta_s$ are not integers, it follows that this equation has no solutions that are rational functions of u. Writing

$$\tilde{y}_1 = \tilde{y}, \ldots, \tilde{y}_r = \frac{d^{r-1}}{du^{r-1}}\tilde{y},$$

we represent this equation in the form of a system of linear inhomogeneous differential equations, and we represent equation (6.22) in the form of the corresponding system of linear homogeneous differential equations.

Further, let us use Theorem 3 in the paper by Nesterenko (1969). Here is its statement:

Let Λ be a field of analytic functions of z that is closed with respect to the differentiation operation and contains the field $\mathbb{C}$. Let a family of analytic functions,

$$f_1(z), \ldots, f_m(z), \quad m \geq 1,$$

form a solution of the system of linear inhomogeneous differential equations

$$y'_k = Q_{k,0} + \sum_{i=1}^{m} Q_{k,i} y_i, \quad k = 1, \ldots, m, \quad Q_{k,i} \in \Lambda.$$

Let every non-trivial solution of the corresponding system of linear homogeneous differential equations

$$y'_k = \sum_{i=1}^{m} Q_{k,i} y_i, \quad k = 1, \ldots, m, \quad Q_{k,i} \in \Lambda$$

consist of the functions algebraically independent over Λ. Then, either all the functions $f_1(z), \ldots, f_m(z)$ are algebraically independent over Λ or all these functions belong to the field Λ.

Note that the statement of this theorem remains valid for formal power series and the corresponding systems of differential equations.

Thus, the series forming the solution of the inhomogeneous system of equations obtained above and corresponding to equation (6.17) are algebraically independent; therefore, series (6.5) and its derivatives up to the order $r - 1$ are algebraically independent.

To prove Theorem 6.1, it remains to use Theorem 3.3.

Chapter 7

Arithmetic Properties of Generalised Hypergeometric Series with Algebraic Irrational Coefficients*

7.1. Generalised Hypergeometric Series with Algebraic Irrational Coefficients Are Not F-Series

In the previous chapter, we considered the series of the form

$$\sum_{n=0}^{\infty} \frac{(\alpha_1)_n \cdots (\alpha_r)_n}{(\beta_1)_n \cdots (\beta_s)_n} z^{(r-s)n},$$

where the numbers $\alpha_1, \ldots, \alpha_r, \beta_1, \ldots, \beta_s$ are rational and none of the numbers $\beta_1, \ldots, \beta_s$ is equal to zero or a negative integer. It was shown that in this case, the series is an F-series.

In this chapter, we consider the case where, among the numbers $\alpha_1, \ldots, \alpha_r, \beta_1, \ldots, \beta_s$, there are algebraic irrational numbers. More precisely, we shall consider series of the form

$$F(\mu_1, \ldots, \mu_m, z) = \sum_{n=0}^{\infty} \frac{(\mu_1)_n \cdots (\mu_m)_n}{n!} z^n, \qquad (7.1)$$

where $(\mu)_0 = 1$, $(\mu)_n = \mu(\mu + 1) \cdots (\mu + n - 1)$, $n \geq 1$.

---

*The results of this chapter are published in Chirskii (2014).

If among the numbers $\mu_1, \ldots, \mu_m$ there are algebraic irrational numbers, then series (7.1) doesn't belong to the class of F-series after the substitution of $z = t^{m-1}$. To see this, we note that if $\sum_{n=0}^{\infty} a_n n! z^n$ is an F-series, then $\sum_{n=0}^{\infty} \frac{a_n}{n!} z^n$ is an E-function (the definition of an E-function is given in the introduction).

Hence, if the series

$$\sum_{n=0}^{\infty} \frac{(\mu_1)_n \cdots (\mu_m)_n}{n!} t^{(m-1)n} \tag{7.2}$$

is an F-series, then the series

$$\sum_{n=0}^{\infty} \frac{(\mu_1)_n \cdots (\mu_m)_n}{n!(((m-1)n)!)^2} t^{(m-1)n} \tag{7.3}$$

is an E-function. Galochkin (1981) established a criterion for a hypergeometric series to be considered an E-function. To state this criterion, we introduce the so-called E-condition.

We say that two systems of algebraic numbers $(a_1, \ldots, a_u)$, $(b_1, \ldots, b_v)$, $v \geq u+1$, satisfy the E-condition if either all of them are rational numbers or one can divide the irrational numbers of these systems into pairs of

$$(a_{i_1}, b_{j_1}), \ldots, (a_{i_w}, b_{j_w})(i_\nu \neq i_\mu \text{ for } \nu \neq \mu, j_\eta \neq j_\lambda \text{ for } \eta \neq \lambda)$$

in such a way that all differences $a_{i_s} - b_{i_s}$, $s = 1, \ldots, w$, are nonnegative integers. We now cite Theorem 1 from Galochkin (1981):

A function

$$\sum_{n=0}^{\infty} \frac{(a_1 + 1)_n \cdots (a_\mu + 1)_n}{(b_1 + 1)_n \cdots (b_v + 1)_n} z^{(v-\mu)n}$$

with complex parameters $(a_1, \ldots, a_\mu), (b_1, \ldots, b_v)$, *different from negative integers and satisfying*

$$a_i \neq b_j, \quad i = 1, \ldots, u, \quad j = 1, \ldots v,$$

is an E-function if and only if the parameters $(a_1, \ldots, a_u), (b_1, \ldots b_v)$ *are algebraic numbers satisfying the E-condition.*

Thus, if at least one of the parameters $\mu_1, \ldots, \mu_m$ is an algebraic irrational number, then the E-condition is not satisfied, and function (7.3) is not an E-function. Based on what was said above, this means that series (7.2) is not an F-series.

Thus, hereinafter, we shall use another method based on the Hermite–Padé approximations.

7.2.   Hermite–Padé Approximations

Many authors have studied the arithmetic properties of the values of generalised hypergeometric series whose parameters include irrational numbers. Sprindzhuk (1968) established that the values of some transcendental functions are irrational. Estimates of linear forms in the values of such functions were obtained by Galochkin (1981, 1976) and Ivankov (1993, 1994, 1995). An important tool in the investigation is provided by the effective construction of the approximating forms. We shall rely heavily on the results of Nesterenko (1995) on the *Hermite–Padé approximations of generalised hypergeometric series.*

We put

$$f_0(z) = F(\alpha_1, \ldots, \alpha_m, z),$$

$$f_1(z) = F(\alpha_1 + 1, \alpha_2, \ldots, \alpha_m, z), \ldots, \tag{7.4}$$

$$f_{m-1}(z) = F(\alpha_1 + 1, \alpha_2 + 1, \ldots, \alpha_{m-1} + 1, \alpha_m, z),$$

where the integers $\alpha_1, \ldots, \alpha_{m-2} \in \mathbb{K}$ are different from negative integers and zero. Suppose that $\alpha_{m-1} = \alpha_m = 1$.

For every valuation v of $\mathbb{K}$ and every integer z from $\mathbb{K}_v$, the sums of all series (7.4) are integers in $\mathbb{K}_v$.

Let N be a positive integer. We define the numbers r, s, t, l by the condition

$$N = ms + r = (m-1)t + l, \quad 1 \le r \le m, \quad 1 \le l \le m - 1, \tag{7.5}$$

where

$$s = \left[\frac{N-1}{m}\right], \quad t = \left[\frac{N-1}{m-1}\right]. \tag{7.6}$$

We put

$$\alpha_N = \alpha_r + s, \tag{7.7}$$

$$f_N(z) = F(\alpha_{N+1}, \ldots, \alpha_{N+m}, z), \tag{7.8}$$

$$u_N(z) = \alpha_1 \ldots \alpha_N z^t f_N(z). \tag{7.9}$$

Lemma 7.1. *For every positive integer N, there are polynomials $P_{N,0}(z), \ldots, P_{N,m-1}(z)$ such that*

$$u_N(z) = P_{N,0}(z)u_0(z) + \cdots + P_{N,m-1}(z)u_{m-1}(z). \tag{7.10}$$

Moreover, the degrees of $P_{N,0}(z), \ldots, P_{N,m-1}(z)$ do not exceed $t - s$, defined by (7.6). The series (7.9) are linearly independent over $\mathbb{K}(z)$. Additionally, the following recurrent equations hold:

$$u_{N+m}(z) = u_{N+1}(z) - \alpha_{N+1}u_N(z) \tag{7.11}$$

if $N = 0$ or $N \geq 1$ and N is not divisible by $m - 1$;

$$u_{N+m}(z) = u_{N+1}(z) - \alpha_{N+1}zu_N(z) \tag{7.12}$$

if $N \geq 1, N$ is divisible by $m - 1$, and for every $i = 0, 1, \ldots, m - 1$;

$$P_{N+m,i}(z) = P_{N+1,i}(z) - \alpha_{N+1}P_{N,i}(z) \tag{7.13}$$

if $N = 0$ or $N \geq 1$ and N is not divisible by $m - 1$;

$$P_{N+m,i}(z) = P_{N+1,i}(z) - \alpha_{N+1}zP_{N,i}(z) \tag{7.14}$$

if $N \geq 1, N$ is divisible by $m - 1$, and for every $i = 0, 1, \ldots, m - 1$. Furthermore, for all non-negative integer values of N, we have

$$\Delta_N(z) = (-1)^{mN}\alpha_1 \ldots \alpha_N z^t, \tag{7.15}$$

where

$$\Delta_N(z) = |P_{N+j,i}|_{\substack{j=0,\ldots,m-1 \\ i=0,\ldots,m-1}}. \tag{7.16}$$

Proof of Lemma 7.1. This lemma is an immediate corollary of Nesterenko's (1995) results. More precisely, it can be obtained in the manner described by Nesterenko (1995) from some of the results proved in Lemma 1, Corollary 2, Theorem 2, and Lemma 2 in his work. For example, formulae (7.11) and (7.12), which are the most important for us, are given by Nesterenko (1995). For completeness, we give the proof of this lemma.

Consider the quantity $u_{N+1}(z) - \alpha_{N+1}u_N(z)$, where $N = 0$ or $N \geq 1$ and N is not divisible by $(m-1)$.

By (7.7)–(7.9), we have

$$u_{N+1}(z) - \alpha_{N+1}u_N(z)$$

$$= \alpha_1 \cdots \alpha_{N+1} z^t F(\alpha_{N+2}, \ldots, \alpha_{N+m+1})$$

$$- \alpha_1 \cdots \alpha_{N+1} z^t F(\alpha_{N+1}, \ldots, \alpha_{N+m})$$

$$= \alpha_1 \cdots \alpha_{N+1} z^t \left(\sum_{n=0}^{\infty} \frac{(\alpha_{N+2})_n \cdots (\alpha_{N+m+1})_n}{n!} z^n \right.$$

$$- \sum_{n=0}^{\infty} \frac{(\alpha_{N+1})_n \cdots (\alpha_{N+m})_n}{n!} z^n \Bigg)$$

$$= \alpha_1 \cdots \alpha_{N+1} z^t \left(\sum_{n=0}^{\infty} \frac{(\alpha_{N+2})_n \cdots (\alpha_{N+m})}{n!} \right.$$

$$\times z^n ((\alpha_{N+m+1})_n - (\alpha_{N+1})_n) \Bigg). \tag{7.17}$$

It follows from (7.7) that

$$\alpha_{N+m+1} = \alpha_{N+1} + 1, \tag{7.18}$$

whence

$$(\alpha_{N+m+1})_n - (\alpha_{N+1})_n = (\alpha_{N+1} + 1)_n - (\alpha_{N+1})_n. \tag{7.19}$$

When $n = 0$, the quantity (7.19) is equal to 0. When $n \geq 1$,

$$(\alpha_{N+1} + 1)_n - (\alpha_{N+1})_n$$
$$= (\alpha_{N+1} + 1) \cdots (\alpha_{N+1} + n - 1)(\alpha_{N+1} + n)$$
$$\quad - \alpha_{N+1}(\alpha_{N+1} + 1) \cdots (\alpha_{N+1} + n - 1)$$
$$= (\alpha_{N+1} + 1) \cdots (\alpha_{N+1} + n - 1)\left((\alpha_{N+1} + n) - \alpha_{N+1}\right)$$
$$= n\,(\alpha_{n+1} + 1) \cdots (\alpha_{N+1} + n - 1)$$
$$= n(\alpha_{N+1} + 1)_{n-1} = n(\alpha_{N+m+1})_{n-1}. \tag{7.20}$$

Moreover, when $n \geq 1$, we obtain in view of (7.18) that

$$(\alpha_{N+k})_n = (\alpha_{N+k} + 1) \cdots (\alpha_{N+k} + n - 1)$$
$$= \alpha_{N+k}\alpha_{N+m+k} \cdots (\alpha_{N+m+k} + n - 2)$$
$$= \alpha_{N+k}(\alpha_{N+m+k})_{n-1}, \quad k = 2, \ldots, m. \tag{7.21}$$

We substitute the resulting values (7.20) of the quantity (7.8) into the right-hand side of (7.17) and use equalities (7.21):

$$\alpha_1 \ldots \alpha_{N+1} z^t \left(\sum_{n=0}^{\infty} \frac{(\alpha_{N+2})_n \cdots (\alpha_{N+m})_n}{n!} z^n ((\alpha_{N+m+1})_n - (\alpha_{N+1})_n) \right)$$

$$= \alpha_1 \ldots \alpha_{N+1} z^t \sum_{n=1}^{\infty} \frac{(\alpha_{N+2})_n \cdots (\alpha_{N+m})_n}{n!} z^n n\, (\alpha_{N+m+1})_{n-1}$$

$$= \alpha_1 \ldots \alpha_{N+1}\alpha_{N+2} \ldots \alpha_{N+m} z^{t+1}$$

$$\times \sum_{n=1}^{\infty} \frac{(\alpha_{N+2+m})_{n-1} \cdots (\alpha_{N+2m})_{n-1}}{(n-1)!} z^{n-1}\, (\alpha_{N+m+1})_{n-1}$$

$$= \alpha_1 \ldots \alpha_{N+m} z^{t+1} F\left(\alpha_{N+m+1}, \ldots, \alpha_{N+2m}\right) = u_{N+m}(z).$$

This proves (7.11).

If $N \geq 1$ and N is divisible by $m - 1$, then it follows from (7.5) that

$$N = (m-1)t + (m-1),$$
$$N+1 = (m-1)t + m = (m-1)(t+1) + 1,$$
$$N+m = (m-1)(t+2) + 1,$$

which, in turn, yields

$$u_{N+1}(z) = \alpha_1 \ldots \alpha_{N+1} z^{t+1} f_{N+1}(z), \tag{7.22}$$

$$u_{N+m}(z) = \alpha_1 \ldots \alpha_{N+m} z^{t+2} f_{N+m}(z). \tag{7.23}$$

Repeating verbatim the argument above and using (7.22) and (7.23), we get (7.12).

It follows from definitions (7.10) that

$$P_{0,0} = 1, \quad P_{0,1} = 0, \ldots, P_{0,m-1} = 0,$$

$$P_{1,0} = 0, \quad P_{1,1} = 1, \quad P_{1,2} = 0, \ldots, P_{1,m-1} = 0, \ldots \tag{7.24}$$

$$\ldots P_{m-1,0} = 0, \ldots, P_{m-1,m-2} = 0, \quad P_{m-1,m-1} = 1.$$

By (7.24), we have

$$\Delta_0(z) = 1. \tag{7.25}$$

The formulae (7.10) now follow from (7.11)–(7.14).

To obtain the formula for $\Delta_{N+1}(z)$ in terms of $\Delta_N(z)$, $N = 0, 1, 2, \ldots$, we consider determinant (7.16):

$$\Delta_N(z) = \begin{vmatrix} P_{N,0}(z) & \cdots & P_{N,m-1}(z) \\ \cdots & \cdots & \cdots \\ P_{N+m-1,0}(z) & \cdots & P_{N+m-1,m-1}(z) \end{vmatrix}.$$

When $N = 0$ or $N \geq 1$ and N is not divisible by $m - 1$, we obtain in view of (7.13) that

$$\Delta_{N+1}(z)$$

$$= \begin{vmatrix} P_{N+1,0}(z) & \cdots & P_{N+1,m-1}(z) \\ \cdots & \cdots & \cdots \\ P_{N+m,0}(z) & \cdots & P_{N+m,m-1}(z) \end{vmatrix}$$

$$= \begin{vmatrix} P_{N+1,0}(z) & \cdots & P_{N+1,m-1}(z) \\ \cdots & \cdots & \cdots \\ P_{N+m-1,0}(z) & \cdots & P_{N+m-1,m-1}(z) \\ P_{N+1,0}(z) - \alpha_{N+1} P_{N,0}(z) & \cdots & P_{N+1,m-1}(z) - \alpha_{N+1} P_{N,m-1}(z) \end{vmatrix}$$

$$
= \begin{vmatrix}
P_{N+1,0}(z) & \cdots & P_{N+1,m-1}(z) \\
\cdots & \cdots & \cdots \\
P_{N+m-1,0}(z) & \cdots & P_{N+m-1,m-1}(z) \\
-\alpha_{N+1}P_{N,0}(z) & \cdots & -\alpha_{N+1}P_{N,m-1}(z)
\end{vmatrix}
$$

$$
= -\alpha_{N+1} \begin{vmatrix}
P_{N+1,0}(z) & \cdots & P_{N+1,m-1}(z) \\
\cdots & \cdots & \cdots \\
P_{N+m-1,0}(z) & \cdots & P_{N+m-1,m-1}(z) \\
P_{N,0}(z) & \cdots & P_{N,m-1}(z)
\end{vmatrix}
$$

$$
= \alpha_{N+1}(-1)^m \Delta_N(z). \tag{7.26}
$$

When $N \geq 1$ and N is divisible by $m-1$, we similarly obtain from (7.14) that

$$
\Delta_{N+1}(z) = \alpha_{N+1}(-1)^m z \Delta_N(z). \tag{7.27}
$$

Equalities (7.25)–(7.27) show that (7.15) holds for all non-negative N. ☐

This lemma will be very useful in the following chapter, too. There, the parameters of the considered series will be transcendental.

7.3. Formulation of the Theorem

For convenience, we recall here the basic facts about the extensions of p-adic valuations.

Let $\mathbb{K}$ be an algebraic number field of finite degree $\varkappa$ over $\mathbb{Q}$. For every prime p, we denote by $\mathbb{Q}_p$ the field of p-adic numbers. For every extension v of the p-adic valuation to $\mathbb{K}$, we denote the corresponding completion of $\mathbb{K}$ by $\mathbb{K}_v$. Then, $\mathbb{K}_v$ is an algebraic extension of $\mathbb{Q}_p$, $[\mathbb{K}_v : \mathbb{Q}_p] = \varkappa_v$, and we have $\sum_v \varkappa_v = \varkappa$, where the sum is taken over all extensions v of the p-adic valuation of $\mathbb{Q}$. The same equality holds for the ordinary absolute value since all these extensions correspond to the fields $\mathbb{K}^{(i)}$ conjugate to $\mathbb{K}$, and we have $\varkappa_v = 1$ if $\mathbb{K}^{(i)} \subset \mathbb{R}$ and $\varkappa_v = 2$ if $\mathbb{K}^{(i)} \not\subset \mathbb{R}$.

All the valuations of $\mathbb{K}$ are extensions of the valuations of $\mathbb{Q}$. We denote the set of Archimedean valuations by V_∞ and the set of non-Archimedean valuations by V_0. It is convenient to consider the following normalised valuations. If v extends the p-adic valuation (which is written as $v \mid p$), then we put

$$|p|_v = p^{-\frac{\varkappa_v}{\varkappa}}.$$

If v extends the Archimedean valuation and corresponds to $\mathbb{K}^{(i)}$, then

$$|x|_v = |x^{(i)}|^{\frac{\varkappa_v}{\varkappa}}$$

(where the superscript i stands for the algebraically conjugate field and algebraically conjugate element, respectively).

The following product formula holds. For every $x \in \mathbb{K}$, $x \neq 0$, we have

$$\prod_{v \in V} |x|_v = 1, \qquad (7.28)$$

where the product is taken over all normalised valuations of $\mathbb{K}$.

For $\alpha \in \mathbb{K}$, let $\overline{|\alpha|}$ be the maximum of the moduli of the numbers conjugate to α.

The aim of this section is to obtain a proof of the following theorem.

Theorem 7.1. *For all integers* $\xi \in \mathbb{K}$, $\xi \neq 0$, $\alpha_1, \ldots, \alpha_{m-2}$, $h_0, \ldots, h_{m-1}$ *in* $\mathbb{K}$ *such that not all* $h_0, \ldots, h_{m-1}$ *vanish, there is an infinite set of valuations* v *of the field* $\mathbb{K}$ *such that*

$$|h_0 f_0(\xi) + \cdots + h_{m-1} f_{m-1}(\xi)|_v \neq 0. \qquad (7.29)$$

7.4.   Estimates for Polynomials $P_{N,0}(z), \ldots, P_{N,m-1}(z)$

Let c_i denote positive constants and S_i positive integers.

For the future, the concept of the size of the polynomial $P(z)$ with coefficients from the field $\mathbb{K}$ will be useful. This value is defined as the largest of the moduli of all numbers conjugate to the coefficients of the polynomial in the field $\mathbb{K}$ and is denoted by $\overline{|P(z)|}$. Let's state the simple properties of the size of the polynomial:

$$\overline{|P_1(z) + P_2(z)|} \leq \overline{|P_1(z)|} + \overline{|P_2(z)|}, \qquad (7.30)$$

and for $\alpha \in \mathbb{K}$,

$$\overline{|\alpha\, P(z)|} \le \overline{|\alpha|}\ \overline{|P(z)|}. \tag{7.31}$$

Inequalities (7.30) and (7.31) immediately follow from the fact that for the numbers γ, δ from the field $\mathbb{K}$, any number conjugate to the sum of these numbers is the sum of the numbers conjugate to them, and any number conjugate to the product of these numbers is the product of the numbers conjugate to them.

Denote

$$c_0 = \left[\max(\overline{|\alpha_1|}, \ldots, \overline{|\alpha_m|})\right] + 1. \tag{7.32}$$

Since $\alpha_{m-1} = 1$, $c_0 \ge 2$.

Lemma 7.2. *The following inequality holds under the hypotheses of the theorem for $N = ms + r$, $1 \le r \le m$ and all $i = 0, \ldots, q$:*

$$\overline{|P_{N,i}(z)|} \le c_0^N (c_0 + 1) \cdots (c_0 + s). \tag{7.33}$$

Proof of Lemma 7.2. The lemma is obvious for $s = 0$ and $1 \le r \le m - 1$, due to equalities (7.24) (with the product $(c_0 + 1) \cdots (c_0 + s)$ being considered equal to 1). When $s = 0, r = m$, the equality $N = m$ is obtained, and

$$P_{m,i}(z) = P_{1,i}(z) - \alpha_1 P_{0,i}(z)$$

follows from (7.24). From (7.24) and (7.32), we get

$$\overline{|P_{m,i}(z)|} \le c_0 + 1.$$

Thus, the lemma is true for $s = 0$, $1 \le r \le m$, that is, for $N \le m$.

Suppose that for some $s, s \ge 1$ and for all $r, i, 1 \le r \le m$, $i = 0, \ldots, m - 1$, the following inequalities are met:

$$\overline{|P_{m(s-1)+r,i}(z)|} \le c_0^{m(s-1)+r} (c_0 + 1) \cdots (c_0 + s - 1). \tag{7.34}$$

For every $i, i = 0, \ldots, m - 1$, one of the following equalities is fulfilled:

$$P_{ms+r,i}(z) = P_{m(s-1)+r+1,i}(z) - \alpha_{m(s-1)+r+1}P_{m(s-1)+r,i}(z), \quad (7.35)$$

$$P_{ms+r,i}(z) = P_{m(s-1)+r+1,i}(z) - z\alpha_{m(s-1)+r+1}P_{m(s-1)+r,i}(z). \quad (7.36)$$

By (7.5), (7.7), and (7.32), we get

$$\overline{|\alpha_{m(s-1)+r+1}|} = \overline{|\alpha_{r+1} + s - 1|} \le c_0 + s. \quad (7.37)$$

If $r \le m - 1$, then both cases (7.35) and (7.36) from (7.27) and (7.30) follow:

$$\overline{|P_{ms+r,i}(z)|} \le \overline{|P_{m(s-1)+r+1,i}(z)|} + \overline{|\alpha_{m(s-1)+r+1}|}\ \overline{|P_{m(s-1)+r,i}(z)|}. \quad (7.38)$$

From (7.34), (7.35), (7.37), and (7.38), we get

$$
\begin{aligned}
\overline{|P_{ms+r,i}(z)|} &\le c_0^{m(s-1)+r+1}(c_0 + 1) \cdots (c_0 + s - 1) \\
&\quad + c_0^{m(s-1)+r}(c_0 + 1) \cdots (c_0 + s - 1)(c_0 + s) \\
&= c_0^{m(s-1)+r}(c_0 + 1) \cdots (c_0 + s - 1)(c_0 + c_0 + s) \\
&\le 2c_0^{m(s-1)+r}(c_0 + 1) \cdots (c_0 + s) \\
&\le c_0^{ms+r}(c_0 + 1) \cdots (c_0 + s), \quad (7.39)
\end{aligned}
$$

since

$$2 \le c_0 \le c_0^m.$$

It remains to consider the case of $r = m$.

For every $i, i = 0, \ldots, m - 1$, we again have one of the equalities

$$P_{m(s+1),i}(z) = P_{ms+1,i}(z) - \alpha_{ms+1} P_{ms,i}(z), \qquad (7.40)$$

$$P_{m(s+1),i}(z) = P_{ms+1,i}(z) - z\alpha_{ms+1} P_{ms,i}(z). \qquad (7.41)$$

It follows from (7.7) and (7.32) that

$$\overline{|\alpha_{ms+1}|} = \overline{|\alpha_1 + s|} \leq c_0 + s. \qquad (7.42)$$

In the cases of (7.40) and (7.41), we obtain in view of (7.30) and (7.31)

$$\overline{|P_{m(s+1),i}(z)|} \leq \overline{|P_{ms+1,i}(z)|} + \overline{|\alpha_{ms+1}|}\,\overline{|P_{ms,i}(z)|}. \qquad (7.43)$$

By what has already been proved, we have

$$\overline{|P_{ms+1,i}(z)|} \leq c_0^{ms+1}(c_0 + 1) \cdots (c_0 + s). \qquad (7.44)$$

Now, by the induction hypothesis,

$$\overline{|P_{ms,i}(z)|} = \overline{|P_{m(s-1)+m}(z)|} \leq c_0^{ms}(c_0 + 1) \cdots (c_0 + s - 1). \qquad (7.45)$$

It follows from (7.42)–(7.45) that

$$\overline{|P_{m(s+1),i}(z)|}$$
$$= c_0^{ms+1}(c_0 + 1) \cdots (c_0 + s) + (c_0 + s)c_0^{ms}(c_0 + 1) \cdots (c_0 + s - 1)$$
$$= c_0^{ms}(c_0 + 1) \cdots (c_0 + s)(c_0 + 1) \leq 2c_0^{ms+1}(c_0 + 1) \cdots (c_0 + s)$$
$$\leq c_0^{m(s+1)}(c_0 + 1) \cdots (c_0 + s). \qquad (7.46)$$

This completes the induction. Inequality (7.33) follows from (7.39) and (7.46). □

7.5. Estimates from Above of the Auxiliary Determinant

In any field $\mathbb{K}$, consider an element of this field having the form

$$L(\xi) = h_1 f_0(\xi) + \cdots + h_m f_{m-1}(\xi).$$

Multiplying it by $\alpha_1 \ldots \alpha_{m-1}$ and taking into account the equalities (7.9), we get the element

$$L^* = b_0 u_0(\xi) + \cdots + b_{m-1} u_{m-1}(\xi), \qquad (7.47)$$

which is a linear form in $u_0(\xi), \ldots, u_{m-1}(\xi)$ with integers from the field $\mathbb{K}$ as coefficients. Since $L(\xi) \neq 0$, the resulting form (7.47) is also non-zero. Denote

$$H = \max_{i=0,\ldots,m-1} \overline{|b_i|}. \qquad (7.48)$$

Consider, for some positive integer N, the linear forms

$$u_N(\xi), \ldots, u_{N+m-1}(\xi) \qquad (7.49)$$

in $u_0(\xi), \ldots, u_{m-1}(\xi)$ defined by (7.10). The determinant $\Delta_N(\xi)$ of the matrix of their coefficients is different from 0 due to equality (7.15) and the conditions of the theorem. Therefore, among the forms in (7.49), there are $m - 1$ forms that are linearly independent with the form L^*. Hence, the determinant of this set of forms

$$\Delta_N^*(\xi) = \begin{vmatrix} b_0 & \cdots & b_{m-1} \\ P_{N_1,0}(\xi) & \cdots & P_{N_1,m-1}(\xi) \\ \cdots & \cdots & \cdots \\ P_{N_{m-1},0}(\xi) & \cdots & P_{N_{m-1},m-1}(\xi) \end{vmatrix}, \qquad (7.50)$$

where

$$N_i \in \{N, \ldots, N + m - 1\}, \qquad (7.51)$$

is non-zero, $\Delta_N^*(\xi) \neq 0$. The value $\Delta_N^*(\xi)$ defined by (7.50) can be considered the value of the polynomial $\Delta_N^*(z)$ at the point $z = \xi$ (where the polynomial $z = \xi$ is defined in an obvious way).

Lemma 7.3. *When $s \geq S_0$, there is an inequality*

$$\prod_{v \in V_0} |\Delta_N^*(\xi)|_v \geq H^{-1} e^{-(m-1)s \ln s - c_1 s}, \qquad (7.52)$$

where the product on the left-hand side of inequality (7.52) is taken over all non-Archimedean valuations of the field $\mathbb{K}$.

Proof of Lemma 7.3. Since $\Delta_N^*(\xi) \neq 0$, in view of the product formula (7.28) and (7.48), it is sufficient to prove the inequality

$$\prod_{v \in V_\infty} |\Delta_N^*(\xi)|_v \leq H \, e^{(m-1)s \ln s + c_1 s}, \qquad (7.53)$$

where the product on the left-hand side is taken over all Archimedean valuations of the field $\mathbb{K}$.

It follows from equality (7.50) that $\Delta_N^*(z)$ is a polynomial with integers from the field $\mathbb{K}$ as coefficients. In addition, in view of (7.51), $N_i \leq N + m - 1$.

Let $\tilde{s}$ and $\tilde{t}$ be defined by the equalities $N + m - 1 = m\tilde{s} + \tilde{r} = (m-1)\tilde{t} + \tilde{l}$. Then,

$$\tilde{s} = \left[\frac{N+m-2}{m}\right], \quad \tilde{t} = \left[\frac{N+m-2}{m-1}\right]. \qquad (7.54)$$

From (7.7) and (7.54), it follows that $s \leq \tilde{s} \leq s + 1$, a $\tilde{t} = t + 1$. Then, using (7.32) with the replacement of s by $s + 1$, we get

$$\overline{|P_{N_i,j}(z)|} \leq c_0^{N+m-1}(c_0 + 1) \cdots (c_0 + s + 1), \quad j = 0, \ldots, m - 1,$$

from which, taking into account (7.47), we have

$$\overline{|\Delta_N^*(z)|} \leq m! H c_0^{(m-1)(N+m-1)}((c_0 + 1) \cdots (c_0 + s))^{m-1}. \qquad (7.55)$$

We represent the product $(c_0 + 1) \cdots (c_0 + s + 1)$ in the form

$$(c_0 + 1) \cdots (c_0 + s + 1) = \frac{\Gamma(c_0 + s + 2)}{\Gamma(c_0 + 1)}, \qquad (7.56)$$

where $\Gamma(x)$ is the Euler gamma function. Further, we use a well-known lemma without proof. $\qquad \square$

Lemma 7.4 (Titchmarsh, 1939). *For every constant a and every* $\delta > 0$ *as* $|s| \to \infty$, *the equality*

$$\ln \Gamma(s + a) = \left(s + a - \frac{1}{2}\right) \ln s - s + \frac{1}{2} \ln 2\pi + O\left(\frac{1}{|s|}\right) \quad (7.57)$$

is performed uniformly in the domain

$$-\pi + \delta \le \arg s \le \pi - \delta.$$

It follows immediately from (7.56) and (7.57) that

$$(c_0 + 1) \cdots (c_0 + s + 1) = \exp\left(s \ln s + O(s)\right) \quad (7.58)$$

as $s \to \infty$. By (7.55) and (7.58), there is a constant c_2 such that

$$\overline{|\Delta_N^*(z)|} \le H \exp((m-1)s \ln s + c_2 s) \quad (7.59)$$

for $s > S_1$. Furthermore, the degree of the polynomial $\Delta_N^*(z)$ does not exceed

$$(m-1)\left(\tilde{t} - \tilde{s}\right) \le (m-1)\left(\left[\frac{N+m-2}{m-1}\right] - \left[\frac{N-1}{m}\right]\right)$$

$$\le (m-1)\left(\frac{N+m-2}{m-1} - \frac{N-1}{m} + 1\right)$$

$$\le \frac{N}{m} + 2(m-1) \le s + 2(m-1). \quad (7.60)$$

Therefore, by (7.59) and (7.60), there is a constant c_3 such that we have

$$\overline{|\Delta_N^*(\xi)|} \le H \exp((m-1)s \ln s + c_3 s) \quad (7.61)$$

for $s > S_2$. Since

$$\prod_{v \in V_\infty} |\Delta_N^*(\xi)|_v \le \prod_{v \in V_\infty} \overline{|\Delta_N^*(\xi)|}^{\frac{\varkappa_v}{\varkappa}} = \overline{|\Delta_N^*(\xi)|},$$

inequality (7.61) yields (7.53) and (7.52) with appropriate S_0, c_1. $\square$

7.6.   The Basic Identity

Determinant (7.50) is transformed as follows: its first column is multiplied by $u_{0,k}(\xi)$, and the other columns multiplied by the corresponding $u_{j,k}(\xi)$, $j = 1,\ldots,m-1$, are added to the result. Taking into account (7.10) and (7.47), we obtain

$$\Delta_N^*(\xi)u_{0,}(\xi) = \begin{vmatrix} L^*(\xi) & \cdots & b_{m-1} \\ u_{N_1}(\xi) & \cdots & P_{N_1,m-1}(\xi) \\ \ddots & \ddots & \ddots \\ u_{N_{m-1}}(\xi) & \cdots & P_{N_{m-1},m-1}(\xi) \end{vmatrix}. \tag{7.62}$$

Similarly, the last column of determinant (7.50) is multiplied by $u_{m-1}(\xi)$, and the other columns multiplied by the corresponding $u_j(\xi)$, $j = 0,\ldots,m-2$, are added to the result. Taking into account (7.10) and (7.47) yields

$$\Delta_N^*(\xi)u_{m-1}(\xi) = \begin{vmatrix} b_0 & \cdots & L^*(\xi) \\ P_{N_1,0}(\xi) & \cdots & u_{N_1}(\xi) \\ \ddots & \ddots & \ddots \\ P_{N_{m-1},0}(\xi) & \cdots & u_{N_{m-1}}(\xi) \end{vmatrix}. \tag{7.63}$$

It follows from (8.12) and (8.19) that

$$u_0(\xi) = 1 + \xi u_{m-1}(\xi). \tag{7.64}$$

Let $\delta_{0,i,j}$ denote the cofactor of the element of determinant (7.62) located at the intersection of row i and column j, and let $\delta_{m-1,i,j}$ denote the cofactor of the corresponding element of determinant (7.63). Then,

$$\Delta_N^*(\xi)u_0(\xi) = L^*(\xi)\delta_{0,1,1} + \sum_{i=2}^m \delta_{0,i,1}u_{N_{i-1}}(\xi), \tag{7.65}$$

$$\Delta_N^*(\xi)u_{m-1}(\xi) = L^*(\xi)\delta_{m-1,1,m} + \sum_{i=2}^m \delta_{m-1,i,m}u_{N_{i-1}}(\xi). \tag{7.66}$$

It follows from (7.64)–(7.66) that

$$\Delta_N^*(\xi) = L^*(\xi)(\delta_{0,1,1} - \xi\delta_{m-1,1,m}) + \sum_{i=2}^{m}(\delta_{0,i,1} - \xi\delta_{m-1,i,m})u_{N_{i-1}}(\xi).$$

(7.67)

7.7. Estimates from Above of the Product of the Valuations of Approximating Forms: End of the Proof of the Theorem

According to (7.8) and (7.9),

$$|u_N(\xi)|_v = |\alpha_1 \ldots \alpha_N|_v \, |\xi^t|_v \, |F(\alpha_{N+1}, \ldots, \alpha_{N+m}, \xi)|_v, \qquad (7.68)$$

and from the hypotheses of the theorem, it immediately follows that

$$|\xi^t|_v \, |F(\alpha_{N+1}, \ldots, \alpha_{N+m}, \xi)|_v \le 1.$$

Thus, from (7.68), we get

$$|u_{N_i}(\xi)|_v \le |\alpha_1 \ldots \alpha_{N_i}|_v \le |\alpha_1 \ldots \alpha_N|_v. \qquad (7.69)$$

Since, by construction, $\delta_{0,i,1}$, $\xi\delta_{m-1,i,m}$, $i = 1, \ldots, m$ are algebraic integers, in view of (7.69), the following inequality holds:

$$\left| \sum_{i=2}^{m}(\delta_{0,i,1} - \xi\delta_{m-1,i,m})u_{N_{i-1}}(\xi) \right|_v \le |\alpha_1 \ldots \alpha_N|_v. \qquad (7.70)$$

From (7.70) and the product formula (7.28), it follows that

$$\prod_{v \in V_0} \left| \sum_{i=2}^{m}(\delta_{0,i,1} - \xi\delta_{m-1,i,m})u_{N_{i-1}}(\xi) \right|_v$$

$$\le \prod_{v \in V_0} |\alpha_1 \ldots \alpha_N|_v = \left(\prod_{v \in V_\infty} |\alpha_1, \ldots, \alpha_N|_v \right)^{-1}. \qquad (7.71)$$

From (7.5) and (7.9), it follows that

$$\alpha_1 \ldots \alpha_N = \alpha_1(\alpha_1 + 1) \cdots (\alpha_1 + s) \cdots \alpha_r(\alpha_r + 1)$$
$$\cdots (\alpha_r + s)\alpha_{r+1}(\alpha_{r+1} + 1) \cdots$$
$$\cdots (\alpha_{r+1} + s - 1) \cdots \alpha_m(\alpha_m + 1) \cdots (\alpha_m + s - 1)$$
$$= \frac{\Gamma(\alpha_1 + s + 1) \cdots \Gamma(\alpha_r + s + 1)\Gamma(\alpha_{r+1} + s) \cdots \Gamma(\alpha_m + s)}{\Gamma(\alpha_1) \cdots \Gamma(\alpha_m)}.$$

$$(7.72)$$

From equations (7.57) and (7.72), we obtain

$$|\alpha_1 \ldots \alpha_N| = \exp(rs \ln s + (m - r)(s - 1) \ln(s - 1) + O(s))$$
$$= \exp(ms \ln s + O(s)) \qquad (7.73)$$

as $s \to \infty$, and equality (7.73) remains true even if we replace all the numbers $\alpha_1, \ldots, \alpha_N$ with their conjugates in the field $\mathbb{K}$.

According to the definitions of normalised valuations,

$$\prod_{v \in V_\infty} |\alpha_1 \ldots \alpha_N|_v = |\text{Norm}(\alpha_1 \ldots \alpha_N)|^{\frac{1}{\varkappa}}, \qquad (7.74)$$

where $\text{Norm}(\alpha)$ denotes the norm of an element $\alpha \in \mathbb{K}$ in $\mathbb{Q}$. By the definition of the norm,

$$\text{Norm}(\alpha_1 \ldots \alpha_N) = \prod_{i=1}^{\varkappa} \alpha_1^{(i)} \ldots \alpha_N^{(i)}, \qquad (7.75)$$

where $\alpha_1^{(i)} \ldots \alpha_N^{(i)}$ are conjugate numbers for $\alpha_1, \ldots, \alpha_N$ in the field $\mathbb{K}$. From (7.73)–(7.75) and (7.57), we get

$$\prod_{v \in V_\infty} |\alpha_1 \ldots \alpha_N|_v = \exp(ms \ln s + O(s)). \qquad (7.76)$$

By the product formula in (7.28), it follows from equality (7.76) that

$$\prod_{v \in V_0} |\alpha_1 \ldots \alpha_N|_v = \exp(-ms \ln s + O(s)). \qquad (7.77)$$

In the infinite product on the left-hand side of equality (7.77), only a finite number of factors are different from 1. They correspond to those $v \in V_0$ for which

$$|\alpha_1 \ldots \alpha_N|_v < 1.$$

These valuations $v \in V_0$ extend the p-adic valuations for those prime numbers p that divide the rational integer

$$\text{Norm}(\alpha_1 \ldots \alpha_N) = \prod_{i=1}^{\varkappa} \alpha_1^{(i)} \ldots \alpha_N^{(i)}.$$

However,

$$\prod_{i=1}^{\varkappa} \alpha_1^{(i)} \ldots \alpha_N^{(i)} = \prod_{i=1}^{\varkappa} \alpha_1^{(i)} \left(\alpha_1^{(i)} + 1\right) \cdots \left(\alpha_1^{(i)} + s\right) \cdots \alpha_r^{(i)} \left(\alpha_r^{(i)} + 1\right)$$

$$\cdots \left(\alpha_r^{(i)} + s\right) \alpha_{r+1}^{(i)} \left(\alpha_{r+1}^{(i)} + 1\right) \cdots \left(\alpha_{r+1}^{(i)} + s - 1\right)$$

$$\cdots \alpha_m^{(i)} \left(\alpha_m^{(i)} + 1\right) \cdots \left(\alpha_m^{(i)} + s - 1\right)$$

$$= \prod_{j=1}^{r} \prod_{k=0}^{s} \prod_{i=1}^{\varkappa} \left(a_j^{(i)} + k\right) \times \prod_{j=r+1}^{m} \prod_{k=0}^{s-1} \prod_{i=1}^{\varkappa} \left(a_j^{(i)} + k\right),$$

and for any fixed j, k, the product

$$\prod_{i=1}^{\varkappa} \left(a_j^{(i)} + k\right) \tag{7.78}$$

is a rational integer. Consequently, the prime numbers p mentioned above do not exceed the largest of the numbers in (7.78). The value of (7.78) is the value of the polynomial

$$\prod_{i=1}^{\varkappa} \left(a_j^{(i)} + x\right)$$

at $x = k$. The polynomial is an increasing function when $x \geq x_0$, where x_0 is some constant. Therefore, at $s \geq S_3$, the largest of the numbers in (7.78) does not exceed the value of $(s + c_0)^{\varkappa}$, where the constant c_0 is defined by equality (7.32).

Hence, at $s \geq S_3$, any prime number p dividing the number $\text{Norm}(\alpha_1 \ldots \alpha_N) = \prod_{i=1}^{\varkappa} \alpha_1^{(i)} \ldots \alpha_N^{(i)}$ satisfies the inequality

$$p \leq (s + c_0)^{\varkappa}. \tag{7.79}$$

Lemma 7.5. *When $s \geq S_4$, the following inequality holds:*

$$\prod_{e^{\sqrt{\ln s}} \leq p \leq (s+c_0)^{\varkappa}} \prod_{v \mid p} |\alpha_1 \ldots \alpha_N|_v < e^{-ms\ln s + ms\sqrt{\ln s} + c_4 s}. \tag{7.80}$$

Proof of Lemma 7.5. To prove this lemma, we need the following statement.   □

Lemma 7.6. *Let $\lambda \in \mathbb{Z}_{\mathbb{K}}$, $\mathbb{K} = \mathbb{Q}(\lambda)$ $s \geq S_5$. Then,*

$$\prod_{p \leq e^{\sqrt{\ln s}}} \prod_{v \mid p} |\lambda(\lambda + 1) \cdots (\lambda + s - 1)|_v \geq e^{-s\sqrt{\ln s} - c_5 s}. \tag{7.81}$$

Proof of Lemma 7.6. The proof of this lemma has much in common with the proofs of the statements in the works of Sprindzhuk (1968) and Galochkin (1981).

Let us denote

$$\omega(s) = \lambda(\lambda + 1) \cdots (\lambda + s - 1).$$

Let $\mathfrak{p}$ be a prime ideal in the ring $\mathbb{Z}_{\mathbb{K}}$, $\mathfrak{p} \mid (\omega(s))$, $\mathfrak{p} \mid (p)$, where p is a prime, with $p \leq e^{\sqrt{\ln s}}$. Since $\mathfrak{p}$ is a prime ideal, $\mathfrak{p} \mid (\lambda + s_1)$ for some $s_1 \in \mathbb{N}$, $s_1 \leq s - 1$. Moreover, if for some t, s_1, $s_2 \in \mathbb{N}$, with $s_1 \leq s - 1$, $s_2 \leq s - 1$, we have $\mathfrak{p}^t \mid (\lambda + s_1)$, $\mathfrak{p}^t \mid (\lambda + s_2)$, then $\mathfrak{p}^t \mid (s_2 - s_1)$. Therefore, the norms of ideals in $\mathbb{K}$ satisfy the inequalities

$$N(\mathfrak{p}^t) \leq N(s_2 - s_1). \tag{7.82}$$

If the ideal $\mathfrak{p}$ has degree f, then we obtain from (7.82) that

$$N(\mathfrak{p}^t) = (N(\mathfrak{p}))^t = p^{ft} \leq N(s_2 - s_1) \leq N(s) = s^{\varkappa}.$$

Hence,

$$p^t \leq p^{ft} \leq s^{\varkappa};$$

moreover,

$$t \leq \frac{\varkappa \ln s}{\ln p}. \tag{7.83}$$

It follows from (7.83) that

$$p^{\frac{t}{\varkappa}} \leq s. \tag{7.84}$$

Inequality (7.84) means that among the ideals $(\lambda), (\lambda+1), \ldots, (\lambda+s-1)$, there are at most

$$\left[\frac{s}{p^{\frac{t}{\varkappa}}}\right] + 1$$

ideals that are divisible by $\mathfrak{p}^t$. By (7.81) and (7.82) for $s \geq S_6$, the power of $\mathfrak{p}$ occurring in the ideal $(\omega(s))$ does not exceed

$$\sum_{1 \leq t \leq c_6 \frac{\ln s}{\ln p}} \left(\left[\frac{s}{p^{\frac{t}{\varkappa}}}\right] + 1\right) \leq s \cdot \frac{1}{p^{\frac{t}{\varkappa}} - 1} + c_6 \frac{\ln s}{\ln p}. \tag{7.85}$$

We first assume that $p \mid \widetilde{D}$, where $\widetilde{D}$ is the discriminant of the field $\mathbb{Q}(\lambda)$.

If the prime ideal $\mathfrak{p}$, corresponding to the valuation v, occurs in the decomposition of (p) to the power e_v, and $\mathfrak{p}$ occurs in the decomposition of the principal ideal (a), $a \in \mathbb{Z}_{\mathbb{K}}$, to the power k, then

$$|a|_v = p^{-\frac{k \varkappa_v}{\varkappa e_v}}.$$

Therefore, in view of (7.53), we get

$$|\omega(s)|_v \geq p^{-\frac{\varkappa_v}{e_v \varkappa}\left(\frac{s}{p^{\frac{1}{\varkappa}} - 1} + c_6 \frac{\ln s}{\ln p}\right)} \geq p^{-\frac{\varkappa_v}{\varkappa}\left(\frac{s}{p^{\frac{1}{\varkappa}} - 1} + c_6 \frac{\ln s}{\ln p}\right)}. \tag{7.86}$$

Since $\sum \varkappa_v = \varkappa$, where the sum is taken over all $v \mid p$, it follows from (7.86) that for $s \geq S_7$, we have

$$\prod_{v \mid p} |\omega(s)|_v \geq \prod_{v \mid p} p^{-\frac{\varkappa_r}{\varkappa}\left(\frac{s}{p^{\frac{1}{\varkappa}} - 1} + c_4 \frac{\ln s}{\ln p}\right)} \geq p^{-\left(\frac{s}{p^{\frac{1}{\varkappa}} - 1} + c_6 \frac{\ln s}{\ln p}\right)} \geq p^{-c_7 s}. \tag{7.87}$$

By the condition $p \mid \tilde{D}$, there are finitely many such primes p, and it follows from (7.87) that

$$\prod_{p \le e^{\sqrt{\ln s}}, p \mid \tilde{D}} \prod_{v \mid p} |\omega(s)|_v \ge e^{-c_6 s}. \qquad (7.88)$$

Now, consider that the prime p, satisfying the inequality $p \le e^{\sqrt{\ln s}}$, does not divide the discriminant $\tilde{D}$ of the field $\mathbb{Q}(\lambda)$. Then, in the ring $\mathbb{Z}_\mathbb{K}$, the decomposition holds: $(p) = \mathfrak{p}_1 \dots \mathfrak{p}_r$.

If $\mathfrak{p}^t \mid (\lambda + s_1), \mathfrak{p}^t \mid (\lambda + s_2)$, then $\mathfrak{p}^t \mid (s_2 - s_1)$ and $\mathfrak{p}^t \mid (s_2 - s_1)$, whence $p^t \le s$. Hence, the ideal $\mathfrak{p}$, dividing (p), enters the decomposition of $(\omega(s))$ with a degree of at most

$$\sum_{1 \le t < c_7 \frac{\ln s}{\ln p}} \left(\left[\frac{s}{p^t} \right] + 1 \right) \le \frac{s}{p-1} + c_8 \frac{\ln s}{\ln p} = \frac{s}{p} + \frac{s}{p(p-1)} + c_8 \frac{\ln s}{\ln p}.$$

Hence,

$$\prod_{v \mid p} |\omega(s)|_v \ge \prod_{v \mid p} p^{-\frac{\varkappa_v}{\varkappa} \left(\frac{s}{p} + \frac{s}{p(p-1)} + c_8 \frac{\ln s}{\ln p} \right)} = p^{-\left(\frac{s}{p} + \frac{s}{p(p-1)} + c_8 \frac{\ln s}{\ln p} \right)}.$$

$$(7.89)$$

Note that we have the following estimates:

$$\sum_{p \le e^{\sqrt{\ln s}}} \left(\frac{s}{p} + \frac{s}{p(p-1)} + c_8 \frac{\ln s}{\ln p} \right) \ln p$$

$$= s \sum_{p \le e^{\sqrt{\ln s}}} \left(\frac{\ln p}{p} \right) + s \sum_{p \le e^{\sqrt{\ln s}}} \left(\frac{\ln p}{p(p-1)} \right) + c_8 \ln s \sum_{p \le e^{\sqrt{\ln s}}} 1$$

$$\le s\sqrt{\ln s} + c_9 s + c_8 \ln s \frac{e^{\sqrt{\ln s}}}{\sqrt{\ln s}} \le s\sqrt{\ln s} + c_{10} s.$$

It follows from (7.78) and (7.89) that for $s \ge S_8$, we have

$$\prod_{p \le e^{\sqrt{\ln s}}, p \nmid \tilde{D}} \prod_{v \mid p} |\omega(s)|_v \ge \prod_{p \le e^{\sqrt{\ln s}}} p^{-\left(\frac{s}{p} + \frac{s}{p(p-1)} + c_8 \frac{\ln s}{\ln p} \right)}$$

$$= e^{-\sum_{p \le e^{\sqrt{\ln s}}} \left(\frac{s \ln p}{p} + \frac{s \ln p}{p(p-1)} + c_8 \ln s \right)}$$

$$\ge e^{-s\sqrt{\ln s} - c_{10} s}. \qquad (7.90)$$

Inequalities (7.88) and (7.90) now yield (7.81), and for $s \geq S_6$, (7.87) follows from (7.81), (7.79), and (7.77). Lemmas 7.5 and 7.6 are proved. $\qquad\square$

Let V_N^* be the set of valuations $v \in V_0$, extending the p-adic valuations for primes p, that satisfy for $s \geq S_8$ the following inequalities:

$$e^{\sqrt{\ln s}} \leq p \leq (s + c_0)^\varkappa. \qquad (7.91)$$

We claim that there is a valuation $v \in V_N^*$, extending the p-adic valuation satisfying (7.91), such that

$$|\Delta_N^*(\xi)|_v > \left| \sum_{i=2}^{m} (\delta_{0,i,1} - \xi \delta_{m-1,i,m}) u_{N_{i-1}}(\xi) \right|_v \qquad (7.92)$$

in the field $\mathbb{K}_v$.

Indeed, by (7.52), we have

$$\prod_{v \in V_0} |\Delta_N^*(\xi)|_v \geq H^{-1} e^{-(m-1)s \ln s - c_1 s}. \qquad (7.93)$$

It follows from (7.77), (7.71), Lemma 7.6, and (7.81) that

$$\prod_{v \in V_N^*} \left| \sum_{i=2}^{m} (\delta_{0,i,1} - \xi \delta_{m-1,i,m}) u_{N_{i-1}}(\xi) \right|_v$$

$$\leq \prod_{v \in V_N^*} |\alpha_1 \dots \alpha_N|_v \leq \exp(-ms \ln s + ms\sqrt{\ln s} + c_4 s).$$

For $s \geq S_4$, we have

$$\ln H < s \ln s - ms\sqrt{\ln s} - (c_1 + c_4)s,$$

whence

$$\ln H + (m-1)s \ln s + c_1 s < ms \ln s - ms\sqrt{\ln s} - c_4 s. \qquad (7.94)$$

It follows from (7.94) that

$$H^{-1} \exp(-(m-1)s \ln s - c_1 s) > \exp(-ms \ln s + ms\sqrt{\ln s} + c_4 s). \qquad (7.95)$$

It follows from (7.93)–(7.95) that

$$\prod_{v \in V_N^*} |\Delta_N^*(\xi)|_v \geq H^{-1} e^{-(m-1)s \ln s - c_1 s}$$

$$> \exp(-ms \ln s + ms\sqrt{\ln s} + c_4 s)$$

$$> \prod_{v \in V_N^*} \left| \sum_{i=2}^{m} (\delta_{0,i,1} - \xi \delta_{m-1,i,m}) u_{N_{i-1}}(\xi) \right|_v.$$

This proves (7.92).

Then, we obtain from (7.71) and (7.92) that

$$0 < |\Delta_N^*(\xi)|_v = |L^* \delta_1|_v \leq |L^*|_v,$$

since δ_1 is an integer in the field $\mathbb{K}$.

To complete the proof of the theorem, it remains to choose a sequence of numbers $s_1, s_2, s_3, \ldots$ that exceed S_8 and satisfy the inequalities

$$(s_k + c_0)^\varkappa < e^{\sqrt{\ln s_{k+1}}}, \quad k = 1, 2, \ldots.$$

Every such interval contains an appropriate prime p, and there is a valuation $v \mid p$ such that inequality (7.29) holds. $\qquad \square$

Chapter 8

Arithmetic Properties of Generalised Hypergeometric Series with Polyadic Transcendental Coefficients*

8.1. Formulation of the Results

Let λ_0 be an arbitrary positive integer. Let's put

$$s_0 = [\exp \lambda_0] + 1.$$

Here and subsequently, the symbol $[a]$ denotes the integer part of the number a. Let $\operatorname{ord}_p a$ denote the degree to which the prime number p enters the prime factorisation of a.

Let λ_1 be an arbitrary positive integer satisfying the following condition: for any prime $p \le s_0 + C_1 \lambda_0^2$, the inequality $\operatorname{ord}_p \lambda_1 \ge m s_0 \ln s_0$ is satisfied, and let $s_1 = [\exp \lambda_1] + 1$. Here and subsequently, C_r, $r = 1, 2, \ldots$, denote some positive constants.

For $k \ge 2$, let λ_k — be an arbitrary positive integer satisfying the following condition: for any prime

$$p \le s_{k-1} + C_1 \lambda_{k-1}^2, \tag{8.1}$$

the inequality

$$\operatorname{ord}_p \lambda_k \ge m s_{k-1} \ln s_{k-1} \tag{8.2}$$

---

*The results of this chapter are published in Chirskii (2022b).

is satisfied, and

$$s_k = [\exp \lambda_k] + 1. \tag{8.3}$$

Let $\mu_{i,0}$, $i = 1, \ldots, m - 1$ be positive integers. Let for all $i = 1, \ldots,$
$m - 1$, $k \geq 1$ the numbers $\mu_{i,k}$ be non-negative integers satisfying

$$\mu_{i,k} \leq \lambda_k. \tag{8.4}$$

Let

$$\alpha_{i,k} = \sum_{l=0}^{k} \mu_{i,l} \lambda_l, \quad i = 1, \ldots, m - 1, \tag{8.5}$$

$$\alpha_i = \sum_{l=0}^{\infty} \mu_{i,l} \lambda_l, \quad i = 1, \ldots, m - 1. \tag{8.6}$$

In what follows, the numbers K_i, $i = 1, 2, \ldots$ are positive integers.
For every $k \geq K_1$, according to (8.1)–(8.5), we have

$$1 \leq \alpha_{i,k} = \sum_{l=0}^{k} \mu_{i,l} \lambda_l \leq 1, 1\lambda_k^2, \quad i = 1, \ldots, m - 1. \tag{8.7}$$

If for every $l \geq K_2$, we have $\mu_{i,l} = 0$, then α_i is a positive integer.

We prove that, otherwise, the series defined by equality (8.6) is
a polyadic Liouville number. This series converges in any field $\mathbb{Q}_p$
according to (8.2) and (8.3), and its sum in this field is a p-adic
integer.

Moreover, the conditions imposed on the positive integers n and
P imply that there exists an integer A such that, for all primes p,
satisfying the inequality $p \leq P$, one has $|\theta - A|_p < |A|^{-n}$. Indeed, for
all prime numbers p satisfying (8.1), for $k \geq K_3$, in view of (8.2)–
(8.7), we have

$$|\alpha_i - \alpha_{i,k}|_p \leq \left| \sum_{l=k+1}^{\infty} \mu_{i,l} \lambda_l \right|_p$$

$$\leq |\lambda_{k+1}|_p \leq p^{-ms_k \ln s_k} \leq s_k^{-ms_k} \leq \alpha_{i,k}^{-ms_k}.$$

For all k, let

$$f_{0,k}(z) = \sum_{n=0}^{\infty} (\alpha_{1,k})_n \cdots (\alpha_{m-1,k})_n z^n,$$

$$f_{m-1,k}(z) = \sum_{n=0}^{\infty} (\alpha_{1,k}+1)_n \cdots (\alpha_{m-1,k}+1)_n z^n, \qquad (8.8)$$

and for $i = 1, \ldots, m-2$, let

$$f_{i,k}(z) = \sum_{n=0}^{\infty} (\alpha_{1,k}+1)_n \cdots (\alpha_{i,k}+1)_n (\alpha_{i+1,k})_n \cdots (\alpha_{m-1,k})_n z^n.$$
$$(8.9)$$

Additionally, we consider the series

$$f_0(z) = \sum_{n=0}^{\infty} (\alpha_1)_n \cdots (\alpha_{m-1})_n z^n,$$

$$f_{m-1}(z) = \sum_{n=0}^{\infty} (\alpha_1+1)_n \cdots (\alpha_{m-1}+1)_n z^n, \qquad (8.10)$$

and for $i = 1, \ldots, m-2$,

$$f_i(z) = \sum_{n=0}^{\infty} (\alpha_1+1)_n \cdots (\alpha_i+1)_n (\alpha_{i+1})_n \cdots (\alpha_{m-1})_n z^n. \qquad (8.11)$$

The coefficients of series (8.8) and (8.9) are positive integers, so these series converge for $|z|_p < p^{\frac{m-1}{p-1}}$ in any field $\mathbb{Q}_p$. Since α_i, $i = 1, \ldots, m-1$ can be treated as p-adic integers, it is true that

$$|(\alpha_i)_n|_p \leq p^{\left(\frac{-n}{p-1}+(p-1)\log_p n\right)}.$$

Indeed, let ω be a p-adic integer. We represent it in the form

$$\omega = a_0 + a_1 p + \cdots + a_n p^n + r_{n+1} = A_n + r_{n+1},$$

where $a_i \in \{0, 1, \ldots, p-1\}$, $i = 0, 1, \ldots, n$, and the number r_{n+1} is a p-adic integer such that $|r_{n+1}|_p < p^{-n-1}$. Then, $(\omega)_n$ can be represented in the form of the sum of $(A_n)_n$ and a finite number of terms, each having a p-adic valuation not greater than p^{-n-1}. Since A_n is a positive integer,

$$|(A_n)_n|_p \leq p^{\left(\frac{-n}{p-1}+C\ln n\right)},$$

with a constant C. Therefore,

$$|(\omega)_n|_p \leq p^{\left(\frac{-n}{p-1}+C\ln n\right)}.$$

This inequality proves the formulated assertion. Therefore, series (8.10) and (8.11) also converge for $|z|_p < p^{\frac{m-1}{p-1}}$.

Note the following identity (important for the subsequent presentation):

$$f_{0,k}(z) = 1 + \alpha_{1,k}\ldots\alpha_{m-1,k}z f_{m-1,k}(z), \qquad (8.12)$$

which easily follows from the definitions in (8.8).

The main results of this chapter are stated as follows. Let M be a positive integer. Consider a reduced system of residues $\mod(M)$. As usual, the number of elements of this system is denoted by $\varphi(M)$, where $\varphi(M)$ is the Euler function. Let $a_1,\ldots,a_\rho$ be ρ distinct elements chosen arbitrarily from this system. Denote by $\boldsymbol{a}_1,\ldots,\boldsymbol{a}_\rho$ the sets of positive integer values taken by the arithmetic progressions $a_i + Mk$, $k \in \mathbb{Z}$. Using the standard notation $\mathbb{P}$ for the set of prime numbers, we denote by $\mathbb{P}(\boldsymbol{a}_1,\ldots,\boldsymbol{a}_\rho)$ the set of prime numbers contained in the union of the sets $\boldsymbol{a}_1,\ldots,\boldsymbol{a}_\rho$.

Theorem 8.1. *Let $m \geq 3$, M, ρ be positive integers. Let*

$$\varphi(M) > m, \qquad \rho m > \varphi(M)(m-1).$$

Then, for any integers $h_0,\ldots,h_{m-1}$, which are not all zero, and for any positive integer ξ, there exist infinitely many prime numbers p from the set $\mathbb{P}(\boldsymbol{a}_1,\ldots,\boldsymbol{a}_\rho)$ such that, in the field $\mathbb{Q}_p$,

$$|L(\xi)|_p = |h_0 f_0(\xi) + \cdots + h_{m-1}f_{m-1}(\xi)|_p > 0. \qquad (8.13)$$

Let ϑ_k be positive integers satisfying the inequality

$$\vartheta_k \leq \lambda_k \qquad (8.14)$$

for any k. Let

$$\Xi = \sum_{l=0}^{\infty} \vartheta_l \lambda_l. \tag{8.15}$$

Theorem 8.2. *Let $m \geq 3$, M, ρ be positive integers. Let*

$$\varphi(M) > m, \quad \rho m > \varphi(M)(m-1).$$

Then, for any integers $h_0, \ldots, h_{m-1}$, which are not all zero, and for the number Ξ defined by equality (8.15) and conditions (8.14), there exist infinitely many prime numbers p from the set $\mathbb{P}(\boldsymbol{a}_1, \ldots, \boldsymbol{a}_\rho)$ such that, in the field $\mathbb{Q}_p$,

$$|L(\Xi)|_p = |h_0 f_0(\Xi) + \cdots + h_{m-1} f_{m-1}(\Xi)|_p > 0. \tag{8.16}$$

Note that the symbols $f_0(\xi), \ldots, f_{m-1}(\xi)$, $f_0(\Xi), \ldots, f_{m-1}(\Xi)$ in (8.13) and (8.16) denote the sums of these series in $\mathbb{Q}_p$.

8.2. Hermite–Padé Approximations

The content of this section repeats the content of the corresponding section in Chapter 7. Only the notation used differs somewhat, since here these constructions are made for an infinite set of values. Therefore, for convenience, we give the wording of the relevant statements and omit their proofs.

For each positive integer k, we consider numbers (8.5) and set $\alpha_{m,k} = 1$. For any $N = ms + r$, where $1 \leq r \leq m$, let

$$\alpha_{N,k} = \alpha_{r,k} + s. \tag{8.17}$$

The number t is defined by $t = \left[\frac{N-1}{m-1}\right]$. With the standard notation

$$_mF_0(\alpha_1, \ldots, \alpha_m, z) = \sum_{n=0}^{\infty} \frac{(\alpha_1)_n \cdots (\alpha_m)_n}{n!} z^n,$$

let

$$f_{N,k}(z) = {}_mF_0(\alpha_{N+1,k}, \ldots, \alpha_{N+m,k}, z). \tag{8.18}$$

Define

$$u_{N,k}(z) = \alpha_{1,k} \ldots \alpha_{N,k} z^t f_{N,k}(z). \tag{8.19}$$

Lemma 8.1. *For any N, there exist polynomials $P_{N,i,k}(z)$, $i = 0, 1$, $\ldots, m - 1$ such that*

$$u_{N,k}(z) = P_{N,0,k}(z)u_{0,k}(z) + \cdots + P_{N,m-1,k}(z)u_{m-1,k}(z). \quad (8.20)$$

Moreover, the degrees of $P_{N,i,k}(z)$, $i = 0, 1, \ldots, m - 1$ are at most $t - s$, the series $f_{0,k}(z), \ldots, f_{m-1,k}(z)$ are linearly independent over $\mathbb{C}(z)$, and the following recurrence relations hold:

$$u_{N+m,k}(z) = u_{N+1,k}(z) - \alpha_{N+1,k}u_{N,k}(z) \quad (8.21)$$

if $N = 0$ or $N \geq 1$ and the number N is not divisible by $m - 1$;

$$u_{N+m,k}(z) = u_{N+1,k}(z) - \alpha_{N+1,k}zu_{N,k}(z) \quad (8.22)$$

if $N \geq 1$, the number N is divisible by $m - 1$, and for every $i = 0, 1, \ldots, m - 1$;

$$P_{N+m,i,k}(z) = P_{N+1,i,k}(z) - \alpha_{N+1,k}P_{N,k}(z) \quad (8.23)$$

if $N = 0$ or $N \geq 1$ and the number N is not divisible by $m - 1$;

$$P_{N+m,i,k}(z) = P_{N+1,i,k}(z) - \alpha_{N+1,k}zP_{N,k}(z) \quad (8.24)$$

if $N \geq 1$ and the number N is divisible by $m - 1$.
  Moreover, for all non-negative integers,

$$\Delta_{N,k}(z) = (-1)^{mN}\alpha_{1,k}\ldots\alpha_{N,k}z^t, \quad (8.25)$$

where

$$\Delta_{N,k}(z) = |P_{N+j,i,k}(z)|_{i,j=0,1,\ldots,m-1}. \quad (8.26)$$

As has been noted already, the proofs of properties (8.20)–(8.25) are presented in Lemma 7.1.
  Note that equality (8.20) implies that

$$P_{0,0,k}(z) = 1, P_{0,1,k}(z) = 0, \ldots, P_{0,m-1,k}(z) = 0,$$
$$P_{1,0,k}(z) = 0, P_{1,1,k}(z) = 1, \ldots, P_{1,m-1,k}(z) = 0,$$

$$\vdots \quad\quad\quad\quad (8.27)$$

$$P_{m-1,0,k}(z) = 0, P_{m-1,1,k}(z) = 0, \ldots, P_{m-1,m-1,k}(z) = 1.$$

8.3.   Estimates of the Coefficients of Approximating Forms

For each k, consider the quantity $\max(\alpha_{1,k}, \ldots, \alpha_{m,k})$. It follows from (8.3) and (8.7) that

$$2 \leq \max(\alpha_{1,k}, \ldots, \alpha_{m,k}) + 1 \leq c_0(k) = C_2(\ln s_k)^2, \tag{8.28}$$

where C_2 is a constant independent of k.

The height $H(P(z))$ of a polynomial $P(z)$ with integer coefficients is defined as the maximum of the absolute values of its coefficients.

Lemma 8.2. *Let $k \in \mathbb{N}$, $k \geq K_4$, $s \in \mathbb{N}$, $s = s_k$, where the number s_k is defined by the equality (8.3). Let $N = ms_k + r$, $1 \leq r \leq m$. Then, the height $H(P_{N,i,k}(z))$ of the polynomial $P_{N,i,k}(z)$, $i = 0, 1, \ldots, m-1$ is at most*

$$\exp(s_k \ln s_k + C_3 s_k \ln \ln s_k). \tag{8.29}$$

Proof of Lemma 8.2. Using an induction argument, we first prove that

$$H(P_{N,i,k}(z)) \leq c_0^N(c_0 + 1) \cdots (c_0 + s). \tag{8.30}$$

The base of induction follows immediately from equalities (8.27) for $r = 1, \ldots, m-1$. For $r = m$, in view of (8.23), we obtain

$$N = m, \quad P_{m,i,k}(z) = P_{1,i,k}(z) - \alpha_{1,k} P_{N,k}(z),$$

and the validity of the assertion follows from inequality (8.28).

The induction hypothesis is as follows: for some s, $s \geq 1$ and for all i, $i = 0, \ldots, m-1$ and all r, $r, \ldots, m$, it is true that

$$H(P_{m(s-1)+r,i,k}(z)) \leq c_0^{m(s-1)+r}(c_0 + 1) \cdots (c_0 + s - 1). \tag{8.31}$$

For each i, $i = 0, 1, \ldots, m-1$, we have either

$$H(P_{ms+r,i,k}(z))$$
$$= H(P_{m(s-1)+r+1,i,k}(z) - \alpha_{m(s-1)+r+1,k} P_{m(s-1)+r,i,k}(z)) \tag{8.32}$$

or

$$H(P_{ms+r,i,k}(z))$$
$$= H(P_{m(s-1)+r+1,i,k}(z) - \alpha_{m(s-1)+r+1,k}zP_{m(s-1)+r,i,k}(z)).$$
$$(8.33)$$

It follows from (8.17) and (8.28) that

$$\alpha_{m(s-1)+r+1,k} = \alpha_{r+1,k} + s - 1 \le c_0 + s. \qquad (8.34)$$

If $r \le m - 1$, then in the case of both (8.32) and (8.33), we obtain

$$H(P_{ms+r,i,k}(z)) \le H(P_{m(s-1)+r+1,i,k}(z))$$
$$+ \alpha_{m(s-1)+r+1,k}H(P_{m(s-1)+r,i,k}(z)). \qquad (8.35)$$

Combining (8.31), (8.34), and (8.35), with $2c_0 \le c_0^m$, we conclude that

$$H(P_{ms+r,i,k}(z)) \le c_0^{m(s-1)+r+1}(c_0 + 1) \cdots (c_0 + s - 1)$$
$$+ c_0^{m(s-1)+r}(c_0 + 1) \cdots (c_0 + s)$$
$$= c_0^{m(s-1)+r}(c_0 + 1) \cdots (c_0 + s - 1)(2c_0 + s)$$
$$\le 2c_0^{m(s-1)+r}(c_0 + 1) \cdots (c_0 + s)$$
$$\le c_0^{ms+r}(c_0 + 1) \cdots (c_0 + s). \qquad (8.36)$$

Consider the case of $r = m$. For each i, $i = 0, 1, \ldots, m - 1$, we have either

$$H(P_{m(s+1),i,k}(z)) = H(P_{ms+1,i,k}(z) - \alpha_{ms+1,k}P_{ms,i,k}(z)) \qquad (8.37)$$

or

$$H(P_{m(s+1),i,k}(z)) = H(P_{ms+1,i,k}(z) - \alpha_{ms+1,k}zP_{ms,i,k}(z)). \qquad (8.38)$$

It follows from (8.17) and (8.28) that

$$\alpha_{ms+1,k} = \alpha_{1,k} + s \le c_0 + s. \qquad (8.39)$$

In the case of both (8.37) and (8.38), we obtain

$$H(P_{m(s+1),i,k}(z)) \le H(P_{ms+1,i,k}(z)) + \alpha_{ms+1,k}H(P_{ms,i,k}(z)).$$
(8.40)

By what was proved above,

$$H(P_{ms+1,i,k}(z)) \le c_0^{ms+1}(c_0 + 1) \cdots (c_0 + s).$$
(8.41)

By the induction hypothesis,

$$H(P_{ms,i,k}(z)) = H(P_{m(s-1)+m,i,k}(z)) \le c_0^{ms}(c_0 + 1) \cdots (c_0 + s - 1).$$
(8.42)

From relations (8.39)–(8.42), it follows that

$$\begin{aligned}
H(P_{m(s+1),i,k}(z)) &\le c_0^{ms+1}(c_0 + 1) \cdots (c_0 + s) \\
&\quad + (c_0 + s)c_0^{ms}(c_0 + 1) \cdots (c_0 + s - 1) \\
&= c_0^{ms}(c_0 + 1) \cdots (c_0 + s)(c_0 + 1) \\
&\le 2c_0^{ms+1}(c_0 + 1) \cdots (c_0 + s) \\
&\le c_0^{m(s+1)}(c_0 + 1) \cdots (c_0 + s)
\end{aligned}$$
(8.43)

since $2 \le c_0 \le c_0^m$.

The induction step is complete, and inequality (8.30) is proved.

The quantity $(c_0 + 1) \cdots (c_0 + s)$ is expressed in terms of values of the Euler gamma function by the equality

$$(c_0 + 1) \cdots (c_0 + s) = \frac{\Gamma(c_0 + s + 1)}{\Gamma(c_0 + 1)}.$$
(8.44)

It is well known that, for any constant a and any $\delta > 0$, uniformly as $|s| \to +\infty$ for $-\pi + \delta \le \arg s \le \pi - \delta$, we have

$$\ln \Gamma(s + a) = \left(s + a - \frac{1}{2}\right) \ln s - s + \frac{1}{2}\pi \ln 2 + O\left(\frac{1}{|s|}\right).$$
(8.45)

It follows from (8.44) and (8.45) that, as $s = s_k \to \infty$ (which is equivalent to $k \to \infty$), the following equality holds:

$$(c_0 + 1) \cdots (c_0 + s_k) = \exp(s_k \ln s_k + O(s_k)).$$
(8.46)

In view of (8.28), $c_0^{m(s_k+1)}$, for $k \geq K_4$, has the upper bound $\exp(C_4 s_k \ln \ln s_k)$.

Combining (8.46) with inequalities (8.43) and (8.36), we complete the proof of estimate (8.29) and the assertion of the lemma.   □

Corollary 8.1. *Let* $\xi \in \mathbb{N}$. *Under the conditions of Lemma 8.2, for all* $k \geq K_5$, *the following inequality holds:*

$$|P_{N,i,k}(\xi)| \leq \exp(s_k \ln s_k + C_5 s_k \ln \ln s_k). \qquad (8.47)$$

Corollary 8.2. *Let* Ξ *be defined by equality (8.15). Let* $\Xi_k = \sum_{l=0}^{k} \vartheta_l \lambda_l$. *Under the conditions of Lemma 8.2, for all* $k \geq K_6$, *the following inequality holds:*

$$|P_{N,i,k}(\Xi_k)| \leq \exp(s_k \ln s_k + C_6 s_k \ln \ln s_k). \qquad (8.48)$$

The proofs of these corollaries are nearly similar. In Corollary 8.1, the point ξ is a fixed positive integer. By Lemma 8.1, the degree of the polynomial $P_{N,i,k}(z)$ is $t - s$, which does not exceed the number $C_7 s_k$; therefore, for $k \geq K_5$,

$$|P_{N,i,k}(\xi)| \leq C_7 s_k \xi^{C_7 s_k} \exp(s_k \ln s_k + C_3 s_k \ln \ln s_k)$$
$$\leq \exp(s_k \ln s_k + C_5 s_k \ln \ln s_k).$$

In Corollary 8.2, the number Ξ_k satisfies the inequality $\Xi_k \leq C_8 (\ln s_k)^2$; therefore, for $k \geq K_6$,

$$|P_{N,i,k}(\Xi_k)| \leq C_7 s_k \Xi_k^{C_7 s_k} \exp(s_k \ln s_k + C_3 s_k \ln \ln s_k)$$
$$\leq \exp(s_k \ln s_k + C_6 s_k \ln \ln s_k). \qquad □$$

8.4. Estimates for Auxiliary Determinants

Together with the form $L(\xi)$, defined by formula (8.13), we consider the form

$$L_k(\xi) = h_0 f_{0,k}(\xi) + \cdots + h_{m-1} f_{m-1,k}(\xi) \qquad (8.49)$$

and the related form

$$l_k(\xi) = \alpha_{1,k} \cdots \alpha_{m-1,k} L_k(\xi) = H_0 u_{0,k}(\xi) + \cdots + H_{m-1} u_{m-1,k}(\xi), \qquad (8.50)$$

where the functions $u_{N,k}(z)$ are defined by equality (8.19). By assumption, the integers $h_0, \ldots, h_{m-1}$ are not all zero. Let $h = \max(h_0, \ldots, h_{m-1})$. Then,

$$H = \max(H_0, \ldots, H_{m-1}) \leq \alpha_{1,k} \ldots \alpha_{m-1,k} h.$$

In view of inequalities (8.28),

$$H \leq C_2^{m-1} (\ln s_k)^{2m-2}. \tag{8.51}$$

Consider the linear forms $u_{N,k}(\xi), \ldots, u_{N+m,k}(\xi)$, defined by (8.19). By Lemma 8.1, equality (8.25) means that determinant (8.26) of the coefficients of these forms computed at the point ξ is non-zero. Therefore, among these forms, $m - 1$ are linearly independent with the form $l_k(\xi)$, defined by (8.50). Suppose that these are the forms

$$u_{N_1,k}(\xi), \ldots, u_{N_{m-1},k}(\xi),$$

where

$$\{N_1, \ldots, N_{m-1}\} \subset \{N, N+1, \ldots, N+m-1\}. \tag{8.52}$$

The determinant of the resulting system of linear forms is given by

$$\Delta_{l,N,k}(\xi) = \begin{vmatrix} H_0 & \cdots & H_{m-1} \\ P_{N_1,0,k}(\xi) & \cdots & P_{N_1,m-1,k}(\xi) \\ \ddots & \ddots & \ddots \\ P_{N_{m-1},0,k}(\xi) & \cdots & P_{N_{m-1},m-1,k}(\xi) \end{vmatrix}, \tag{8.53}$$

which, according to what was stated above, is a non-zero integer.

Consider forms (8.49) and (8.50) at the point $\Xi_k = \sum_{l=0}^{k} \vartheta_l \lambda_l$. Among the forms $u_{N,k}(\Xi_k), \ldots, u_{N+m,k}(\Xi_k)$, there are $m - 1$ linear forms that are linearly independent with the form $l_k(\Xi_k)$. Suppose that these are the forms

$$u_{N_1,k}(\Xi_k), \ldots, u_{N_{m-1},k}(\Xi_k),$$

where

$$\{N_1, \ldots, N_{m-1}\} \subset \{N, N+1, \ldots, N+m-1\}.$$

Now, the choice of the numbers $N_1, \ldots, N_{m-1}$ can be different from (8.52), which corresponded to determinant (8.53). Nevertheless, for the determinant of the newly obtained system of m linearly independent forms, we retain the same notation: $\Delta_{l,N,k}(\Xi_k)$.

In what follows, the numbers K_i depend on the number h, which is constant for the considered form $L_k(\xi)$.

Lemma 8.3. *For any $k \geq K_7$, it is true that*

$$|\Delta_{l,N,k}(\xi)| \leq \exp((m-1)s_k \ln s_k + C_9 s_k \ln \ln s_k), \qquad (8.54)$$

$$|\Delta_{l,N,k}(\Xi_k)| \leq \exp((m-1)s_k \ln s_k + C_{10} s_k \ln \ln s_k). \qquad (8.55)$$

Proof of Lemma 8.3. By Corollary 8.1 to Lemma 8.2 and inequality (8.47), for $k \geq K_5$,

$$
\begin{aligned}
|\Delta_{l,N,k}(\xi)| &\leq m! H \max |P_{N,i,k}(\xi)| \\
&\leq m! H \exp((m-1)s_k \ln s_k + C_5 s_k \ln \ln s_k) \\
&\leq H \exp((m-1)s_k \ln s_k + C_{11} s_k \ln \ln s_k). \qquad (8.56)
\end{aligned}
$$

Combining inequalities (8.51) and (8.56) yields (8.54).

The proof of inequality (8.55) is almost identical to that of (8.54), with the only difference being that Corollary 8.1 is replaced by Corollary 8.2 to Lemma 8.2 (inequality (8.48)). Here, K_7 is the largest of the numbers K_5 and K_6. □

8.5.   Upper Bounds for the Product of the Valuations of Approximating Forms

Lemma 8.4. *Let $k \in \mathbb{N}$, $k \geq K_8$, where K_8 is an effective constant, $s \in \mathbb{N}$, and $s = s_k$. Let M, ρ be positive integers. Let*

$$\rho > \frac{\varphi(M)(m-1)}{m}.$$

Then, for any $N_i \in \{N, N + 1 < \cdots, N + m - 1\}$, *it is true that*

$$\prod_p \max_i |u_{N_i,k}(\xi)|_p \leq \exp\left(-\frac{\rho m}{\varphi(M)} s_k \ln s_k + C_{12} s_k \sqrt{\ln s_k}\right),$$

$$(8.57)$$

where the product on the left-hand side is taken over all prime numbers p from the set $\mathbb{P}(\boldsymbol{a}_1, \ldots, \boldsymbol{a}_\rho)$ *that satisfy*

$$\exp\sqrt{\ln s_k} \leq p \leq s_k + C_1 (\ln s_k)^2.$$

$$(8.58)$$

Lemma 8.5. *Let* $k \in \mathbb{N}$, $k \geq K_9$, *where* K_9 *is an effective constant,* $s \in \mathbb{N}$, *and* $s = s_k$. *Let* M, ρ *be positive integers. Let*

$$\rho > \frac{\varphi(M)(m-1)}{m}.$$

Then, for any $N_i \in \{N, N + 1 < \ldots, N + m - 1\}$, *it is true that*

$$\prod_p \max_i |u_{N_i,k}(\Xi_k)|_p \leq \exp\left(-\frac{\rho m}{\varphi(M)} s_k \ln s_k + C_{13} s_k \sqrt{\ln s_k}\right),$$

$$(8.59)$$

where the product on the left-hand side is taken over all prime numbers p from the set $\mathbb{P}(\boldsymbol{a}_1, \ldots, \boldsymbol{a}_\rho)$ *that satisfy* (8.58).

Proof of Lemma 8.4. It follows from (8.18) and (8.19) that

$$u_{N_i,k}(\xi) = \alpha_{1,k} \ldots \alpha_{N_i,k} F(\alpha_{N_i,k} + 1, \ldots, \alpha_{N_i,k} + m, \xi).$$

Since for any prime p the quantity $F(\alpha_{N_i,k} + 1, \ldots, \alpha_{N_i,k} + m, \xi)$ is a p-adic integer, we have

$$\max_i |u_{N_i,k}(\xi)|_p \leq |\alpha_{1,k} \ldots \alpha_{N_i,k}|_p \leq |\alpha_{1,k} \ldots \alpha_{N,k}|_p.$$

Because all $\alpha_{l,k}$ are integers and in view of (8.52), $N_i \geq N$. Hence,

$$\prod_p \max_i |u_{N_i,k}(\xi)|_p \leq \prod_p |\alpha_{1,k} \ldots \alpha_{N,k}|_p, \qquad (8.60)$$

where the products are taken over any set of prime numbers p.

According to (8.17),

$$\alpha_{1,k} \cdots \alpha_{N,k}$$

$$= \alpha_{1,k}(\alpha_{1,k}+1)\cdots(\alpha_{1,k}+s_k)\cdots\alpha_{r,k}(\alpha_{r,k}+1)\cdots(\alpha_{r,k}+s_k)$$

$$\cdot\,\alpha_{r+1,k}(\alpha_{r+1,k}+1)\cdots(\alpha_{r+1,k}+s_k-1)\cdots\alpha_{m,k}(\alpha_{m,k}+1)$$

$$\cdots(\alpha_{m,k}+s_k-1). \tag{8.61}$$

It follows from (8.28) and (8.3) that the prime factorisation of $\alpha_{1,k}\cdots\alpha_{N_i,k}$ includes only primes p that satisfy

$$p \le s_k + C_1\lambda_k^2 \le s_k + C_1(\ln s_k)^2. \tag{8.62}$$

Let us estimate from above the product

$$\prod_p |\alpha_{1,k}\cdots\alpha_{N,k}|_p, \tag{8.63}$$

taken over all prime numbers p from $\mathbb{P}(\boldsymbol{a}_1,\ldots,\boldsymbol{a}_\rho)$ that satisfy inequalities (8.62).

Consider the product

$$\alpha_{i,k}(\alpha_{i,k}+1)\cdots(\alpha_{i,k}+s_k) = \frac{(\alpha_{i,k}+s_k)!}{(\alpha_{i,k}-1)!}, \quad i=1,\ldots,r. \tag{8.64}$$

The prime p in product (8.64) is raised to the power

$$\mathrm{ord}_p(\alpha_{i,k}(\alpha_{i,k}+1)\cdots(\alpha_{i,k}+s_k))$$

$$= \frac{\alpha_{i,k}+s_k-S_{\alpha_{i,k}+s_k}}{p-1} - \frac{\alpha_{i,k}-1-S_{\alpha_{i,k}-1}}{p-1}$$

$$= \frac{s_k - S_{\alpha_{i,k}+s_k} + S_{\alpha_{i,k}-1}+1}{p-1}, \tag{8.65}$$

where the symbol S_x with a positive integer x denotes the sum of the digits in the p-adic expansion of this number, and hence

$$1 \le S_x \le (p-1)([\log_p x]+1) \le (p-1)(\log_p x+1).$$

Therefore,

$$\frac{s_k + 2}{p - 1} - \log_p(\alpha_{i,k} + s_k) - 1 \le \frac{s_k - S_{\alpha_{i,k}+s_k} + S_{\alpha_{i,k}-1} + 1}{p - 1}$$

$$\le \frac{s_k}{p - 1} + \log_p(\alpha_{i,k} - 1).$$

Thus, in view of (8.28), power (8.65) for $k \ge K_{10}$ lies in the range

$$\frac{s_k}{p - 1} - C_{14} \log_p s_k \le \mathrm{ord}_p(\alpha_{i,k}(\alpha_{i,k} + 1) \cdots (\alpha_{i,k} + s_k))$$

$$\le \frac{s_k}{p - 1} + C_{15} \log_p \ln s_k. \tag{8.66}$$

Consider the product

$$\alpha_{i,k}(\alpha_{i,k} + 1) \cdots (\alpha_{i,k} + s_k - 1) = \frac{(\alpha_{i,k} + s_k - 1)!}{(\alpha_{i,k} - 1)!},$$

$$i = r + 1, \ldots, m. \tag{8.67}$$

For each of the products in (8.67), by analogy with (8.66), we conclude that, for $k \ge K_{11}$, the prime p appears in this product in the power satisfying the inequalities

$$\frac{s_k}{p - 1} - C_{16} \log_p s_k \le \mathrm{ord}_p(\alpha_{i,k}(\alpha_{i,k} + 1) \cdots (\alpha_{i,k} + s_k - 1))$$

$$\le \frac{s_k}{p - 1} + C_{17} \log_p \ln s_k. \tag{8.68}$$

Thus, combining (8.66), (8.68), and (8.61), we conclude that, for all the considered primes p and $k \ge K_{12}$, the following inequality is true:

$$|\alpha_{1,k} \cdots \alpha_{N,k}|_p \le \exp\left(-\frac{\ln p}{p - 1} m s_k + C_{18} \ln s_k\right). \tag{8.69}$$

Inequality (8.69) means that

$$\prod_p |\alpha_{1,k} \cdots \alpha_{N,k}|_p \le \exp\left(\sum_p \left(-\frac{\ln p}{p - 1} m s_k + C_{18} \ln s_k\right)\right), \tag{8.70}$$

where the product and the sum are taken over all prime numbers p from $\mathbb{P}(\boldsymbol{a}_1, \ldots, \boldsymbol{a}_p)$ that satisfy inequalities (8.62).

For these p, we have the estimate

$$C_{18} \sum_p \ln s_k \leq C_{19} s_k. \tag{8.71}$$

We use the equality

$$\sum_p \frac{\ln p}{p-1} = \sum_p \frac{\ln p}{p} + \sum_p \frac{\ln p}{p(p-1)} = \sum_p \frac{\ln p}{p} + C_{20} \tag{8.72}$$

and the well-known estimate

$$\sum_{p \leq x} \frac{\ln p}{p} = \frac{1}{\varphi(M)} \ln x + O(1), \quad x \to +\infty,$$

where the sum extends over all prime numbers $p \leq x$ belonging to the set $\boldsymbol{a}_i$ of values of any of the considered arithmetic progressions. Therefore, in view of (8.62) and (8.72), for such p and with $k \geq K_{13}$, we have

$$\sum_p \frac{\ln p}{p-1} = \frac{1}{\varphi(M)} \ln(s_k + C_1(\ln s_k)^2) m s_k + C_{21} s_k$$

$$\geq \frac{1}{\varphi(M)} m s_k \ln s_k + C_{22} s_k.$$

Consequently, in view of (8.70) and (8.71),

$$\prod_p |\alpha_{1,k} \cdots \alpha_{N,k}|_p \leq \exp\left(-\frac{m\rho}{\varphi(M)} s_k \ln s_k + C_{23} s_k\right), \tag{8.73}$$

where the product is taken over all prime numbers p from $\mathbb{P}(\boldsymbol{a}_1, \ldots, \boldsymbol{a}_\rho)$ that satisfy inequalities (8.62).

Let us estimate from below product (8.63), taken over all prime numbers p, satisfying the inequality

$$p \leq \exp \sqrt{\ln s_k}. \tag{8.74}$$

In view of (8.66) and (8.68), this product satisfies the lower bound

$$\prod_p |\alpha_{1,k} \ldots \alpha_{N,k}|_p \geq \exp\left(\sum_p \left(-\frac{\ln p}{p-1} m s_k - C_{24} \ln s_k\right)\right), \quad (8.75)$$

where the product and the sum are taken over all prime numbers p satisfying inequality (8.74). Using relation (8.72) and the estimate

$$\sum_p \ln s_k \leq C_{25} \sqrt{\ln s_k} \exp \sqrt{\ln s_k},$$

which follows from (8.74), we estimate the sum

$$\sum_p \left(\frac{\ln p}{p-1} m s_k + C_{24} \ln s_k\right)$$

$$\leq m s_k \sum_p \frac{\ln p}{p} + C_{20} m s_k + C_{25} \sqrt{\ln s_k} \exp \sqrt{\ln s_k}. \quad (8.76)$$

It is well known that

$$\sum_{p \leq x} \frac{\ln p}{p} = \ln x + O(1), \quad x \to +\infty,$$

where the sum extends over all primes $p \leq x$. Therefore, for $k \geq K_{14}$, it is true that

$$m s_k \sum_p \frac{\ln p}{p} \leq m s_k \sqrt{\ln s_k} + C_{26} s_k. \quad (8.77)$$

Hence, in view of inequalities (8.74)–(8.77), for $k \geq K_{15}$, we have

$$\sum_p \left(\frac{\ln p}{p-1} m s_k + C_{24} \ln s_k\right) \leq m s_k \sqrt{\ln s_k} + C_{27} s_k. \quad (8.78)$$

From (8.75) and (8.78), it follows straightforwardly that

$$\prod_p |\alpha_{1,k} \ldots \alpha_{N,k}|_p \geq \exp\left(-m s_k \sqrt{\ln s_k} - C_{27} s_k\right), \quad (8.79)$$

where the product and the sum are taken over all prime numbers p satisfying inequality (8.74).

Combining (8.73) and (8.79) yields, for $k \geq K_{16}$,

$$\prod_p |\alpha_{1,k} \cdots \alpha_{N,k}|_p$$

$$\leq \exp\left(-\frac{m\rho}{\varphi(M)} s_k \ln s_k + C_{23}s_k + ms_k\sqrt{\ln s_k} + C_{27}s_k\right)$$

$$\leq \exp\left(-\frac{m\rho}{\varphi(M)} s_k \ln s_k + C_{28}s_k\sqrt{\ln s_k}\right),$$

where the product is taken over all prime numbers p from $\mathbb{P}(a_1, \ldots, a_\rho)$ that satisfy inequalities (8.58).

However, assuming that K_8 is the largest of the numbers $K_1, K_2, \ldots, K_{16}$, $k \geq K_8$, and $C_{12} = C_{28}$, it then follows from (8.60) that inequality (8.57) holds, i.e.,

$$\prod_p \max_i |u_{N_i,k}(\xi)|_p \leq \exp\left(-\frac{\rho m}{\varphi(M)} s_k \ln s_k + C_{12}s_k\sqrt{\ln s_k}\right),$$

where the product is taken over all prime numbers p from $\mathbb{P}(a_1, \ldots, a_\rho)$ that satisfy inequalities (8.58).

Lemma 8.4 is proved.   □

The proof of inequality (8.59) in Lemma 8.5 repeats word for word the proof of Lemma 8.4.   □

8.6.   Lower Bounds for $|l_k(\xi)|_p$, $|l_k(\Xi_k)|_p$, $|L_k(\xi)|_p$, $|L_k(\Xi_k)|_p$

Determinant (8.53) is transformed as follows: its first column is multiplied by $u_{0,k}(\xi)$, and the other columns multiplied by the corresponding $u_{j,k}(\xi)$, $j = 1, \ldots, m - 1$ are added to the result. Taking into account (8.20) and (8.50), we obtain

$$\Delta_{l,N,k}(\xi)u_{0,k}(\xi) = \begin{vmatrix} l_k(\xi) & \cdots & H_{m-1} \\ u_{N_1,k}(\xi) & \cdots & P_{N_1,m-1,k}(\xi) \\ \ddots & \ddots & \ddots \\ u_{N_{m-1},k}(\xi) & \cdots & P_{N_{m-1},m-1,k}(\xi) \end{vmatrix}. \qquad (8.80)$$

Similarly, the last column of determinant (8.53) is multiplied by $u_{m-1,k}(\xi)$, and the other columns multiplied by the corresponding

$u_{j,k}(\xi)$, $j = 0, \ldots, m-2$ are added to the result. Taking into account (8.20) and (8.50) yields

$$\Delta_{l,N,k}(\xi)u_{m-1,k}(\xi) = \begin{vmatrix} H_0 & \cdots & l_k(\xi) \\ P_{N_1,0,k}(\xi) & \cdots & u_{N_1,k}(\xi) \\ \ddots & \ddots & \ddots \\ P_{N_{m-1},0,k}(\xi) & \cdots & u_{N_{m-1},k}(\xi) \end{vmatrix}. \quad (8.81)$$

It follows from (8.12) and (8.19) that

$$u_{0,k}(\xi) = 1 + \xi u_{m-1,k}(\xi). \quad (8.82)$$

Let $\delta_{0,i,j}$ denote the cofactor of the element of determinant (8.80) located at the intersection of row i and column j, and let $\delta_{m-1,i,j}$ denote the cofactor of the corresponding element of determinant (8.81). Then,

$$\Delta_{l,N,k}(\xi)u_{0,k}(\xi) = l_k(\xi)\delta_{0,1,1} + \sum_{i=2}^{m}\delta_{0,i,1}u_{N_{i-1},k}(\xi), \quad (8.83)$$

$$\Delta_{l,N,k}(\xi)u_{m-1,k}(\xi) = l_k(\xi)\delta_{m-1,1,m} + \sum_{i=2}^{m}\delta_{m-1,i,m}u_{N_{i-1},k}(\xi). \quad (8.84)$$

It follows from (8.82)–(8.84) that

$$\Delta_{l,N,k}(\xi) = l_k(\xi)(\delta_{0,1,1} - \xi\delta_{m-1,1,m})$$

$$+ \sum_{i=2}^{m}(\delta_{0,i,1} - \xi\delta_{m-1,i,m})u_{N_{i-1},k}(\xi). \quad (8.85)$$

Lemma 8.6. *Let $k \in \mathbb{N}$, $k \geq K_{17}$. Then, there exists a prime number p_k satisfying inequalities (8.58) such that*

$$|l_k(\xi)|_{p_k} \geq \exp(-(m-1)s_k \ln s_k - C_{29}s_k\sqrt{\ln s_k}), \quad (8.86)$$

$$|L_k(\xi)|_{p_k} \geq \exp(-(m-1)s_k \ln s_k - C_{30}s_k\sqrt{\ln s_k}). \quad (8.87)$$

Proof of Lemma 8.6. First, we prove the existence of a prime number p_k satisfying inequalities (8.58) for which (8.86) holds. Assume the opposite, i.e., for all primes p, satisfying inequalities (8.58), we have

$$|l_k(\xi)|_p < \exp(-(m-1)s_k \ln s_k - C_{29}s_k \sqrt{\ln s_k}). \tag{8.88}$$

In (8.85), the coefficient multiplying $l_k(\xi)$ is an integer. Determinant (8.53) is non-zero. For a non-zero integer A, it is true that

$$|A|_p \geq \frac{1}{|A|},$$

which, in view of (8.54), implies that, for all the considered prime numbers p,

$$|\Delta_{l,N,k}(\xi)|_p \geq \exp(-(m-1)s_k \ln s_k - C_9 s_k \ln \ln s_k).$$

Combining this inequality with (8.88) yields

$$|l_k(\xi)|_p < |\Delta_{l,N,k}(\xi)|_p \tag{8.89}$$

for $\geq K_{18}$. Then, according to the well-known properties of the p-adic valuation, (8.85) and (8.89) mean that for all the considered primes p, it is true that

$$|\Delta_{l,N,k}(\xi)|_p = \left| \sum_{i=2}^{m} (\delta_{0,i,1} + \xi \delta_{m-1,i,m}) u_{N_{i-1},k}(\xi) \right|_p. \tag{8.90}$$

Since the numbers $(\delta_{0,j,1} + \xi \delta_{m-1,j,m-1})$ are integers, we obtain

$$|\delta_{0,j,1} + \xi \delta_{m-1,j,m-1}|_p \leq 1$$

for all the considered primes p, and equality (8.90) implies that

$$\left| \sum_{i=2}^{m} (\delta_{0,i,1} + \xi \delta_{m-1,i,m}) u_{N_{i-1},k}(\xi) \right|_p \leq \max_j |u_{N_j,k}(\xi)|_p.$$

Therefore,

$$|\Delta_{l,N,k}(\xi)|_p \leq \max_j |u_{N_j,k}(\xi)|_p. \tag{8.91}$$

It follows from (8.91) that, for any subset $\mathbb{P}_0$ of the prime number set $\mathbb{P}$, the following inequality holds:

$$\prod_{p \in \mathbb{P}_0} |\Delta_{l,N,k}(\xi)|_p \leq \prod_{p \in \mathbb{P}_0} \max_j |u_{N_j,k}(\xi)|_p. \tag{8.92}$$

By Lemma 8.4, inequality (8.57) holds, i.e.,

$$\prod_p \max_i |u_{N_i,k}(\xi)|_p \leq \exp\left(-\frac{\rho m}{\varphi(M)} s_k \ln s_k + C_{12} s_k \sqrt{\ln s_k}\right),$$

where the product is taken over all prime numbers p from $\mathbb{P}(a_1, \ldots, a_\rho)$ that satisfy inequalities (8.58).

Given a rational number $A \neq 0$, the following product formula is true:

$$\prod_p |A|_p = \frac{1}{|A|}.$$

Therefore, it follows from (8.54) that

$$\prod_p |\Delta_{l,N,k}(\xi)|_p \geq \exp(-(m-1)s_k \ln s_k - C_9 s_k \ln \ln s_k), \qquad (8.93)$$

where the product is taken over all prime numbers p from $\mathbb{P}(a_1, \ldots, a_\rho)$ that satisfy inequalities (8.58).

In view of the inequality

$$\frac{\rho m}{\varphi(M)} > m - 1,$$

estimates (8.92), (8.57), and (8.93) contradict each other for $k \geq K_{19}$. For $K_{17} = \max(K_{18}, K_{19})$, this refutes the assumption made and proves inequality (8.86) for some p_k satisfying (8.58). Since $l_k(\xi)$, defined by (8.50), differs from $L_k(\xi)$, defined by (8.49), only by the factor $\alpha_{1,k} \ldots \alpha_{m-1,k}$, inequalities (8.86) and (8.51) yield (8.87), and Lemma 8.6 is proved.   □

Lemma 8.7. *Let $k \in \mathbb{N}$, $k \geq K_{20}$. Then, there exists a prime number p_k, satisfying inequalities (8.58), such that*

$$|l_k(\Xi_k)|_{p_k} \geq \exp(-(m-1)s_k \ln s_k - C_{31} s_k \sqrt{\ln s_k}), \qquad (8.94)$$

$$|L_k(\Xi_k)|_{p_k} \geq \exp(-(m-1)s_k \ln s_k - C_{32} s_k \sqrt{\ln s_k}). \qquad (8.95)$$

The proof of inequalities (8.94) and (8.95) repeats word for word the proof of Lemma 8.6, with the only difference being that inequality (8.57) is replaced by (8.59) and inequality (8.54) is replaced by (8.55).   □

8.7. Completion of the Proofs of Theorems 8.1 and 8.2

Let p_k be a prime number satisfying (8.58).

Consider the linear form

$$L(\xi) = h_0 f_0(\xi) + \cdots + h_{m-1} f_{m-1}(\xi).$$

As was noted above, this is an integer p_k-adic number, so the difference of the forms,

$$L(\xi) - L_k(\xi) = h_0 f_0(\xi) + \cdots + h_{m-1} f_{m-1}(\xi) - (h_0 f_{0,k}(\xi) + \cdots$$
$$+ h_{m-1} f_{m-1,k}(\xi)), \tag{8.96}$$

is also a p_k-adic integer. According to (8.5), (8.6), (8.8), and (8.10),

$$f_0(\xi) - f_{0,k}(\xi) = \sum_{n=0}^{\infty} ((\alpha_1)_n \cdots (\alpha_{m-1})_n - (\alpha_{1,k})_n \cdots (\alpha_{m-1,k})_n) \xi^n,$$
$$\tag{8.97}$$

$$f_{m-1}(\xi) - f_{m-1,k}(\xi) = \sum_{n=0}^{\infty} ((\alpha_1 + 1)_n \cdots (\alpha_{m-1} + 1)_n$$
$$- (\alpha_{1,k} + 1)_n \cdots (\alpha_{m-1,k} + 1)_n) \xi^n. \tag{8.98}$$

Furthermore, for $i = 1, \ldots, m - 2$, in view of (8.9) and (8.11),

$$f_i(\xi) - f_{i,k}(\xi) = \sum_{n=0}^{\infty} (\alpha_1 + 1)_n \cdots (\alpha_i + 1)_n (\alpha_{i+1})_n \cdots (\alpha_{m-1})_n \xi^n$$

$$- \sum_{n=0}^{\infty} (\alpha_{1,k} + 1)_n \cdots (\alpha_{i,k} + 1)_n (\alpha_{i+1,k})_n \cdots (\alpha_{m-1,k})_n \xi^n$$

$$= \sum_{n=0}^{\infty} ((\alpha_1 + 1)_n \cdots (\alpha_i + 1)_n (\alpha_{i+1})_n \cdots (\alpha_{m-1})_n$$

$$- (\alpha_{1,k} + 1)_n \cdots (\alpha_{i,k} + 1)_n (\alpha_{i+1,k})_n \cdots (\alpha_{m-1,k})_n) \xi^n. \tag{8.99}$$

Consider the quantities

$$(\alpha_1)_n \cdots (\alpha_{m-1})_n - (\alpha_{1,k})_n \cdots (\alpha_{m-1,k})_n, \qquad (8.100)$$

$$(\alpha_1 + 1)_n \cdots (\alpha_{m-1} + 1)_n - (\alpha_{1,k} + 1)_n \cdots (\alpha_{m-1,k} + 1)_n, \quad (8.101)$$

and, for $i = 1, \ldots, m - 2$,

$$(\alpha_1 + 1)_n \cdots (\alpha_i + 1)_n (\alpha_{i+1})_n \cdots (\alpha_{m-1})_n$$
$$- (\alpha_{1,k} + 1)_n \cdots (\alpha_{i,k} + 1)_n (\alpha_{i+1,k})_n \cdots (\alpha_{m-1,k})_n. \qquad (8.102)$$

For $n = 0$, all differences (8.100)–(8.102) vanish.

For $n \geq 1$, difference (8.100) is represented in the form

$$(\alpha_1)_n \cdots (\alpha_{m-1})_n - (\alpha_{1,k})_n \cdots (\alpha_{m-1,k})_n$$
$$= (\alpha_1)_n \cdots (\alpha_{m-1})_n - (\alpha_{1,k})_n (\alpha_2)_n \cdots (\alpha_{m-1})_n$$
$$+ (\alpha_{1,k})_n (\alpha_2)_n \cdots (\alpha_{m-1})_n - (\alpha_{1,k})_n (\alpha_{2,k})_n \cdots (\alpha_{m-1})_n + \cdots$$
$$+ (\alpha_{1,k})_n (\alpha_{2,k})_n \cdots (\alpha_{m-2,k})_n (\alpha_{m-1})_n - (\alpha_{1,k})_n \cdots (\alpha_{m-1,k})_n.$$
$$(8.103)$$

Consider the following quantities involved in (8.103):

$$(\alpha_{1,k})_n (\alpha_{2,k})_n \cdots (\alpha_{j-1,k})_n (\alpha_{j,k})_n (\alpha_{j+1})_n \cdots (\alpha_{m-1})_n$$
$$- (\alpha_{1,k})_n (\alpha_{2,k})_n \cdots (\alpha_{j-1,k})_n (\alpha_j)_n (\alpha_{j+1})_n \cdots (\alpha_{m-1})_n$$
$$= ((\alpha_{j,k})_n - (\alpha_j)_n)(\alpha_{1,k})_n (\alpha_{2,k})_n \cdots (\alpha_{j-1,k})_n (\alpha_{j+1})_n \cdots (\alpha_{m-1})_n.$$
$$(8.104)$$

The quantity

$$(\alpha_{j,k})_n - (\alpha_j)_n$$
$$= \alpha_{j,k}(\alpha_{j,k} + 1) \cdots (\alpha_{j,k} + n - 1) - \alpha_j(\alpha_j + 1) \cdots (\alpha_j + n - 1)$$
$$(8.105)$$

can be treated as the difference between the values of the polynomial $x(x + 1) \cdots (x + n - 1)$ at the points $\alpha_{j,k}$ and α_j. Since, according to

(8.5) and (8.6),

$$\alpha_{j,k} - \alpha_j = -\sum_{l=k+1}^{\infty} \mu_{j,l}\lambda_l,$$

it follows from inequality (8.2) that, for any $j = 1, \ldots, m-1$,

$$|\alpha_{j,k} - \alpha_j|_{p_k} \leq p_k^{-ms_k \ln s_k}.$$

Therefore, the quantity defined by equality (8.105) satisfies the relation

$$|(\alpha_{j,k})_n - (\alpha_j)_n|_{p_k} \leq p_k^{-ms_k \ln s_k}. \tag{8.106}$$

Since $(\alpha_{1,k})_n(\alpha_{2,k})_n \cdots (\alpha_{j-1,k})_n(\alpha_{j+1,k})_n \cdots \cdots (\alpha_{m-1})_n$ is a p_k-adic integer, it follows from (8.106) that the p_k-adic valuation of (8.104) does not exceed the number

$$p_k^{-ms_k \ln s_k}. \tag{8.107}$$

Therefore, quantity (8.100) represented in the form of sum (8.103) satisfies the inequality

$$|(\alpha_1)_n \cdots (\alpha_{m-1})_n - (\alpha_{1,k})_n \cdots (\alpha_{m-1,k})_n|_{p_k} \leq p_k^{-ms_k \ln s_k}. \tag{8.108}$$

It is easy to see that difference (8.101) satisfies the same estimate. This means that all terms of the convergent p_k-adic series (8.97) and (8.98) are estimated from above by (8.107). Therefore,

$$|f_0(\xi) - f_{0,k}(\xi)|_{p_k} \leq p_k^{-ms_k \ln s_k},$$
$$|f_{m-1}(\xi) - f_{m-1,k}(\xi)|_{p_k} \leq p_k^{-ms_k \ln s_k}. \tag{8.109}$$

Representing difference (8.100) in a form similar to (8.101), we note that

$$|(\alpha_{j,k} + 1) - (\alpha_j + 1)|_{p_k} = |\alpha_{j,k} - \alpha_j|_{p_k} \leq p_k^{-ms_k \ln s_k}.$$

Proceeding by analogy with (8.103)–(8.109) yields

$$|f_i(\xi) - f_{i,k}(\xi)|_{p_k} \leq p_k^{-ms_k \ln s_k} \tag{8.110}$$

for any $i = 1, \ldots, m - 2$. It follows from equality (8.96) that

$$L(\xi) - L_k(\xi) = h_0(f_0(\xi) - f_{0,k}(\xi)) + \cdots$$
$$+ h_{m-1}(f_{m-1}(\xi) - f_{m-1,k}(\xi)). \qquad (8.111)$$

Therefore, (8.109) and (8.110) yield

$$|L(\xi) - L_k(\xi)|_{p_k} \leq p_k^{-ms_k \ln s_k}. \qquad (8.112)$$

In Lemma 8.6, it was proved that for any $k \geq K_{17}$, inequality (8.87) holds, i.e.,

$$|L_k(\xi)|_{p_k} \geq \exp(\ -(m - 1)s_k \ln s_k - C_{30}s_k \sqrt{\ln s_k}\ ).$$

Combined with (8.110), this inequality yields

$$|L(\xi)|_{p_k} = |L_k(\xi)|_{p_k} \geq \exp(\ -(m - 1)s_k \ln s_k - C_{30}s_k \sqrt{\ln s_k}\ ) > 0.$$
$$(8.113)$$

This is the inequality that was to be proved.

Consider the linear form

$$L(\Xi) = h_0 f_0(\Xi) + \cdots + h_{m-1} f_{m-1}(\Xi).$$

As was noted above, it is a p_k-adic integer.

Consider the difference

$$L(\Xi) - L_k(\Xi_k) = L(\Xi) - L_k(\Xi) + L_k(\Xi) - L_k(\Xi_k). \qquad (8.114)$$

For $L(\Xi) - L_k(\Xi)$, we follow the line of reasoning used for difference (8.111). Although ξ is a positive integer and Ξ is a polyadic number, in deriving estimate (8.112), we used only the fact that ξ is a p_k-adic integer, so the replacement of ξ by Ξ does not affect the validity of the estimate. In other words,

$$|L(\Xi) - L_k(\Xi)|_{p_k} \leq p_k^{-ms_k \ln s_k}. \qquad (8.115)$$

Consider the quantity $L_k(\Xi) - L_k(\Xi_k)$. It has the form

$$L_k(\Xi) - L_k(\Xi_k) = \sum_{i=0}^{m-1} h_i(f_{i,k}(\Xi) - f_{i,k}(\Xi_k)). \tag{8.116}$$

Representing each of the differences $f_{i,k}(\Xi) - f_{i,k}(\Xi_k)$ in the form

$$f_{i,k}(\Xi) - f_{i,k}(\Xi_k)$$
$$= \sum_{n=0}^{\infty} (\alpha_{1,k} + 1)_n \cdots (\alpha_{i,k} + 1)_n (\alpha_{i+1,k})_n \cdots (\alpha_{m-1,k})_n (\Xi^n - \Xi_k^n),$$
$$\tag{8.117}$$

we note that $\Xi^n - \Xi_k^n$ is equal to the product of the number $\Xi - \Xi_k$ by a p_k-adic integer; therefore, according to (8.2) and (8.15),

$$|\Xi^n - \Xi_k^n|_{p_k} \leq |\Xi - \Xi_k|_{p_k} \leq p_k^{-ms_k \ln s_k}. \tag{8.118}$$

Equalities (8.116) and (8.117) and inequality (8.118) mean that

$$|L_k(\Xi) - L_k(\Xi_k)|_{p_k} \leq p_k^{-ms_k \ln s_k}. \tag{8.119}$$

Combining (8.114) with (8.115) and (8.119) yields

$$|L(\Xi) - L_k(\Xi_k)|_{p_k} \leq p_k^{-ms_k \ln s_k}. \tag{8.120}$$

Inequalities (8.95) in Lemma 8.7 and (8.120) show that, for $k \geq K_{21}$, we have

$$|L(\Xi)|_{p_k} = |L_k(\Xi_k)|_{p_k} > 0;$$

in other words, inequality (8.16) is proved.

To complete the proofs of the theorems, we need to check whether the inequality $p_k < p_{k+1}$ holds for $k \geq K_{22}$. For this purpose, in view of (8.58), it suffices to prove that for $k \geq K_{22}$, the following inequality is true:

$$s_k + C_2(\ln s_k)^2 < \exp \sqrt{\ln s_{k+1}}.$$

According to (8.2) and (8.3), $s_{k+1} \geq \exp\lambda_{k+1}$ and

$$\lambda_{k+1} \geq \prod_{p \leq s_k + 2(\ln s_k)^2} \exp(\ln p(ms_k \ln s_k))$$

$$= \exp\left(\sum_{p \leq s_k + 2(\ln s_k)^2} \ln p(ms_k \ln s_k)\right) \geq \exp s_k^2.$$

(Here, we used the rough estimate $\sum_{p \leq s_k + 2(\ln s_k)^2} \ln p \geq \frac{s_k}{2 \ln s_k}$.)
Thus, in view of (8.3),

$$s_{k+1} \geq \exp \lambda_{k+1},$$

$$\ln s_{k+1} \geq \lambda_{k+1} \geq \exp s_k^2, \quad \sqrt{\ln s_{k+1}} \geq \exp \frac{1}{2} s_k^2;$$

therefore,

$$\exp \sqrt{\ln s_{k+1}} \geq \exp\left(\exp \frac{1}{2} s_k^2\right) > s_k + C_2(\ln s_k)^2,$$

as required.

Thus, we have proved that, for arbitrary linear forms $L(\xi)$ and $L(\Xi)$, there exist infinitely many numbers k and primes p_k for which $|L(\xi)|_{p_k} > 0$ and $|L(\Xi)|_{p_k} > 0$, as was stated in the theorems.   □

8.8.   Concluding Remarks

The theorems proved do not imply that the p-adic numbers $f_0(\xi), \ldots, f_{m-1}(\xi)$ and, accordingly, $f_0(\Xi), \ldots, f_{m-1}(\Xi)$ are linearly independent for the given prime p. However, they allow us to prove the transcendence of one of the p-adic numbers $f_0(\xi), f_{m-1}(\xi)$ in infinitely many p-adic fields, which is described in the following chapter.

Chapter 9

Transcendence of p-Adic Values of Generalised Hypergeometric Series with Transcendental Polyadic Parameters*

9.1. Basic Identity

In the previous chapters, we considered series of the form

$$F_0(z) = \sum_{n=0}^{\infty} \frac{(\alpha_1)_n \cdots (\alpha_r)_n}{(\beta_1)_n \cdots (\beta_s)_n} z^n,$$

$$F_1(z) = \sum_{n=0}^{\infty} \frac{(\alpha_1 + 1)_n \cdots (\alpha_r + 1)_n}{(\beta_1 + 1)_n \cdots (\beta_s + 1)_n} z^n.$$

We recall that $(\gamma)_n$ is defined by the equalities $(\gamma)_0 = 1$, $(\gamma)_n = \gamma(\gamma + 1) \cdots (\gamma + n - 1)$ for $n \geq 1$. Of course, each β_j, $j = 1, \ldots, s$, is not equal to zero or a negative integer.

We further consider the following basic identity:

$$F_0(z) = 1 + \frac{\alpha_1 \ldots \alpha_r}{\beta_1 \ldots \beta_s} z F_1(z). \tag{9.1}$$

This identity implies that if the numbers $\alpha_1, \ldots, \alpha_r, \beta_1, \ldots, \beta_s$, and ξ are algebraic, and if the series $F_0(\xi), F_1(\xi)$ converge, then either

---

*The results of this chapter are quite recent. They are published in Chirskii (2022b, 2023).

both numbers, $F_0(\xi), F_1(\xi)$, are algebraic or they are both transcendental. If the number $\frac{\alpha_1...\alpha_r}{\beta_1...\beta_s}\xi$ is transcendental and if $F_1(\xi) \neq 0$, then it follows from (9.1) that at least one of the numbers $F_0(\xi), F_1(\xi)$ is transcendental. These assertions are valid in the case of the field of complex numbers as well as in the case of any algebraic extension of the field of p-adic numbers. So, if we prove that in the field of p-adic numbers the inequality $F_1(\xi) \neq 0$ holds, then equation (9.1) implies that at least one of the p-adic numbers $F_0(\xi), F_1(\xi)$ is transcendental.

9.2. Formulations of Theorems

Here, we consider the series

$$\Psi_0(z) = \sum_{n=0}^{\infty} (\alpha_1)_n \cdots (\alpha_m)_n z^n,$$

$$\Psi_1(z) = \sum_{n=0}^{\infty} (\alpha_1 + 1)_n \cdots (\alpha_m + 1)_n z^n,$$

(9.2)

and using the results from the previous chapter, we shall prove that if $\alpha_1, \ldots, \alpha_m$ are polyadic Liouville numbers and if ξ is a positive integer or Ξ is a polyadic Liouville number, then there are infinitely many primes p such that in the field of p-adic numbers, at least one of the p-adic numbers $\Psi_0(\xi), \Psi_1(\xi)$ (correspondingly, $\Psi_0(\Xi), \Psi_1(\Xi)$) is transcendental.

Earlier, we noted that studies of the arithmetic nature of the values of hypergeometric series in the complex domain are of great interest. A large number of significant results have been obtained by many authors. We have already mentioned several wonderful books (e.g., Shidlovskii, 1989; Nesterenko and Feldman, 1998) addressing these problems and presented numerous references. But the situation is different for p-adic domains. Previously, it was only possible to prove the infinite transcendence or infinite linear independence of the values of such series (see Bertrand *et al.*, 2004; Chirskii, 2019, 2020, 2021). The infinite transcendence of the series means that for any non-zero polynomial $P(x)$ with integer coefficients, there is an infinite set of primes p such that when substituting the values of the series in question into this polynomial in the field of p-adic numbers, a non-zero p-adic number is obtained. However, this does not mean that the value of this series is irrational in a particular field of p-adic numbers. So, the theorems proved in the following seem to represent the first result on the transcendence of the values of generalised hypergeometric series in the field of p-adic numbers.

Let's move on to the exact wording. The result is a consequence of the theorems proved in Chapter 8 using the Hermite–Padé approximations from Nesterenko (1995), so we recall the notation used and the basic definitions.

Let λ_0 be an arbitrary positive integer. Put $s_0 = [\exp\lambda_0]+1$. Let λ_1 be an arbitrary positive integer, such that for any prime $p \leq s_0 + 2\lambda_0^2$, the inequality $\operatorname{ord}_p \lambda_1 \geq ms_0 \ln s_0$ holds, and let $s_1 = [\exp \lambda_1] + 1$.

For $k \geq 1$, let λ_{k+1} be an arbitrary positive integer, such that for any prime $p \leq s_k + 2\lambda_k^2$, the inequality

$$\operatorname{ord}_p \lambda_{k+1} \geq ms_k \ln s_k \tag{9.3}$$

holds, and let

$$s_{k+1} = [\exp \lambda_{k+1}] + 1. \tag{9.4}$$

Let $\mu_{i,0}$, $i = 1, \ldots, m$ be positive integers. Let, for all $i = 1, \ldots, m$, $k \geq 1$, the numbers $\mu_{i,k}$ be non-negative integers, satisfying

$$\mu_{i,k} \leq \lambda_k. \tag{9.5}$$

Let $\alpha_i = \sum_{l=0}^{\infty} \mu_{i,l}\lambda_l$, $i = 1, \ldots, m$. If for some k and for all $l \geq k$, we have $\mu_{i,l} = 0$, then α_i is a positive integer. In the other case, they are polyadic Liouville numbers. Let at least one of the numbers α_i, $i = 1, \ldots, m$, be a polyadic Liouville number.

Let M be a positive integer. Consider a reduced residue system $\mod(M)$. As usual, the number of elements in this system is denoted by $\varphi(M)$, where $\varphi(M)$ is the Euler function. Let ρ various elements $a_1, \ldots, a_\rho$ of this reduced system of residues be arbitrarily chosen. We denote by $\boldsymbol{a_1}, \ldots, \boldsymbol{a_\rho}$ sets of positive integer values taken by progressions $a_i + Mk$, $k \in \mathbb{Z}$, $i = 1, \ldots, \rho$. Using the standard notation $\mathbb{P}$ for the set of primes, we denote by $\mathbb{P}(\boldsymbol{a_1}, \ldots, \boldsymbol{a_\rho})$ the set of primes included in the union of the sets $\boldsymbol{a_1}, \ldots, \boldsymbol{a_\rho}$. Let $\varphi(M) > m + 1$.

Theorem 9.1. *Let M, ρ be positive integers. Let $(m+1)\rho > \varphi(M)m$. Then, for any positive integer ξ, there exists an infinite set of primes*

p from the set $\mathbb{P}(\boldsymbol{a_1}, \dots, \boldsymbol{a_\rho})$ such that in the field $\mathbb{Q}_p$, at least one of the numbers $\Psi_0(\xi), \Psi_1(\xi)$ is transcendental.

Let the positive integers μ_k for any k satisfy $\mu_k \leq \lambda_k$.
Let $\Xi = \sum_{l=0}^{\infty} \mu_l \lambda_l$.

Theorem 9.2. *Let M, ρ be positive integers. Let $(m+1)\rho > \varphi(M)m$. Then, there exists an infinite set of primes p from the set $\mathbb{P}(\boldsymbol{a_1}, \dots, \boldsymbol{a_\rho})$ such that in the field $\mathbb{Q}_p$, at least one of the numbers $\Psi_0(\Xi), \Psi_1(\Xi)$ is transcendental.*

9.3.   Proofs of Theorems

The **proofs of both theorems** are practically identical. In Chapter 8, we considered the series

$$f_0(z) = \sum_{n=0}^{\infty} (\alpha_1)_n \cdots (\alpha_m)_n z^n, \quad i = 1, \dots, m,$$

$$f_i(z) = \sum_{n=0}^{\infty} (\alpha_1 + 1)_n \cdots (\alpha_i + 1)_n (\alpha_{i+1})_n \cdots (\alpha_m)_n z^n,$$

$$i = 1, \dots, m.$$

In this chapter, we denoted

$$f_0(z) = \Psi_0(z), \quad f_m(z) = \Psi_1(z).$$

So, (9.1) takes the form

$$\Psi_0(z) = 1 + \alpha_1, \dots, \alpha_m z \Psi_1(z).$$

For convenience, we recall the formulation of Theorem 8.1.
  Let $m \geq 3$, M, ρ be positive integers. Let

$$\varphi(M) > m, \quad \rho m > \varphi(M)(m-1).$$

Then, for any integers $h_0, \dots, h_{m-1}$ that are not all zero and for any positive integer ξ, there exist infinitely many prime numbers p from the set $\mathbb{P}(\boldsymbol{a_1}, \dots, \boldsymbol{a_\rho})$ such that, in the field $\mathbb{Q}_p$,

$$|L(\xi)|_p = |h_0 f_0(\xi) + \cdots + h_{m-1} f_{m-1}(\xi)|_p > 0.$$

(Theorem 8.2 is quite similar and concerns the case of the point Ξ.)

From Theorems 8.1 and 8.2, it follows that for points ξ and Ξ, there are infinitely many primes p such that in the field $\mathbb{Q}_p$, inequality $\Psi_1(\xi) \neq 0$ (respectively, $\Psi_1(\Xi) \neq 0$) holds. Indeed, it is enough to consider the linear form $L(\xi) = f_m(\xi)$ (respectively, $L(\Xi) = f_m(\Xi)$) and apply the above theorems.

According to the remarks made above, it remains only to prove that the number $\alpha_1 \ldots \alpha_m \xi$ (respectively, $\alpha_1 \ldots \alpha_m \Xi$) is a polyadic Liouville number. To do this, consider the positive integers

$$\alpha_{i,k} = \sum_{l=0}^{k} \mu_{i,l} \lambda_l, \quad i = 1, \ldots, m. \tag{9.6}$$

Put $A = \alpha_{1,k} \ldots \alpha_{m,k} \xi$. Then,

$$\alpha_1 \cdots \alpha_m \xi - \alpha_{1,k} \cdots \alpha_{m,k} \xi$$
$$= (\alpha_1 - \alpha_{1,k})\alpha_2 \cdots \alpha_m \xi + \alpha_{1,k}(\alpha_2 - \alpha_{2,k})\alpha_3 \cdots \alpha_m \xi$$
$$+ \alpha_{1,k}\alpha_{2,k}(\alpha_3 - \alpha_{3,k})\alpha_4 \cdots \alpha_m \xi + \cdots$$
$$+ \alpha_{1,k} \cdots (\alpha_{m-1} - \alpha_{m-1,k})\alpha_m \xi + \alpha_{1,k} \cdots (\alpha_m - \alpha_{m,k})\xi.$$

Since all the numbers $\alpha_1, \ldots, \alpha_m, \alpha_{1,k}, \ldots, \alpha_{m,k}, \xi$ are p-adic integers,

$$|\alpha_1 \ldots \alpha_m \xi - \alpha_{1,k} \ldots \alpha_{m,k} \xi|_p \leq \max_{i=1,\ldots,m} |\alpha_i - \alpha_{i,k}|_p. \tag{9.7}$$

For given numbers n, P, we choose the number K_0 such that for $k \geq K_0$, we have

$$P \leq s_k + 2\lambda_k^2, \quad n \leq \frac{ms_k \ln s_k}{\ln 2(\ln \xi + m \ln k + 2m \ln \ln s_k)}. \tag{9.8}$$

Then, for any prime p such that $p \leq P$, in view of (9.3), the inequality

$$\max_{i=1,\ldots,m} |\alpha_i - \alpha_{i,k}|_p \leq p^{-\mathrm{ord}_p \lambda_{k+1}} \leq p^{-ms_k \ln s_k} \tag{9.9}$$

holds. In turn, from (9.4), (9.5), and (9.6), it follows that

$$|\alpha_{i,k}| \leq k\lambda_k^2 \leq k \ln^2 s_k.$$

Therefore,

$$|A| = |\alpha_{1,k} \ldots \alpha_{m,k}\xi| \leq |\xi| k^m \ln^{2m} s_k = p^{\frac{\ln \xi + m \ln k + 2m \ln \ln s_k}{\ln p}}$$

$$\leq p^{\frac{\ln \xi + m \ln k + 2m \ln \ln s_k}{\ln 2}}. \qquad (9.10)$$

From (9.7)–(9.10), the inequality

$$|\alpha_1 \ldots \alpha_m \xi - A|_p \leq A^{-n}$$

follows. The reasoning for the numbers $\alpha_1 \ldots \alpha_m \Xi$ is quite similar.

Note once again that checking the inequality $\Psi_1(\xi) \neq 0$ (respectively, $\Psi_1(\Xi) \neq 0$) for a given prime p allows us to assert that at least one of the numbers $\Psi_0(\xi), \Psi_1(\xi)$ (respectively, $\Psi_0(\Xi), \Psi_1(\Xi)$) is transcendental p-adic.

9.4.   Example

Let us turn back to the above-mentioned assertion and consider the simplest case of $p = 2$.

Let $\lambda_0 = 1$. Put $s_0 = [\exp \lambda_0] + 1 = 3$. Let λ_1 be an arbitrary positive integer such that for any prime

$$p \leq s_0 + 2\lambda_0^2 = 5,$$

the inequality $\mathrm{ord}_p \lambda_1 \geq 2s_0 \ln s_0$ holds, and let $s_1 = [\exp \lambda_1] + 1$.

For $k \geq 1$, let λ_{k+1} be an arbitrary positive integer, such that for any prime $p \leq s_k + 2\lambda_k^2$, the inequality

$$\mathrm{ord}_p \lambda_{k+1} \geq 2s_k \ln s_k$$

holds, and let

$$s_{k+1} = [\exp \lambda_{k+1}] + 1.$$

As was noted above, the series

$$\lambda = \sum_{k=0}^{\infty} \lambda_k \qquad (9.11)$$

converges. Further, let

$$f_0(z) = \sum_{n=0}^{\infty} (\lambda)_n z^n, \quad f_1(z) = \sum_{n=0}^{\infty} (\lambda+1)_n z^n.$$

These series converge in the field $\mathbb{Q}_2$. This can be easily proved by analogy with the proof of the convergence of series (9.8)–(9.11).

Theorem 9.3. *At least one of the 2-adic numbers $f_0(1), f_1(1)$ is a transcendental 2-adic number.*

Proof of Theorem 9.3. As we noted above,

$$f_0(z) = 1 + \lambda z f_1(z).$$

We put here $z = 1$ and get

$$f_0(1) = 1 + \lambda f_1(1).$$

We will prove that in the field $\mathbb{Q}_2$ the inequality $f_1(1) \neq 0$ holds. To do this, let us consider the expansion

$$f_1(1) = \sum_{n=0}^{\infty} (\lambda+1)_n = 1 + (\lambda+1) + (\lambda+1)(\lambda+2) + \cdots . \quad (9.12)$$

Note that for $n \geq 1$, the inequality $\operatorname{ord}_2(\lambda+1)_n \geq 1$. Indeed, $\lambda_0 = 1$, $\operatorname{ord}_2 \lambda_1 \geq 6 \ln 3 \geq 6$. Moreover, the values $\operatorname{ord}_2 \lambda_k$ rapidly increase. Therefore, from (9.11), we obtain

$$\operatorname{ord}_2(\lambda+1) \geq 1.$$

Since the series $f_1(1)$ converges, we see that it is a sum of two numbers, namely, 1 and the 2-adic number

$$(\lambda+1) + (\lambda+1)(\lambda+2) + \cdots ,$$

and

$$\operatorname{ord}_2((\lambda+1) + (\lambda+1)(\lambda+2) + \cdots) \geq 1.$$

This proves that $f_1(1) \neq 0$. Therefore, we have proved Theorem 9.3.

□

Bibliography

Adams W. (1966). Transcendental numbers in the p-adic domain. *Am. J. Math.*, 88, pp. 279–307.

André Y. (1996). *G-Functions and Geometry*. Vieweg and Teubner Verlag, Wiesbaden.

Bertrand D., Chirskii V. and Yebbou J. (2004). Effective estimates for global relations on Euler-type series. *Ann. Fac. Sci. Toulouse Math.* (6) 13(2), pp. 241–260.

Beukers F., Brownawell W. D. and Heckman G. (1988). Siegel normality. *Ann. Math. Ser. II*, 127, pp. 279–308.

Bombieri E. (1981). On G-functions. *Recent Progress in Analytic Number Theory*, Vol. 2. Academic Press, London, pp. 1–67.

Chirskii V. G. (2014). On the arithmetic properties of generalized hypergeometric series with irrational parameters. *Izv. Math.*, 78(6), pp. 1244–1260.

Chirskii V. G. (2017). Arithmetic properties of polyadic series with periodic coefficients. *Izv. Math.*, 81(2), pp. 444–461.

Chirskii V. G. (2019). Product formula, global relations and polyadic integers. *Russ. J. Math. Phys.*, 26(3), pp. 286–305.

Chirskii V. G. (2020a). Arithmetic properties of generalized hypergeometric F-series. *Russ. J. Math. Phys.*, 27(2), pp. 175–184.

Chirskii V. G. (2020b). Arithmetic properties of Euler-type series with a Liouvillean polyadic parameter. *Dokl. Math.*, 102(2), pp. 412–413.

Chirskii V. G. (2021). Arithmetic properties of an Euler-type series with polyadic Liouvillean parameter. *Russ. J. Math. Phys.*, 28(3), pp. 294–302.

Chirskii V. G. (2022a). Polyadic Liouville numbers. *Dokl. Math.*, 106(S2), S137–S141.

Chirskii V. G. (2022b). Arithmetic properties of the values of generalized hypergeometric series with polyadic transcendental parameter. *Dokl. Math.*, 106(2), pp. 386–397.

Chirskii V. G. (2023). Transcendence of p-adic values of generalized hypergeometric series with transcendental polyadic parameter. *Dokl. Math.*, 107(2), pp. 109–111.

Chudnovsky G. V. (1985). On applications of Diophantine approximations. *Proc. Natl. Acad. Sci. USA*, 81, pp. 7261–7265.

Ernvall-Hytönen A.-M., Matala-aho T. and Seppälä L. (2019). Euler's divergent series in arithmetic progressions. *J. Integer Seq.*, 22, Article 19.2.2.

Ernvall-Hytönen A.-M., Matala-aho T. and Seppälä L. (2023). Euler's factorial series, Hardy integral, and continued fractions. *J. Number Theory*, 244, pp. 224–250.

Euler L. (1760). De seriebus divergentibus. *Novi. Commentii Acad. Sci. Petrop.*, 5, pp. 205–237; *Opera Omnia, Series 1*, 14, pp. 585–617.

Flicker Y. (1977). On p-adic G-functions. *J. London Math. Soc.*, 15(3), pp. 395–402.

Fomin A. A. (1999). Some mixed abelian groups as modules over the ring of pseudo-rational numbers. *Abelian Groups and Modules*. Trends in Mathematics. Birkhäeuser, Basel, pp. 87–100.

Galochkin A. I. (1970). The algebraic independence of values of E-functions at certain transcendental points. *Moscow Univ. Math. Bull.*, 25(5), pp. 41–45.

Galochkin A. I. (1974). Estimates from below of polynomials in the values of analytic functions of a certain class. *Math. USSR-Sb.*, 24(3), pp. 385–407.

Galochkin A. I. (1981). A criterion for hypergeometric Siegel function to belong to the class of E-functions. *Math. Notes*, 29, pp. 3–8.

Hata M. (1992). Irrationality measures of the values of hypergeometric functions. *Acta Arith.*, 60, pp. 335–347.

Hermite C. (1873). Sur la function exponentielle. *C. R. Acad. Sci. Ser. A (Paris)*, 77, pp. 18–24, 74–79, 226–233, 285–293.

Ivankov P. L. (1993). On linear independence of the values of entire hypergeometric functions with irrational parameters. *Siberian Math. J.* 34(5), pp. 839–847.

Koblitz N. (1984) [1977]. *p-Adic Numbers, p-Adic Analysis, and Zeta-Functions*, 2nd edn. Graduate Texts in Mathematics, Vol. 58. Springer-Verlag, New York.

Korobov A. N. (1981). Estimates for certain linear forms. *Moscow Univ. Math. Bull.*, 36(6), pp. 45–49.

Kurepa D. (1971). On the left factorial function $!n$. *Math. Balkanica*, 1, pp. 147–153.

Lindemann F. (1882). Über die zahl π. *Math. Ann.*, 20, pp. 213–225.

Liouville J. (1844). Sur des classes très étendues de quantités don't la valeur n'est ni algébrique, ni même réductible à des irrationelles algébriques. *C. R. Acad. Sci. Sér. A (Paris)*, 18, pp. 883–885.

Mahler K. (1935). Über transzendente p-adische zahlen. *Compos. Math.* 2, pp. 259–275.

Matala-aho T. and Zudilin W. (2018). Euler's factorial series and global relations. *J. Number Theory*, 186, pp. 202–210.

Nesterenko Y. V. (1969). On the algebraic independence of the values of E-functions satisfying nonhomogeneous differential equations. *Math. Notes*, 5(5), pp. 352–358.

Nesterenko Y. V. (1995). Hermite-Padé approximants of generalized hypergeometric functions. *Russ. Acad. Sci. Sb. Math.* 83, pp. 189–219.

Novosyolov E. V. (1961). Integrating on a bicompact ring and number-theoretical applications. *Izv. Vuzov. USSR Math.*, 32(22), pp. 66–79 (Russian).

Postnikov A. G. (1988). *Introduction to Analytic Number Theory*. Translations of Mathematical Monographs, Vol. 68. American Mathematical Society, Providence.

Salikhov V. K. (1989). Algebraic independence of values of hypergeometric E-functions. *Dokl. Math.*, 40(1), pp. 71–74.

Salikhov V. K. (1990). Irreducibility of hypergeometric equations, and algebraic independence of values of E-functions. *Acta Arith.*, 53, pp. 453–471.

Salikhov V. K. (1991). A criterion for the algebraic independence of a class of hypergeometric E-functions. *Math. USSR-Sb.*, 69(1), pp. 203–226.

Shidlovskii A. B. (1954). On transcendence and algebraic independence of the values of certain classes of entire functions. *Dokl. Akad. Nauk SSSR*, 96, pp. 697–700 (Russian).

Shidlovskii A. B. (1955). A criterion for algebraic independence of the values of a class of entire functions. *Dokl. Akad. Nauk SSSR*, 100, pp. 977–980 (Russian).

Shidlovskii A. B. (1989). *Transcendental Numbers*. W. de Gruyter, Berlin, New York.

Siegel C. L. (1929). Über einige anwendungen diophantischer approximationen. *Abh. Preuss. Akad. Wiss. Phys.-Math. Kl.*, 1, pp. 1–70.

Siegel C. L. (1949). *Transcendental Numbers*. Princeton University Press, Princeton.

Vladimirov V. S. (2002). Left factorials, Bernoulli numbers, and the Kurepa conjecture. *Publ. Inst. Math. Beograd (N.S.)*, 72(86), pp. 11–22.

Weierstrass K. (1885). Zu Lindemann's abhandlung: Über die Ludolph'sche zahl. *Sitzungsher. Preuss. Akad. Wiss.*, pp. 1067–1085.

Index